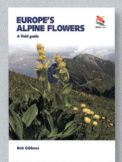

Other **WILD**Guides titles

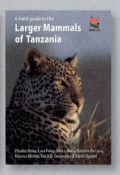

press.princeton.edu/wildguides/

TREES
of Britain and Ireland

Jon Stokes

Published by Princeton University Press,
41 William Street, Princeton, New Jersey 08540
99 Banbury Road, Oxford OX2 6JX
press.princeton.edu

Copyright © 2025 by Jon Stokes
All illustrations are © Stuart Jackson-Carter (SJC Illustration)

Copyright in the photographs remains with the individual photographers.

GPSR Authorized Representative: Easy Access System Europe - Mustamäe tee 50,
10621 Tallinn, Estonia, gpsr.requests@easproject.com

All rights reserved. Princeton University Press is committed to the protection of
copyright and the intellectual property our authors entrust to us. Copyright promotes
the progress and integrity of knowledge created by humans. Thank you for supporting
free speech and the global exchange of ideas by purchasing an authorized edition of this
book. If you wish to reproduce or distribute any part of it in any form, please obtain
permission.

Requests for permission to reproduce material from this work should be sent to
permissions@press.princeton.edu

British Library Cataloging-in-Publication Data is available

Library of Congress Control Number 2023948257
ISBN 978-0-691-22416-9
Ebook ISBN 978-0-691-22417-6

Editorial: Jacqui Sayers and Rob Still
Design: Rob Still
Cover Design: Rob Still
Production: Ruthie Rosenstock
Publicity: Caitlyn Robson-Iszatt and William Pagdatoon

Cover image: Treescape, County Durham by Jon Stokes
Title page image: Birches *Betula* spp. by Jon Stokes

Printed in Italy

10 9 8 7 6 5 4 3 2 1

Contents

Acknowledgements	6
Introduction	7
What is a native species?	8
What is the difference between a tree and a shrub?	9
Using this book	10
How does a tree 'work'?	11
Life cycle of a tree	16
Tree shapes	18
Tree shapes in the wild	18
Managed tree shapes	18
Tree habitats	24
Scrub	24
Hedgerows	28
Orchards, Wood Pasture and Parkland	30
Woodland	31
Types of woodland	33
History of our treescapes and habitats	37
Trees and wildlife	44
Trees as characters – the fascination of old trees	47
Tree identification	51
Getting started	51
One-step ID	52
Key to leaves	58
Key to flowers	68
Key to fruit	74
Key to winter twigs	78

THE SPECIES ACCOUNTS

Introduction to conifers	88
Key to conifers	90
Conifers with long needles	96
Conifers with short needles	104
Conifers with needles in rosettes	106
Conifers with broad, triangular needles	111
Conifers with short, thin needles in 3s	112
Conifers with flat needles	114
Conifers with scale-like needles	124

Introduction to broadleaves 130

  Introduction to the rose family 131

    Introduction to the whitebeams 133

    Plums, cherries, apples & pears 170

    Hawthorns & cotoneasters 186

  Buckthorns and Sea-buckthorn 192

  Walnut 196

  Introduction to the elms 197

  Introduction to Beech, oaks & chestnuts 213

  Introduction to willows & poplars 226

    Identification of willows & poplars 227

    Willows 232

    Aspen and poplars 256

  Introduction to birches, alders, hazels & Hornbeam 265

    Introduction to birches 266

    Alders, hazels & Hornbeam 272

  Spindles 280

  Planes 282

  Maples & Sycamore 284

  Limes 292

  Ashes 298

  False Acacia 300

  Strawberry-tree 301

Widespread native shrubs 302

Native woody peas 318

Currants 322

Widespread introduced shrubs and small trees 326

Park, street & garden trees and shrubs 338

References and further reading 345

Appendix 1: Measurements of optimal short shoot leaves of elm species 346

Appendix 2: Taxonomic list of trees and shrubs mentioned in the text 348

Appendix 3: Non-tree species mentioned in the text 352

Photographic and artwork credits 355

Index 356

Foreword

The Tree Council, the national charity that brings everyone together for the love of trees, is delighted in its 50th year to celebrate this stunning new Tree ID Guide by Jon Stokes, our Director of Trees, Science & Research.

In this guide, Jon draws on his more than three decades' professional experience roaming every corner of the UK and Ireland driven by his enormous enthusiasm and desire to research, understand and share his knowledge of trees and the natural world.

Trees are a vital part of our landscape and culture, they provide habitat for thousands of species, absorb carbon to help reduce the impacts of climate change, and their beauty and colourful rustling contribute immeasurably to our health and wellbeing.

This comprehensive collection is a tribute to these incredible organisms, sparking interest and curiosity, setting us off with all the right tools on a lifelong journey of arboreal learning. Being a tree detective means slowing down and getting up close with the detail – and the more you delve and compare, the more connected you become.

With bounteous facts and detailed illustrations to help experts and amateurs alike, this book draws together 113 British and Irish native trees and shrubs and tells a wealth of stories about our natural heritage. Another 190 introduced species and hybrids paint a wider picture of our treescape and show the richness of the history of our trees, to create a never-ending journey of awe.

We hope you will find Jon's expert compendium an invaluable aid on your own journeys into the field and that it will help you too get wonderfully waylaid by the world of trees.

Sara Lom
The Tree Council CEO

Acknowledgements

Writing this book has been a dream of mine for many years, and I couldn't have written it without the support of my family, friends and expert colleagues.

First, I would like to thank my family, especially my wife Alison and children, Amy and Philip, for their unending support and for never complaining when being dragged up the side of a mountain on the hunt for the perfect photograph! Additional thanks to Amy for research in developing this book.

I would also like to thank my friends and colleagues at The Tree Council for their advice and input when drafting this book. Clare Bowen, Sara Lom, Jess Allan, Sam Village, Will Fitzpatrick, Jackie Shallcross and Ian Turner have given invaluable feedback which has helped me immensely.

The author having just found Net-leaved Willow on the slopes of Ben Lawers during the preparation of this book.

Furthermore, I would like to extend my appreciation to the BSBI for the use of the maps and distribution data; Louise Hill from Defra, Emma Dear from Natural England, and Rob Wolton of the Devon Hedge Group who all provided helpful feedback; Glyn Marsh for access to his fantastic tree collection; Stuart Jackson-Carter for his stunning illustrations and to Rob Still of **WILD**Guides for his support and photographs.

The elm section of the book was developed with invaluable support from Alex Prendergast from Natural England, Brian Eversham of the Bedfordshire, Cambridgeshire and Northamptonshire Wildlife Trust and Cicely Marshall of Cambridge University.

There are also so many of you who have shaped my thoughts and views on trees over the years and to all of you – Caroline, Ted, Derek, John, Archie, Jill, Oliver, Tony, Esmond, Neil, Lisa, Guy, James, Glyn, Bianca, Gary, Nigel, Chris, Hugh, Andy, Pauline, Jonathan, Jeremy, Paul, David, Dave and Simon plus Margaret (to name but a few) – thank you all!

I would also like to acknowledge and thank the many Tree Warden volunteers and other professional tree people whom I have had the joy and privilege of working alongside throughout my career. I thank you all for your friendship and hope that this book proves to be a valuable resource for you and the many Tree Wardens and tree lovers to come. Your efforts to plant and care for trees in your communities have always inspired me.

As with many things, this book is an evolving resource, and there will be new discoveries and research findings that require updates to be made, and therefore my request to all is to share updates and comments with me, which can be considered in future editions.

I would like to thank you, the reader, for choosing my book as your tree guide. This book has been a labour of love, and I hope it brings you as much joy as it has brought me over the last four years.

Finally to Mum, Dad, Mervyn and Mollie. Thank you!

Jon Stokes

Introduction

In developing this book, we set out with a clear concept – to 'describe and celebrate all of Britain & Ireland's native trees and shrubs'. Britain & Ireland have several unusual and uncommon native tree and shrub species, ranging from a diverse number of rare whitebeams to the dwarf willows and birches, plus we even have a unique cotoneaster. However, these species have rarely been covered in previous tree books.

For me as a professional tree conservationist of 35 years standing, I felt it was important to pull together all these plants into one place, to help to improve the understanding of our native tree flora. Its diversity and complexity is surprising and there is still much to learn about our wide range of tree species and subspecies.

However, it also became clear that to stick to the original vision would result in some species that people would widely encounter (like Horse Chestnut) not being included in the book, as they are not native. Therefore, the number of species included in the book grew, to include a range of non-native trees.

To determine which species to include, we consulted the Botanical Society of Britain & Ireland (BSBI) *Plant Atlas 2020* and have included any tree or shrub which has become naturalized (with a self-sustaining population) and occurs in more than 5% of hectads (a hectad being a notional 10 × 10 km square, in a grid which covers these islands). We have also included a range of trees that are widely encountered in parks, gardens and streets.

However, to keep true to the original 'native species' concept, the introduced species have been treated differently, with some only having short descriptions to enable their separation from the natives.

Jon Stokes

January 2025

Wistman's Wood on Dartmoor – an amazing example of an ancient, upland oak woodland

INTRODUCTION

What is a native species?

In this book, we define a native tree or shrub as a species that colonized Britain & Ireland naturally after the last ice age (±15,000 years ago) and before these islands were cut off from the rest of Europe by rising sea levels ±8,000 years ago (see *p. 37*).

Along with native species, we have also included trees and shrubs which are known as 'archaeophytes', which are plants introduced to Britain & Ireland, either intentionally or accidentally, by humans (predominantly farmers) at the start of the Neolithic period (±6,500 years ago) and the end of the Medieval period (1500 CE) and which have since become naturalized into the landscape.

Species introduced after *c.*1500 are known as 'neophytes'. The 16th century is used as the transition period from 'archaeophyte' to 'neophyte' because it includes the period of discovery of the 'New World', and the rise of plant hunters who began importing trees and shrubs from across the globe for commercial and ornamental purposes.

Key dates of introduction to Britain & Ireland of some common non-native trees

16th century or earlier	Norway Spruce, Sycamore, Black Mulberry, White Mulberry, Evergreen Oak, Oriental Plane, Tamarisk, Judas Tree, Laburnum
17th century	European Larch, Tulip-tree, False Acacia, Horse Chestnut, Norway Maple, Swamp Cypress (*c.*1638), Cedar of Lebanon (1638), Red Maple (1656), London Plane (1663).
18th century	Indian Bean Tree (1726), Turkey Oak (*c.*1740), Ginkgo (1754), Corsican Pine (1759), Rhododendron (*ponticum*) (1763), Irish Yew (*c.*1770), Monkey-puzzle (1795)
19th century	Douglas Fir (1827), Noble Fir (1830), Grand Fir (1831), Deodar (1831), Sitka Spruce (1831), Monterey Pine (1833), Coast Redwood (1843), Giant Redwood (1853), Katsura Tree (1865), Leyland Cypress (*natural hybrid found in Welshpool in 1888*).
20th century	Dawn Redwood (1948)

The introduced Sycamore (*left*) is perhaps a more familiar tree than the closely related native Field Maple (*right*).

INTRODUCTION

What is the difference between a tree and a shrub?

Trees are the tallest and largest free-standing organisms on earth. It is thought that there are between 60,000 and 80,000 tree and shrub species in the world, the difference reflecting the difficulty of defining a 'tree'.

Simplistically, a tree is a 'large' woody organism. That definition, however, is inadequate in the case of small willows, which are 'small' woody organisms yet can be described as 'trees'. Shrubs are often described as 'small-to-medium-sized perennial woody plants', with persistent woody stems above the ground.

Some tree books have created the distinction that a tree has a single stem or 'trunk' and an elevated head of branches, whereas a shrub branches from the base, with no obvious trunk. Unfortunately, the way 'a tree' is managed can make it appear to be 'a shrub', or 'a shrub' that has not been cut could look like 'a tree' – see the pictures (*right*) of a hawthorn 'tree' and 'shrub'.

In this book, we have therefore used the definition that a tree 'is regarded as a plant that has a self-supporting perennial woody stem', which covers everything from a tiny Dwarf Willow to the tallest Coast Redwood, and we do not worry about the distinction between shrubs and trees. We have, however, chosen not to cover some woody plants like the heathers, and climbers like roses and ivy, as they would typically never be described as a tree.

Hawthorn can be a 'tree' (*top*) or a 'shrub' (*bottom*), as a result of grazing or management. Mature 'tree' hawthorns are rarely seen, but are beautiful small trees.

Some woody plants such as brambles and roses (left) and Heather (right) are not covered as they are rarely regarded as shrubs.

Using this book

This book aims to provide the tools for anyone interested in trees, whatever their level of expertise or experience to improve their knowledge and enjoyment of Britain & Ireland's trees and shrubs.

The focus is on identification of trees throughout the year and, as far as possible, technical botanical terms have been avoided – any that are used are illustrated and described.

INTRODUCTORY SECTIONS (*pp. 11–50*) | cover how a tree 'works'; their life cycle, shape, habitats, landscape history, character and importance to wildlife.

TREE IDENTIFICATION (*pp. 51–87, 90–95*) | provides an introduction to tree identification and annotated visual keys to leaves, flowers, fruit, deciduous winter twigs, and coniferous leaves and cones which reference species groups and individual species accounts.

THE SPECIES ACCOUNTS (*pp. 51–344*) | contain a concise description of each species together with photos that show the species in its habitat; close-up photos and illustrations of significant identification features; and information covering similar species, interesting facts pertaining to the species and ecological notes/associated non-tree species where relevant.

Introductions and further keys to particular species groups, together with information on how to identify are given where relevant.

TAXONOMY, ENGLISH AND SCIENTIFIC NAMES | The accounts are not presented in strict taxonomic order, but instead are pragmatically grouped by similarity. Scientific names follow those found in the BSBI's *PlantAtlas 2020*; as is the case for the vast majority of English names with a few minor exceptions due to the author's preference. See *p. 348* for a taxonomic list of the tree and shrub species included.

MAPS AND PHENOLOGY | The accounts feature a distribution map and phenology chart based upon the BSBI *PlantAtlas 2020* data; coded as below.

SPECIES STATUS (see *p. 8* for definitions) | A colour coded strip at the top of the page designates whether the main species on that page are **native**, an **archaeophyte**, a neophyte or a **hybrid** as per the key below.

CONSERVATION STATUS CODES | IUCN Red List status is given for the world (*The IUCN Red List of Threatened Species*, Version 2024-2, IUCN. (2024). iucnredlist.org); Great Britain (*GB Red List for Vascular Plants*, BSBI (2019) – dark outline) and/or Ireland (*Ireland Red List No. 10: Vascular Plants*, NWPS (2016) – green outline) as per the key below.

EDIBILITY AND TOXICITY | Icons (as shown below) after each species name indicate whether the fruit of a species is toxic and should not be eaten; edible if cooked; or edible when raw. The lack of an icon indicates that the fruit should not be eaten but is not significantly harmful if inadvertently ingested.

Key to codes used in the book

RED LIST CONSERVATION STATUS CODES

MAPS AND PHENOLOGY
- in flower
- in fruit (unripe)
- in fruit (ripe)
- in leaf

SPECIES STATUS
- Native
- Archaeophyte
- Neophyte
- Hybrid

EDIBILITY AND TOXICITY
Do not eat (icon absent)
- Edible raw
- Edible if cooked/processed
- **Toxic – do not eat**

J F M A M J J A S O N D

IUCN		GB	Ireland
CR	Critically Endangered	CR	CR
EN	Endangered	EN	EN
VU	Vulnerable	VU	VU
NT	Near Threatened	NT	NT
LC	Least Concern	LC	
DD	Data Deficient	DD	
NE	Not Evaluated	NE	
	No Data		

How does a tree 'work'?

A tree consists of four key parts:

Roots | These anchor the trunk and crown of the tree into the ground and make them stable. They also take up water and nutrients from the soil and serve as a store of energy in the form of carbohydrates.

Trunk and branches | These give a tree height and width so its leaves can capture as much light energy as possible. The trunk of a tree acts as the highway via which nutrients and water are distributed around the tree. Water travels from the roots up to the leaves through xylem (a series of vascular vessels which also provide physical support to a tree). When water reaches the leaves, it enables photosynthesis (see p. 13) to occur.

Leaves | The tree's food producers which use water, carbon dioxide and sunlight to create energy (in the form of glucose) for the tree.

Flowers and fruit (or cones…) | Coniferous trees do not have flowers. Instead they produce both ♂ and ♀ cones; ♂ cones release pollen, fertilising a separate ♀ cone – the seeds developing within a protective structure that eventually matures into the familiar 'pine cone'.

In contrast; angiosperm (flowering plant) trees, once mature, produce flowers – although these are inconspicuous in many species. Flowering trees reproduce either sexually (via the exchange of pollen between ♂ and ♀ reproductive systems), or asexually (by vegetative reproduction). Once pollinated, the ovary of ♀ flower expands to form a fruit that protects the seed. Fruits are found in a wide range of forms – from nuts, to winged seeds, to berries –depending on the species.

How does a tree grow?

Trees have three points of growth:

1) At the tip of the **roots**, allowing the roots to extend through the soil
2) At the **buds**, where growth lengthens the branches and increases the size of the tree's crown (the name given to the complete canopy of leaves and branches)
3) Within a layer of the **trunk** called the cambium which is situated between the outer layers of bark and inner layers of wood. The cambium consists of active cells which divide and thicken the trunk of the tree.

The amazing root spread of an old Beech, by a stream in the New Forest.

The trunk gives a tree height and width, so its leaves can capture as much light energy as possible.

Leaves come in a wide range of shapes and sizes – from large 'broadleaves' to very thin 'needles'.

The diversity of tree seed is striking – both in shapes and colours.

How do roots function?

There is a common misconception that the roots of a tree reflect its trunk and branches. Instead, a tree's root system is often surprisingly shallow, dominated by long, relatively thin lateral roots usually spreading out close to the soil surface rather than by a deeply penetrating taproot. The best description is that the tree resembles a wine glass, sitting on a dinner plate.

When a seed germinates, it has a taproot which initially grows vertically downward, providing that soil conditions are suitable and the ground is not too hard. In the first two or three years, the taproot grows rapidly downward, but its growth slows with the tree's age, and it eventually declines.

However, side-roots start to form almost immediately and spread out to form the main anchor for the tree, providing essential structural support. Bending and swaying of young trees in the wind encourages the growth of these fibrous side-roots, increasing the strength of support.

Most roots are found close to the soil surface, with 90% or more of all roots located in the top 60 cm of the soil. Whilst the typical depth of tree roots is usually overestimated, root spread is often underestimated as it is commonly believed that roots extend only to the spread of the branches (the drip line). However, roots can grow beyond the branch spread, in some cases extending outwards for a distance equivalent to at least (and, in some cases, up to three times) the height of the tree.

Even when meeting obstacles and physical barriers in the soil, such as kerbs or building foundations, roots will tend to grow around them, taking the line of least resistance, often resuming their original direction of growth once the object has been bypassed.

This fallen oak shows the wide variety of root shapes and sizes and the often shallow depth of the roots.

Bark behaves like a waterproof skin around the trunk of the tree and its shape and form can be helpful in the identification of the species.

What is bark?

Bark consists of the two outermost layers (**bark** and **phloem**) of five that make up a tree trunk: **bark** ❶, the outermost layer, behaves like a waterproof skin; **phloem** ❷, the adjacent layer, transports the sugars created by photosynthesis around the tree. Part of the outer bark layer is filled with air, tannin, and/or waxy substances. This is the **phelloderm** ⓐ and it provides protection against damage, pests and diseases, as well as dehydration and extreme temperatures. In some trees the phelloderm has become substantially thicker (*e.g.* Cork Oak, Giant Redwood), providing increased protection against the challenges of their habitat. These bark and phloem are separated by the **cambium** ❸ from the two innermost wood layers (**sapwood** ❹ and **heartwood** ❺) – see diagram (*right*).

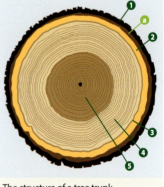

The structure of a tree trunk

What is wood?

Wood is the two layers of hard, fibrous material in the centre of the trunk surrounded by the cambium and bark. The innermost wood layer (**heartwood**) provides physical support to the stem; the outermost wood layer (**sapwood**) conducts water around a tree. Heartwood is not found in all trees but is the material that is usually sought for timber.

What are tree rings?

Trees grow outwards by the addition of new wood in the cambium layer immediately under the bark. The cambium is a very thin layer which actively produces the cells which create more wood and bark. Each year, during the tree's active growth period, visible growth rings are created. Differences in ring width indicate varying year-on-year growth rates. Because the growth season is annual, rings can be counted to determine the age of the tree and used to date wood taken from trees in the past in a practice known as dendrochronology.

Each year, trees produce new wood immediately under the bark. These visible rings indicate the annual growth of the tree.

Why are leaves green?

Leaves are normally green in colour because of chlorophyll which enables a tree to produce its own food. Chlorophyll is an amazing compound that absorbs the red and blue wavelengths of light to produce the chemical reactions needed to turn carbon dioxide and water into sugars and oxygen. This results in the yellow to green wavelengths being reflected – giving leaves their colour. This process is called photosynthesis and only takes place in leaves in specialist leaf cells called chloroplasts. Photosynthesis is a set of complex reactions, usually simplified as:

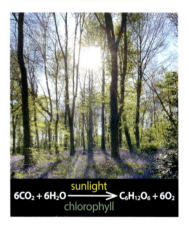

$$6CO_2 + 6H_2O \xrightarrow[\text{chlorophyll}]{\text{sunlight}} C_6H_{12}O_6 + 6O_2$$

Carbon dioxide + Water + Light energy [*in the presence of Chlorophyll*] → **Glucose (sugar) + Oxygen**

Simply put, a tree takes in carbon dioxide in order to produce sugars for itself and, as a waste product, emits oxygen into the atmosphere. It is truly astonishing, that a tree can create its amazing structures from gas in the atmosphere combined with water and nutrients from the ground.

Once created the sugars pass through another set of vessels – the phloem tubes – to the growing areas of the tree where their energy is needed.

Why do leaves change colour in the autumn?

As autumn approaches, a tree reabsorbs the chlorophyll from its leaves revealing various other pigments, such as carotenoids and anthocyanins, that were previously masked by the chlorophyll. As the leaves die, red anthocyanins are produced from any remaining sugars in the leaf. For strong, deep autumn leaf colours to occur there needs to be a high quantity of these sugars present and a weather pattern that gives a combination of warm, bright days and cold nights. These optimal weather conditions only sporadically occur in Britain & Ireland although are frequent on the east coast of the USA, resulting in the well-documented vivid autumn leaf colours that are found there.

How do leaves fall off in the autumn?

With deciduous trees, the mechanism to shed leaves is called abscission. Leaf fall is not a random process, but is a deliberate sequence, triggered by decreasing daylight and reduced air temperature. Across the base of the leaf is the abscission zone which breaks as the autumn approaches, causing the leaf to fall from the tree. The breaking of this zone is caused by increased levels of the plant hormones – ethylene, abscisic acid and/or auxin. A leaf scar remains on the twig at the site of the shed leaf.

Although leaves grow to a finite size in one season, the twigs that they grow from continue to expand into branches year on year, so over time all leaves (even needles) are forced to drop off.

As a leaf falls from the twig, it leaves a leaf scar which can be distinctive, like the large 'Y'-shaped scar found on a Walnut twig.

What is a bud?

A bud is a collection of tiny leaves, stems or flowers (usually contained within scales) which develop usually at the base of a leaf during the summer months in preparation for the following year. In spring, as a tree produces new leaves after the winter, the bud scales fall off and the leaves and/or flowers and twigs begin to expand and grow.

Do trees have flowers?

Technically, coniferous trees do not have flowers but instead have small ♂ and ♀ cones (strobili, *singular* strobilus). These lack many of the features of the flowers found in broadleaved trees but still perform the same function – transferring pollen from ♂ to ♀ reproductive organs. Most non-coniferous (or angiosperm) trees, as with all flowering plants, produce flowers and seeds.

Different species of tree (both coniferous and flowering) are either **dioecious** or **monoecious**.

Monoecious trees have ♂ and ♀ reproductive organs contained within flowers that are found on the same plant. Some species have ♂ and ♀ organs in the same bisexual flower (*e.g.* cherries and maples); others have separate unisexual ♂ and ♀ flowers (*e.g.* pines, oaks and hornbeam).

Dioecious trees have either unisexual (single sex) ♂ or ♀ flowers on separate plants (*e.g.* willows and Yew).

Sexual reproduction, known as pollination, relies on pollen from the ♂ reproductive organs reaching the female's. In conifers, pollination occurs as a result of the wind-borne transfer of pollen grains from a ♂ pollen cone to a ♀ seed cone. In Britain & Ireland's flowering trees this process (from the ♂ anther to the ♀ stigma) occurs by wind dispersal or by transfer by insects.

Ash buds: dormant and budding with, in this example, leaves sprouting from the top buds and male flowers from the side-buds (*inset*).

HOW A TREE WORKS

A range of tree flowers and 'not flowers': male and female catkins of a Hornbeam (*top left*); strobili (the reproductive parts of a conifer) of a Monterey Pine (*top right*); male and female catkins of a willow (*bottom left*); and the bisexual monoecious early spring flowers of Cherry Plum (*bottom right*).

As a result of pollination, a seed is produced which will be dispersed. If the environmental conditions are right it will germinate to become a seedling with a genetic make-up that is a combination of both its parents.

How are seeds dispersed?

Once a seed has developed, it may by dispersed by the wind (such as in birches or the winged seeds of maples and elms), drop to the ground (such as in Horse Chestnut and Beech) or be eaten off the tree by birds and animals (such as Blackbirds eating Rowan berries or squirrels transporting acorns).

Some seeds may take root if they fall close to the parent tree. Other fallen seeds may be eaten by birds and animals and spread further afield as they move on and defecate.

Wind dispersed Field Maple 'helicopter' seeds (*left*); Beech mast which drops to the ground containing the seeds (*centre*); Waxwing eating a Rowan berry which will be dispersed by the bird (*right*).

Life cycle of a tree

As a tree grows it passes through various stages of its life, from a sprout to a seedling, then a sapling, and onwards to become a mature tree and, if it is fortunate, it will become an ancient tree. To reach old age, a tree needs the combination of a suitable space with enough water, nutrients, and sunlight. It also requires other factors such as wind, drought, insect pests, fungal diseases and competition from other trees to be favourable, as any of these could end a tree's life prematurely.

Sprout

When a tree begins to grow, it is called a sprout. The first task of these tiny trees is to produce a root, using energy stored within the seed. The root heads down into the soil, seeking out water and nutrients, and then an embryonic trunk begins to develop and head towards the light. Reaching the light, the baby tree will produce its first leaves, needles, or scales which allow the young tree to make its own food through photosynthesis.

A germinating tree seedling called a 'sprout' with its developing roots which help the young tree to establish.

Seedling

As the young tree develops, it becomes a seedling – a risky stage when the tree can easily be eaten by slugs and snails, small mammals, rabbits or deer. As it grows, the seedling begins to develop woody characteristics with a hardening of the stem and the development of a thin early bark. It also further develops its covering of leaves. The roots continue to grow, developing particularly in the upper soil, where there are the water, nutrients and oxygen the young tree needs. If the tree can survive this stage, it becomes a sapling.

The first 'seed leaves' (or cotyledons) of a young Small-leaved Lime seedling.

Sapling

As the young tree develops into a sapling, it begins to develop a flexible trunk but is not mature enough to produce flowers or fruit. The young bark is often a different colour and texture from its more mature form. Growth can be rapid during this stage of life, with a thickening trunk as the tree begins to mature.

Mature tree

The mature stage of a tree's life develops when it starts to produce flowers and fruits. During this phase of its life, a tree will grow as large as its genetics and the site conditions will allow. The bark changes, often becoming darker and thicker with more cracks. The age at which a tree reaches maturity and starts to produce flowers and fruit depends on the species. Trees such as birch and Hazel start to produce flowers after about 10 years, whilst Sweet Chestnut, Wych Elm and oaks only start to produce flowers at about 30–40 years old. Individual

A young Ash sapling begins to develop a flexible trunk and its first leaves.

trees then continue to produce their maximum number of flowers and seeds, for the duration of their mature phase, which again differs between species. Shorter-lived species like birches continue to produce large numbers of seeds until they are about 70 years old, whilst oaks produce large numbers of acorns until they are more than 300 years old. Even as a tree moves into the ancient phase of its life, it can still produce flowers and seeds, just not in the quantity that it did during the mature phase.

Ancient trees

As a tree ages, it passes from its most productive mature phase to become ancient, the definition of which depends on the individual species (see *p. 48*). During this phase seed productivity declines, the trunk can hollow, and the canopy starts to reduce – lessening the distances over which water and nutrients are transported and thereby prolonging the survival of the tree. These are all natural processes and are not necessarily an indication of ill health or decline. As a tree ages, the dead wood in the ancient tree creates conditions that fungi and insects can take advantage of, which in turn provides food and shelter for birds and bats who use cracks and hollows within the trunk and branches. Even larger species such as the rare Pine Marten make their dens in holes in ancient trees.

Decaying trees

In the final stage of a tree's existence, it can be dying or even standing dead. As the branches break off and fall to the ground, they return nutrients to the soil. The decaying tree can also be very helpful to a whole range of other species which use the dead and decaying wood, such as insects or fungi. Research has even shown that up to a fifth of woodland species depend on dead or dying wood. This is why it is vital to understand the value of dead wood, both standing or lying on the ground, and not 'tidy it up'. Eventually the tree will fall, and the rotting hulk will be recycled by fungi and insects, returning more nutrients to the ground for the next generation of trees to use.

The mature stage of a tree begins when it starts to produce flowers and fruits. The age of maturity depends on the species, with Hazel producing flowers after 10 years and oaks at about 30–40.

Ancient trees produce less seed, the canopy starts to reduce and the trunk usually hollows.

As a tree decays it creates valuable habitat for a wide range of insects and other species, which is why deadwood shouldn't be 'tidied up'.

Tree shapes

Tree shapes in the wild

The primary factors determining the shape of a tree are its species and its genetics. However, although every tree species tends to have a characteristic shape, the specific shape of any individual tree develops in response to the environmental factors around it, such as competition for light, wind, drought and soil type.

The interplay between a tree's genetics and the environment where it grows results in myriad shapes and forms for individual trees. For example, if the soil conditions are favourable and there is nothing competing with it for light, a tree will usually develop the characteristic shape of the species. However, competition for light (*e.g.* in woodland) may cause the tree to develop a taller, thinner form, or to grow sideways and on one side, *e.g.* if growing on the edge of a wood.

This adaptability to conditions creates for us a unique opportunity with trees – a fascinating, almost archaeological ability to look at a tree's shape and work out the factors in the past that led to the tree's current appearance. Once learnt, you will never look at a tree in the same way again!

The best time to start looking at trees in this way is in winter, when a tree's shape and form is unobscured by leaves. Obviously, there are a countless range of factors that can influence a tree's shape and therefore a vast range of ways it can grow. Following are a few frequently seen examples.

Managed tree shapes

The way a tree's natural shape and form can be changed by the world around it allowed humans to exploit these natural processes to produce specific products – like thin whippy poles for building fences, or curved bits of wood for a ship.

Whilst the trunk size increases significantly over time, a tree's canopy shape can remain consistent. This allows an open grown tree to show a 'characteristic' shape for a species. In the case of these oaks, they have a distinctive 'domed' shape. However management and the environment can significantly change the shape of a tree.

Throughout history, this type of usage led to the creation of tree management techniques which provided the main forms of wood products people needed and resulted in trees acquiring new 'managed' shapes. The two most common treatments – coppicing and pollarding – involved the removal of the branches or larger limbs, which then produced specific growth forms in the trees. Being able to spot these forms also opens a window into a tree's life.

TREE SHAPES

Natural shapes I

When a tree grows in a woodland, the proximity of neighbouring trees can cause the tree to grow straight and tall. This creates a longer and straighter trunk as shown by these Beech trees.

In some dense woodlands, a tree may grow sideways towards the available light. This can happen when there is a clearing, caused by another tree falling.

The Beech in the centre of this picture has all its branches on the right-hand side. This is because the tree grew with a large tree on its left (which reduced the light and prevented branch formation), whilst on the right-hand side there was light, which allowed branches to develop.

As a tree gets older it may lose parts of its upper canopy and the remaining trunk may hollow. This has the effect of making the tree appear a bit like a traffic cone, with a narrow top and a wide base. This is a very stable shape and helps the tree to avoid further significant wind damage.

Most trees in exposed situations are shaped by the prevailing wind, often leading to their canopies becoming highly streamlined.

TREE SHAPES

Natural shapes II

If a tree falls over it does not necessarily result in its death – if sufficient roots remained unbroken. These trees can grow horizontally for many decades and even develop new canopies and are known as 'phoenix' trees as they have 'risen again'.

Branches can cross and fuse together in a process called 'inosculation' which can lead to complex structures. More uncommonly, whole trees, usually of the same species, can touch and apparently fuse together. Occasionally as here, two different species are involved – known as a 'twin tree'.

Browsing by animals affects tree shape. Here in the New Forest, the lower branches have been grazed off by deer and ponies as high as they can reach, leaving a 'browse line' under the canopy.
This restricts the tree's ability to produce branches lower down the trunk, which may have damaging long-term effects for the health of the tree.

Some species just have unusual natural shapes. This example of the native subspecies of Black Poplar has the characteristic shape of the species – a natural lean to the trunk. What the values are of developing a leaning trunk is unclear, although there has to be some evolutionary advantage to it, or Black Poplars would simply grow straight.

As trees get older, their lower branches can often droop down to, and eventually touch, the ground. It has been suggested that these touching lower branches provide the old tree with additional stability – rather like a walking stick for a human. In some species these touching branches can then grow new roots in a process called 'layering'.

If a tree branch 'layers', it can begin to grow into a new tree. Beyond the point of contact with the ground, the branch begins to thicken thanks to new roots, as the new tree begins to develop. Yew often produces low hanging, layering branches from which can grow new trees, leading to the ancient belief that the Yew was immortal.

What is coppicing?

Coppicing is a traditional method of woodland management in which young trees are cut down to near the ground. Before the introduction of chainsaws, cutting would have been by billhook, and the stool (the remaining cut-off trunk of the tree) would have been cut ±50 cm from the ground. Regrowth takes the form of many new shoots, which are allowed to grow for future harvest as poles or firewood. After a number of years, the shoots are re-cut and the cycle begins again.

Traditionally coppice woodland was cut in sections, rotationally, ensuring that a crop was available each year. A side effect of the rotational cutting was that a rich variety of habitats were created, as the woodland would always have a range of differently aged regrowth in it. This is beneficial for the woodland's wildlife. The length of the cutting cycle depended on the species and the local conditions, combined with the use to which the product was put. Hazel was usually cut on a seven to nine-year cycle whereas oak was coppiced on a 20-year cycle.

However, it was not only woodland that was coppiced, and a traditional method of reinvigorating a hedge was to coppice it. This system involved cutting the hedge back to a series of low stumps which would then regrow. The hedge could then be cut on a short rotation of between 7 and 15 years to provide small wood, usually for fuel.

Historically in parts of Essex and Suffolk, and probably elsewhere in the country, the timing of hedge coppicing needed to be integrated into the overall management of the farm. During the first years after coppicing the hedge, when the hedge was cut to the ground, the adjacent fields were turned to an arable crop. Then as the hedge regrew and became or was made stock proof, the land was turned to grass and stock kept in the fields until the next coppicing.

Unfortunately, as coppiced hedges are not stock proof for several years after cutting, they needed reinforcing, usually using material taken from the hedge and banged into the ground to fill the gap. Alternatively, a newly coppiced hedge can be fenced off to provide effective stock proofing until the hedge regrows.

The management of trees by coppicing – the act of cutting them down and letting them regrow – may have been learnt by early humans observing what happens to a tree after it is felled by a Beaver. This natural ability of many tree species to regrow was then developed by humans to produce a wide range of wooden products.

When cut back to the ground, coppiced trees regrow a new canopy by producing a large number of new shoots from all the available buds. These new young shoots develop a characteristic shape and form as seen here.

The effect of regular coppicing is to allow light periodically to the woodland floor. This has significant impacts on the ground flora – with many wildflower species adapting to the pattern of light every few years. One species particularly adapted to these conditions is the now rare Oxlip – here growing in Hayley Wood, Cambs.

TREE SHAPES

If a tree had been coppiced but then left to grow again uncut, a distinctive form develops called an 'overstood' coppice. These trees have multiple stems that have developed from a single root plate and often grow in a circular pattern related to the original cut tree. If left to grow, they can produce amazing old trees like these ancient coppiced Small-leaved Limes in Piles Coppice in Warwickshire.

What is pollarding?

Historically, pollarding is the rotational cutting of trees to encourage the growth of branches by cutting off a tree's stem, usually 2 m or so above ground level. The repeat cutting led to an expanded 'fist-shape' (a bolling) either at the top of the trunk or at the end of the cut limbs. Multiple shoots would grow from the bolling point.

Pollarding was common in pasture woodland where grazing animals were present. Pollarding above head height protected valuable timber or poles from being damaged by browsing animals such as cattle, horses or deer.

The other use of pollarding was in hedgerows. Historically in tenanted farms, the wood from a tree was the property of the farm owner, but the arisings from a tree – the branches and other produce – were the property of the tenant. Therefore, most tenants pollarded all their hedge trees to ensure that they obtained maximum 'arisings' from each tree for fodder or firewood. This practice makes coppiced hedges a curious hybrid between the two types of woodland management – coppice and wood pasture (see *p. 30*).

This ancient Beech is halfway between a coppice and a pollard and could be described either as a 'high cut coppice' or a 'low cut pollard'!

TREE SHAPES

Where old pollards are a long time out of their cutting rotation they are called lapsed pollards, and professional advice is always advisable before attempting restoration cutting of these trees.

Why are urban trees pollarded?

Pollarding is often used in urban areas to reduce a tree's size or for health and safety reasons. It prevents the tree from becoming too large for its space and reduces the risk of falling branches that could harm property and people. Once cut, the tree will grow a new crown.

Low pollard: These beautiful low-cut pollards in Wales were cut with the aim of having sheep feeding between them. Therefore they have been cut about 1 metre from the ground, which has formed these delightful short pollarded trees.

This is a 'shred' – a tree that has had all the side branches regularly removed to produce an odd form of vertical coppice or pollard. Not commonly seen these days, although some are still visible in eastern England and Gloucestershire – and more widely so in Normandy.

This is a beautiful pollarded tree in Normandy – where a single stem has been left on the cut pollard to act as a sap riser. This has the effect of increasing the pull on the rising sap in the spring after cutting – increasing the speed of regrowth. In French this process is called 'Tire sève' – literally 'pull sap'.

Urban pollard

Tree habitats

The treescapes of Britain & Ireland are diverse and varied. Almost anywhere could theoretically have a covering of trees, ranging from oak woodlands in southern Ireland, to large expanses of Scots Pine forest in Scotland, dark Yew woodlands on the chalk of southern England or dwarf willow scrub on mountainsides. Historically, these treescapes have been described as the 'climax community' – the plant community that, through ecological succession, will eventually establish on an area of land and remain largely unchanging in their species composition.

All our treescapes have, however, been shaped and moulded by humans, often to perform different tasks over millennia – from hedges that keep cows and horses in fields, to orchards that produce food. Some tree habitats are more natural than others, like scrub, which is often the first natural colonisation of an open area by the local trees and shrubs, to woodland which has become established over centuries but where the species have often been selected for their timber or other qualities.

However they have arisen, it is the diversity of our treescapes that gives much of the history, texture and interest to our landscape. What follows is a brief review of the main types of tree habitat.

Scrub

Scrub as a habitat is hard to define as it covers a wide range of treescapes. Sitting ecologically on the spectrum from open ground to woodland, it is the 'in between' habitat. This characteristic led historically to scrub often being dismissed by some environmentalists as not being important or even as a problem, as it colonizes more valued habitats like chalk downland. However, scrub is valuable. It provides breeding opportunities for many species of birds and insects, and fruits as food for many more. The issue with scrub habitats is that the shrubs that make up scrub will over time be replaced by larger tree species, eventually evolving into woodlands. This means that to retain scrub habitats often requires management, to favour the species that are found there.

Ecologically, scrub communities have been divided into three main types: coastal, lowland and upland scrubs.

A great example of shingle scrub with Gorse and Blackthorn in West Sussex.

TREE HABITATS

Coastal scrub
Coastal scrub in Britain & Ireland develops on shingle, sea cliffs and sand dunes.

Shingle scrub
If sediment accumulates in shingle, miniature Blackthorn and Broom can develop, with brambles, Elder and gorses, plus occasionally Juniper. Where the soil is deep enough, stunted oaks and Holly trees can develop.

Sea cliff scrub
Sea cliff scrub is typically composed of brambles, Blackthorn, Hazel, gorses and Wild Privet – all usually stunted by wind and salt spray. On limestone cliffs, Juniper can occur, and the Great Orme in north Wales is the only site for Wild Cotoneaster.

Sea cliff scrub at The Dizzard in Cornwall, where there are also stunted oaks.

Dune scrub
Dune scrub develops in stable areas of sand dunes, with Sea-buckthorn and Creeping Willow often being the dominant species. In wetter areas, Alder, Bog Myrtle and Grey Willow scrub can develop with Blackthorn, Hawthorn, Elder and Privet.

Lowland scrub communities
Lowland scrub occurs widely as two main types – **lowland wet scrub** and **lowland dry scrub**. The latter is itself subdivided by soil type – acidic soils (often on gravels and grits), neutral soils (neither particularly acid or alkaline) and alkaline or calcareous soils (such as chalky or limestone soils).

Lowland wet scrub
Lowland wet scrub is dominated by Grey Willow, often with Downy Birch. Sometimes there can be Alder and Pedunculate Oak, with Hawthorn and brambles forming the understory. Depending on the soil type there can also be Alder Buckthorn, Bog Myrtle, currants, Dog Rose and Elder.

Lowland wet scrub filled with willows, Alder and bramble.

TREE HABITATS

Neutral soil scrub dominated by large Hawthorns.

Calcareous scrub on chalk with prominent Juniper and Common Whitebeam

Lowland dry scrub
Lowland dry scrub communities are split into three main types based on the soil types: acidic, neutral and calcareous.

Acidic soil scrub
Scrub on acidic soils is dominated by Gorse and Broom with occasional bramble. If the soil is disturbed or burnt, young birches, Holly, pines and oaks can develop into scrub with scattered Crab Apple and Alder Buckthorn on occasions.

Neutral soil scrub
On neutral soils Hawthorn scrub dominates, with Blackthorn, Elder and elms. Elder scrub develops on moist disturbed soils, alone or with brambles and Grey Willow, conditions which also support the growth of the introduced Butterfly-bush. On the coast, Blackthorn dominates neutral soils, because it can tolerate salt and is effectively the climax community, as few other species can tolerate the wind and salt spray.

Lowland scrub in the New Forest where Gorse and birches (both Silver and Downy) are prominent, with large oaks visible in the distance.

Calcareous soil scrub
In the warmer, drier parts of the country, calcareous soils have a diverse lowland dry scrub community composed mainly of Hawthorn, Blackthorn and brambles, occasionally with Box, Dogwood, Hazel, Privet, Juniper, Wayfaring-tree, Common Whitebeam and Yew. In some places on these lime-rich soils in the west and north of Britain, whitebeams have evolved, leading to areas like the Bristol and Avon gorges becoming home to collections of rare endemic trees. Around the coast, stunted Hazel also grows on calcareous soils, which can be covered in rare mosses and lichens.

Upland scrub communities
Upland scrub occurs in mountainous areas and in northern coastal Scotland. There are five main types:

Wet upland scrub
In this scrub habitat the lowland willow species are replaced by other willows with more northerly distributions, such as Bay Willow, Dark-leaved Willow, Eared Willow and Tea-leaved Willow.

Dry upland scrub
Dry upland scrub is dominated by Hawthorn with Blackthorn, Crab Apple, Grey Willow, Hazel and Rowan. There can also be Aspen, birches, oaks, Rowan and Scots Pine. Often stunted by the soil and weather, the trees can be effectively miniaturised by the conditions, even when they are several centuries old.

Upland Juniper scrub
Juniper scrub occurs on both calcareous and acid upland soils, and at elevations of up to 650 m, where it can grow with Downy Birch. On lower slopes it also provides cover for the development of birches, oaks and Scots Pine woodland.

Sub-arctic willow scrub on the slopes of Ben Lawers in Scotland (see *p. 28*).

TREE HABITATS

Upland Juniper scrub

Dwarf Birch scrub

Dwarf Birch scrub
Dwarf Birch scrub grows on bogs and on upland heaths. Highly susceptible to grazing, this form of scrub is a national rarity and grows mostly in the Scottish Highlands.

Sub-arctic willow scrub
In one of Britain & Ireland's rarest habitats, Sub-arctic willow scrub grows on wet alpine areas, where grazing pressure is low. Downy Willow is the most widespread species, with Mountain, Woolly and Whortle-leaved Willow. Widely overgrazed, few areas still exist, with only five sites having more than 100 Woolly Willow plants and a total estimated population of only 2,100 plants of this species across Scotland.

Hedgerows
Hedgerows are effectively managed forms of scrub, designed to deliver specific benefits to the land manager – keeping animals in a field, for example, or providing timber or food close to home. They are collectively one of the largest treescapes in our islands.

A healthy elm hedge in Essex.

Hedges and their trees are extremely important places for wildlife because they offer a wide diversity of ecological niches in a relatively small area. Hedges can provide shelter, breeding opportunities, nesting sites, song posts, hiding places and ecologically friendly links between habitats.

Many of the trees in our hedges are mature, veteran or ancient and they contribute significantly to the landscape character of an area. In the past these hedgerow trees were an extremely valuable sustainable source of wood fuel through pollarding (see *p. 22*), and where they are found, they represent an important part of our living cultural heritage, *e.g.* the landscape of south Norfolk.

The commonest woody species in hedgerows are Hawthorn, which appear in 90% of Britain's hedges, with Blackthorn and Elder occurring in 47% and 35% of hedges, respectively. As full trees growing above the hedgerow, overall, Ash is the commonest tree with oaks and Field Maple next. However, Hazel also occurs in more than 25% of hedges in England and nearly 75% of those in Wales. Rowan was commonest in Welsh hedges, with Holly the remaining relatively abundant tree.

In Ireland, Hawthorn and Blackthorn are common, with Gorse, Holly, Dog-rose and brambles also abundant. Ash and Beech are widespread as hedgerow trees along with Hazel, Elder, elms and willows.

A wide range of other woody species occur in hedges, including (in decreasing order of abundance) Grey Willow, Dogwood, Goat Willow, Crab Apple, Wild Cherry, Alder, Midland Hawthorn, Buckthorn, Spindle, Bird Cherry, Hornbeam, Wayfaring-tree, Guelder-rose, Yew, Aspen, Eared Willow, Wild Service-tree, Large-leaved Lime, Alder Buckthorn, Common Whitebeam and Small-leaved Lime.

There are also a series of local variations in hedge types. This has led to a significant number of regionally distinctive hedge types, including:

Elm hedges: Usually single species or dominated by suckering shoots of *Ulmus* species, particularly English Elm.

Holly hedges: Commonly as 5–10 metre sections, and also whole field or medieval deer park boundaries as the sole dominant species.

Beech hedges: Often on banks, either close clipped or now grown into a line of mature trees. Pioneered as a hedge shrub during the late Enclosure period (see *p. 41*).

Hazel-dominated hedges: Forming small field patterns at forest margins.

Fruit hedges: Dominated by Damson, Cherry Plum and Blackthorn, Wild Plum, Crab Apple and other *Prunus* species.

Windswept hedges composed of various species: Usually Gorse, Hawthorn, Blackthorn or elms.

Tamarisk hedges: A non-native shrub planted because of its salt tolerance.

Pine hedges: Planted and managed as hedges during the Enclosure period then left unmanaged, resulting in contorted shapes.

Fuchsia hedges: Fuchsia, an introduced shrub, is a common component of hedgerows in parts of the south and west of Ireland.

Pine hedges – a feature of Norfolk and Suffolk's Breckland.

Fuchsia hedges, Co. Kerry, Ireland.

Orchards, Wood Pasture and Parkland

These three treescapes are ecologically and structurally very similar: open-grown trees in grasslands. The differences are the mixture of tree species, and the density of the trees.

Traditional Orchards

Orchards are treescapes composed primarily of apples, cherries, pears, plums and nuts (hazelnuts and walnuts). One important feature of traditional orchards is the wide range of fruit varieties they can contain, for example, 101 varieties of Perry Pears were found in Gloucestershire orchards.

A traditional orchard in Herefordshire

These traditional orchards are a widely distributed and often long-established habitat which is vital for local biodiversity and landscape character. The trees are often densely planted – the spacing of trees in traditional orchards ranging from one tree every 3 m to one every 20 m. They are also usually in relatively small areas, with the trees being managed to maximise fruit production. This was achieved usually by grafting specific fruit varieties onto rootstocks and then managing the tree's shape to increase flowering and therefore fruit production.

The minimum size of a traditional orchard is defined as five trees with crown edges less than 20 m apart.

Grassland is also an integral part of a traditional orchard, as is scrub usually in the form of hedgerows which border the orchard. The hedges play a vital ecological role, providing shelter and breeding sites for pollinators like bumble bees. A healthy traditional orchard will have trees of differing ages and a wide range of veteran and ancient characteristics in some of the trees (see *p. 48*), including dead wood, which boosts their value to wildlife.

Traditional orchards are found throughout Britain & Ireland, with the majority found in England, but they are now one of the rarer treescapes.

Wood Pasture and Parkland

Like orchards, wood pasture systems are a mixture of grassland and trees. Wood pasture historically occurred naturally along the edges of woodlands and forest, where the grazing wild animals strayed from the darker forests to graze on the open grassland. This can still be seen occurring naturally in Romania, where the Wild Boar, deer and even bears come to feed on the grass and fruit from the trees and shrubs around the forest edge. Sheep and cows also fed in this transition zone from grassland to forest, and the trees provided the land managers with wood fuel, especially if pollarded at a height which kept the regrowth of the tree away from the browsing animals.

As wood pasture is a complex mosaic, this allowed multiple land uses in a single site and created a rich and varied habitat with great biodiversity importance. Where they remain in Britain & Ireland, the remnants of these wood pasture systems are still vital to our wildlife and landscape, often housing many ancient trees and their associated wildlife. They also became formalised and celebrated for their ancient trees as 'Parkland', especially by some of the great landscape designers like Humphry Repton and 'Capability' Brown in the 18th and early 19th centuries (see *p. 41*).

The value of these wood pasture systems lies in the fact that the trees grow in their natural open grown shape (see *p. 18*) until they are pollarded. The pollarding then creates new ecological niches and prolongs the life of the trees, creating dead wood which boosts biodiversity (see *p. 45*). This creates opportunities for species such as fungi, not only in the dead and decaying wood, but also around the tree roots, often in mutually beneficial partnerships with the trees (known as mycorrhizal relationships). In the open grassland, fungi including waxcaps (the 'colourful orchids' of the fungal world) also develop. Other wildlife that benefit from these wood pasture systems are specialist insects on dead wood known as saproxylic invertebrates (see *p. 49*), and those that use highly specialist niches, for example, water-filled holes or sap runs. This diversity of niches leads to birds and bats finding food and breeding opportunities, increasing the importance of the habitat.

From the 19th century onwards, trees were planted into these landscapes, including some non-native species like the Cedar of Lebanon (see *p. 108*) which are themselves becoming rare in their native habitats.

Some of our wood pasture systems make for our finest treescapes – such as the New Forest and Epping Forest – whilst others may only be small remnants comprising a few veteran or ancient oaks. However, many of the wood pasture systems that still exist are of international importance for their biodiversity, landscape or history, such as Windsor Great Park, the Hamilton High Parks, South Lanarkshire and Crom, Northern Ireland.

Dinefwr Park in Carmarthenshire is a stunning example of this habitat, with its White Cattle and deer mingling amongst amazing open grown trees.

Estimates of wood pasture systems are difficult to obtain, but it is thought that very little of the original wood pasture systems are left across these islands.

Woodland

Left to develop naturally from scrubland, or planted deliberately for timber or other uses, woodland is our last major natural treescape.

As with all our treescapes, there are many complications in defining our woodlands. Whole books have been written on the types of woodland habitats in Britain & Ireland, and it is difficult in a few paragraphs to explain many of the complexities, but two of the key factors used to categorise our woodlands are (1) the length of time the area has been wooded, and (2) the species that are the dominant trees present.

Wood-pasture and parkland – one of Britain's most beautiful treescapes as the open-grown trees develop their full canopy.

TREE HABITATS

Ancient Woodland
In Britain & Ireland, the oldest and most important woodlands are described as Ancient Woodlands. These woodlands are defined as an 'irreplaceable habitat', places which are important natural assets for (amongst other things) their wildlife, soils, carbon capture and storage abilities, plus their contribution to the genetic diversity of our trees. They are also called 'ancient semi-natural woodland', reflecting the fact that these woodlands have usually been managed by humans, altering their natural ecology.

In the 1980s, when determining what might be considered 'ancient woodland', a decision was made in England to use the date of 1600, because detailed estate maps (and other sources) began to appear in the 16th and 17th centuries, and there was a practical need to set a cut-off point, to draw the distinction between the longest-lived woodlands and those planted more recently. Therefore, if a woodland was on the first maps, it was an 'ancient woodland', and these are deemed our most important woodlands. The differences in defining ancient woodland dates – 1650 in Ireland, 1750 in Scotland and 1830 in Wales – represent the availability of good maps in those countries.

Whatever the date, what these 'ancient woodlands' represent are places that were woodland when the first maps were drawn up and the wooded status of which may have existed for decades, centuries or potentially even millennia before the maps were created. These woodlands are the 'crown jewels' of our woodland heritage.

Secondary Woodland
Woodlands that have been created on land that was at some point cleared of trees (and therefore not shown as woodland on the first maps) can still have important assemblages of plants and animals, particularly if they are adjacent to an existing ancient wood. Those that have developed naturally are termed 'secondary' woodlands, to distinguish from 'planted' woodlands (*below*).

Planted Woodland
Created deliberately to increase the tree cover of an area, these may use native species (both conifers and broadleaves) or non-native species such as Corsican Pine or Douglas Fir. Planted non-native coniferous woodlands are often referred to as plantations.

Ancient Woodland in Oxfordshire with mature oaks, and Bluebells carpeting the woodland floor.

Types of woodland

Woodlands have different species compositions; the dominant native trees in a woodland are generally determined by the soil type, the elevation and the availability of water.

The two major woodland types in Britain & Ireland are **Wet Woodland** and **Dry-land Woodland**. The latter is subdivided into Lowland Mixed Deciduous Woodlands, Lowland Beech and Yew Woodlands, Native Pine Woodlands, Upland Mixed Ashwoods, Upland Birchwoods, and Upland Oakwoods. In addition to these, woodlands along the Atlantic coast have also recently been dubbed the **Atlantic Rainforest**.

Wet Woodland consisting of Alder.

Wet Woodland

Wet woodlands are found on floodplains, along streams, in peaty hollows and as a stage in the development of fens, mires and bogs. The main tree species are usually Alder, birch and willows, but there can be Ash, oak, pine and Beech. Wet woodlands can develop alone, or as the wetter edges of other woodland types, *e.g.* oak woodland.

Alder woodland often has a long management history, usually of coppicing, and some of the best wet woodlands are found in the East Anglian Broads and in Cheshire and Shropshire. Wet woodlands along riverside floodplains have become rare, but the New Forest still has some excellent examples.

In Ireland, Almond Willow woodland grows on river islands and banks of lowland rivers. Species-rich woodland dominated by Alder, willow (mostly Grey Willow) and Ash also occur on wet, fertile soils. Birch and occasionally Hawthorn usually dominate the shrub layer.

Dry-land Woodland

Native Pine Woodlands

Native Scots Pine forest used to cover large areas of Scotland but now cover only 1% of the area where they once occurred. Sizeable areas, however, still occur mainly in the Grampians and the Highlands of Scotland. They grow on poor soils and are not as rich in biodiversity as other forms of woodland but do support some specific rare and unusual species like Scottish Crossbill and Twinflower. Often other tree species can be found growing in these pinewoods, including Alder, birches, Bird Cherry, Rowan and various willows, with an understory of Aspen, Hazel, Holly and Juniper.

Upland Birchwoods

In Scotland, some upland areas are dominated by woodlands of Downy and/or Silver Birch. There can also be Rowan (when the soil is acidic), willows, Juniper and Aspen. At higher elevations Downy Birches can be of the subspecies *tortuosa* (also known as Mountain Birch) (see *p. 270*) with occasional Dwarf Birch.

In Ireland, birch woodlands develop on raised bogs and in peaty hollows and on mineral soils. Downy Birch is usually the dominant species, with Holly, oaks, Rowan, Scots Pine and willows.

Upland and Lowland Mixed Ashwoods

This type of woodland occurs on alkaline soils in the north and west of the country, where Ash is the major species, but where oaks, birches, elms, Small-leaved Lime and Hazel are often part

TREE HABITATS

Yew Woodland

Native Pine Woodland

Upland Birchwood

Mixed Ashwood

Upland Oakwood

Lowland Beech Woodland

of the woodland structure. Although the name suggests this type of woodland is at higher elevations, it can also be found down to sea level.

Good examples of this type of woodland can be found in the Derbyshire Dales, although with the arrival of a fungal disease called Ash Dieback (which is killing many Ash trees), these woodlands have been changing rapidly over the last 10 years.

Historically, much of the woodland was coppiced for wood fuel, which helped to create a wildlife-rich habitat that is now threatened by the decline of Ash. Bluebells and Wild Garlic are common in Ash woods in the south of England, and the rare Large-leaved Lime tree can occasionally be found growing in the same places. With Ash bark often covered in lichens, the loss of these trees and the decline of these woodlands will have a significant impact on biodiversity in many areas of the country.

In Ireland, Ash and Hazel woods grow on dry, lime-rich soils in the lowlands. Ash and occasionally Pedunculate Oak form the canopy, with Hazel dominating the shrub layer combined with Blackthorn, Hawthorn and Spindle.

Upland Oakwood

Another upland woodland type is the upland oakwood, which is often mainly Sessile Oak, although Pedunculate Oak can occur. Birch, Holly, Rowan and Hazel often grow as the understory in these woods, and there are usually abundant mosses, ferns and lichens. The more open structure of the Sessile Oak woodland also supports bird like Redstart and Pied Flycatcher. Good examples of upland oakwoods can be found in Argyll, Cornwall, Cumbria, Devon and Gwynedd.

Lowland Mixed Deciduous Woodland

This woodland type grows on a wide range of soils and includes most of the semi-natural woodland in southern and eastern England, lowland Wales and Scotland. The various sites differ in species composition, with either or both oaks and a wide mixture of any other native tree species,

Lowland Mixed Deciduous Woodland

depending on location and woodland history. Other main components of the woodland can be Field Elms, Field Maple, Hornbeam and Small-leaved Lime.

In Ireland, Sessile Oak dominates the woodland canopy, with Holly, Downy Birch and Rowan usually forming the shrub layer. On more lime-rich soils, Ash and Hazel grow with Bluebells, and if the humidity is high, then lichens and ferns often grow in abundance on the trees. These woodlands are found particularly in the south and south-west, as well as in Connemara, Co. Galway, and in Cos. Wicklow, Mayo, Donegal and elsewhere.

Lowland Beech and Yew Woodland

Lowland Beech and Yew woodlands occur on a range of soils, although they are widely found on calcareous (chalk and limestone) soils. They can occur separately – Beech woodlands or Yew woodlands – or as a combination of both species. Also present can be Ash, Sycamore and Common Whitebeam, with oaks especially on less calcareous soils. On more acidic soils, Beech woodland can develop with a Holly understory and the occasional Yew. Good examples of this type of woodland can be found in the High and Low Wealds of Kent and Sussex, the New Forest, the Cotswolds and the Wye Valley.

In Ireland, Yew woods develop often with a poorly developed shrub and herb flora, but they often develop a luxuriant moss layer. They can be found on limestone soils in the south-west.

Atlantic (Temperate) Rainforest

Over recent years the ancient Ash, oak, birch, pine and Hazel woodlands which grow along the Atlantic coast of Britain & Ireland have been recognized as a Temperate Rainforest. Here in the wet, often humid and mild climate, carpets of moss and lichens have grown on our coastal trees, especially those in steep valleys and river corridors where the humidity is increased. Unfortunately, the extent of this habitat has reduced, with our remaining fragments found mostly in the south-west of England, in Wales and Northern Ireland, along the west coast of Scotland and in Cork, Clare, Donegal, Galway, Kerry and Mayo in Ireland.

Atlantic Rainforest – dripping in mosses and lichens – these wet woodlands of the Atlantic coasts are extremely special places.

History of our treescapes and habitats

In Britain there are an estimated 3 billion trees in total, with an estimated 123 million trees outside woodland. The non-woodland trees can be found as in-field trees especially in parklands and wood pastures and in our extensive hedgerow network, plus many trees in our parks, gardens and streets. Our woodlands cover nearly 13·2% of Britain, and our trees outside woodlands have recently been estimated to account for another 3·2% of Britain's landmass.

By comparison, in Ireland woodland cover had fallen to below 20% of the country by the 16th century, but over the next 300 years this percentage dropped even further, heading towards only 1% of the country being wooded by 1900. However, over the last 30 years, significant tree planting grants to farmers have significantly increased Ireland's tree cover, which currently stands around 11%, with an aspiration of 18% over the next few decades.

By comparison, 66% of Finland is wooded, 37% of Spain and 32% of France, Germany and Italy. Indeed, in Europe, only Malta and Iceland have less woodland cover than found in Britain & Ireland.

From the mountains of Scotland to the marshes of the English south coast, all the way to the lakes and mountains of County Kerry, the treescape of Britain & Ireland is rich and diverse and controlled by a mixture of landform, local biodiversity, land use and human history. Even within small areas huge changes can be seen, for example in coastal Hampshire, a 40-mile drive to the north travels from salty coastal marshes where Blackthorn, Hawthorn and elms dominate, through chalk downland of Ash, Yew and Common Whitebeam, to clay vales where oaks dominate, finally reaching birch-dominated heathlands.

History of the treescape

Our varied treescape resulted from the history of these islands. At the end of the last ice age, about 15,000 years ago, the landscape of Britain & Ireland was largely covered in snow and ice, and only a few alpine plants lived in the cold summers. As the snow and ice retreated, the land was colonized by plants, creating a landscape like the arctic tundra of modern-day Greenland. At that time (about 9,000 years ago), Britain was still connected to mainland Europe and to Ireland, allowing the free movement of pioneering plants and animals, including humans, over the whole continent.

As the climate warmed, the first tree and shrub species began to arrive. These included birches, Hazel, willows and Juniper, which arrived first. Pollen analysis shows that the early trees arriving back led eventually to the creation of a well-wooded landscape, often referred to as the 'Wildwood'. Research by Frans Vera in the Netherlands has suggested that at this time Europe's large herbivores opened – or kept open – grassy patches in forests and it is now thought that the Wild Ox or Auroch, (the native ancestor of domestic cattle), along with Elk, the Irish Elk which became extinct about 10,000 years ago, Beaver and Wild Boar influenced the structure and ecology of these original wild forests. The woods and forests consisted of a variety of species, including Hazel, oaks, elms and Ash, with alders and willows in the wetter areas and Scots Pine in Scotland and around the south-west coasts of Kerry.

Between 8,000 and 7,500 years ago, the sea level rose sufficiently to sever the land bridge between Britain, Ireland and mainland Europe. The countryside would have become largely wooded, with the clear areas being a result of poor soils, coastal location, animal clearings, bogs and open water. The early human settlers of these islands at this time were thought to be largely coastal dwellers and beachcombers,

Once extinct, Wild Boar again roam wild in Britain.

who maintained a 'wandering' lifestyle, making a living wherever they could find food.

However, evidence from Portland in Dorset, of an 8,000-year-old midden (a rubbish dump consisting of limpets and other molluscs – all in a matrix of clayey loam mixed with charcoal), hearths and a cooking/storing pit raises the possibility that even these early people led a settled lifestyle, possibly all year-round. A limited analysis of charcoal pieces from this midden has produced the remains of Wild Cherry, Crab Apple, pear and Hazel nuts, showing that these early people had already begun to use Britain's tree products.

Zennor, Cornwall – ancient field systems from the Bronze Age.

During the Neolithic period (6,000–4,000 years ago), arriving settlers brought with them crops and animals and began farming Britain. They used a slash and burn system, killing the trees by ring barking (removing a strip of bark from around a tree) and then burning the land. They converted tracts of the countryside to farmland and established settled farmsteads by clearing land with stone tools. In Ireland, this change of land use coincided with a change in the climate, which became much wetter, and there was a build-up of thick peat and blanket bogs, which spread to cover large areas of Ireland. This process of burning the trees, followed by the build-up of peat, explains the regular discovery of dead tree stumps in the bogs.

Between 4,000 and 3,000 years ago, Bronze Age systems of farming became well developed. The wildwood had been progressively felled and probably covered only about 50% of the country by this time. Complex field systems were created in some parts of the country, and the Land's End peninsula still provides a clear image of this early countryside. At Zennor, for example, small irregular fields have retained their original prehistoric shape and boundaries, with Cornish 'hedges' comprising massive granite block walls and earth banks, made from rocks cleared from the fields. In Ireland during this period, the percentage of elm in the woodlands decreased (as it did across the whole of Northern Europe) while the Ash and oak percentage increased.

Whilst the Zennor and Dartmoor early field systems have survived intact, the evidence that living hedges existed around early fields is obviously harder to obtain. Nevertheless, in Suffolk, Norfolk, Essex, Kent and Hertfordshire, hedged, often parallel-sided, ancient field systems are still apparent, and the boundaries of these fields may have remained largely unchanged for millennia.

Unfortunately, much of this period of our landscape history inevitably remains vague as the small human population, and absence of records leaves us with only tantalising glimpses into the past. However, it is amazing that in certain parts of our landscape, the woodlands, field systems and boundaries established more than 6,000 years ago remain and influence the nature of the modern countryside.

Roman trees and hedges

Soon after the Romans arrived in Britain, they brought some of their most useful trees and other plants to support their existence in these cold northern climes. They are credited with the introduction of Sweet Chestnut – valued for its nutritious nuts and useful coppice wood – and Walnut, which was eaten and often pressed for its oil. They also introduced Stone Pine *Pinus pinea*, presumably for pine nuts with their salads.

The Romans built fine villas, and to augment their grand designs they planted formal gardens to remind them of home. There are records of incredible Box topiary in the gardens of Rome and, because Box was a readily available native tree in Britain, it is reasonable to suggest that British villas were similarly adorned.

The reconstructed Roman gardens at Fishbourne Palace in Sussex show the use of Box, elm pergolas for vines, and trellises for fruit trees.

The Romans also brought orchard culture to these shores. They had discovered wonderful sources of fine fruit tree varieties across the Caucasus. They planted groves of apples, pears (Pliny the Elder, writing in the first century, knew of 41 varieties) and plums. The seeds of Britain's great gardening culture had been sown.

With their innovative ideas the Romans improved ploughs and scythes, thereby raising their productivity and so helping to feed the increasing population, some of whom began to move to the new towns. By this time the woodlands had been reduced to an estimated national cover of approximately 11% (similar to today) with little remaining wildwood. The residual woodland was managed, especially through coppicing, with hedges and fences enclosing fields close to settlements and farmsteads.

By AD 400 there was a significant urban population engaged in trades and crafts and no longer directly involved in agriculture. This period also saw the extensive development of Roman villas, properties owned by local magnates and often surrounded by thousands of acres of land. As a result of these changes there was a corresponding increase in new hedges and fences to enclose larger rectangular fields or to mark boundaries, some of which can still be seen in south-east Essex.

Anglo Saxon and Medieval times

During the Saxon period (5th–11th centuries) many of the changes created by the Romans were reversed, with town dwellers moving back to the countryside and re-adopting subsistence farming methods. This coincided with a large-scale depopulation of Britain because of war, disease and emigration. The effect of the population decline was that areas of the countryside reverted to woodland and scrub.

In Ireland the arrival of the Normans in the early 12th century led to large scale clearance of forests, but between 1315 and 1350, Britain & Ireland were struck by famine and the plague, which reduced the population by 50%, resulting in abandonment of agriculture from many marginal farming lands. Within decades, scrub replaced previously cultivated and grazed land and new woodlands formed, many of which remain today as ancient semi-natural woodland.

By 1400 much of the population was still engaged in subsistence agriculture, which occupied nearly all usable land. Through the 15th and 16th centuries, rising demand for firewood and charcoal placed increasing pressure on woodlands, although their area remained broadly stable at around 10%. This period also saw the beginnings of the practice of enclosure that would continue through several centuries, creating tens of thousands of miles of new hedge. As the total length of hedge continued to grow, the number of hedge trees also began to increase as a source of wood fuel.

An early painting of the English treescape at Henley-on-Thames, Oxfordshire by Jan Siberechts (1627–c.1703).

The 16th and 17th centuries

Prior to 1500, little information was available about tree planting or selection of species, but after the Roman withdrawal, very few new trees arrived in Britain & Ireland. It is thought that the Sycamore, an excellent colonizer, arrived from northern Europe sometime after the late 13th century, although no specific date can be identified. Equally, the arrival of the Horse Chestnut from Turkey around 1600 was the beginning of an entertaining relationship with a characterful tree.

It was not until 1523, when Fitzherbert published his *Book of Husbandry*, that a written work explained exactly how people might plant and manage trees. Further valuable plant information became accessible in the early English herbals. William Turner published his illustrated *Herbal* in three parts between 1551 and 1568. This was followed in 1597 by John Gerard's *Herball*.

In the following century John Parkinson published his *Paradisi in sole Paradisus terrestris* in 1629. He was an enthusiastic gardener and plant collector who was keen to pass on his wealth of experience and knowledge. However, it was with the publication of John Evelyn's *Sylva – A Discourse on Forest Trees* in 1662 that a dedicated work for tree planters became available. Evelyn went into the minutest detail of how to nurture various species successfully and, with the commercial forester in mind, set forth the virtues of the timber and expectations of their harvest potential. So comprehensive was this work that it went to several editions for more than a century.

By the 17th century in Ireland, woodland cover had dropped significantly to somewhere between about 12% and 2% (the estimates differing as the real data are poor), because of trees being felled for timber which was often exported to Britain for ships, house building and barrel making. The remaining woodland was often coppiced to produce charcoal for smelting iron ore.

Introduction of the London Plane

Meanwhile the plant hunters had begun to visit distant shores, returning with all manner of trees and plants. In the mid-16th century, the Eastern Plane arrived from the Caucasus and began to grow alongside the Western Plane, one of the earliest arrivals from the Americas in the 1620s. The result was a hybrid between the two – named the London Plane.

The first hybrid is generally thought to date from 1663 in Britain, and it is highly likely that the cross occurred in the Lambeth Garden of the famous 17th-century gardener and plant hunter John Tradescant, as it is known that both species were growing there. It is now estimated that around 50% of trees in the streets and squares of Britain's capital are London Planes, the oldest, in Berkeley Square, being more than 200 years old.

In 1638 the first Cedar of Lebanon seed was brought to Britain from Syria by Dr Edward Pococke, and tradition has it that he planted one in his rectory garden at Childrey in Oxfordshire in 1646. Stately homes, country mansions and municipal parks then spread them around the country.

Throughout the 17th century, new trees continued to cross the Atlantic from eastern and central North America. One of the most handsome was the Tulip-tree, which was planted as an amenity tree with striking golden yellow autumnal colour.

In the late 17th and early 18th centuries the influence of Dutch garden design during the reign of William and Mary brought a renewed passion for topiary, which had fallen out of fashion. John Evelyn, in 1662, promoted Yew for topiary as an alternative to the benchmark species, Box, and in 1694 a grand Yew design was created at Levens Hall in Cumbria, which remains the finest topiary garden in Britain.

In 1638, the first Cedar of Lebanon seed was brought to Britain. This tree was planted by Dr Edward Pococke in Oxfordshire from those first seeds.

The 18th and 19th centuries

Almost all the maps of the 17th century show parishes across England having some length of hedge. By the 18th century, reforms of agricultural practices paved the way for the Industrial Revolution and had two major effects: i) increasing the output of farms and ii) reducing the number of people who needed to work on the land. This resulted in a movement of people from the countryside to the towns, where there was work available in the emerging industries.

Enclosure Acts

To allow farmers to farm in the new ways, they needed large, consolidated plots of land instead of the scattered 'strip' system that had previously been practised in the open field system. As a result, Parliament enacted a series of laws (mostly between 1760 and 1830) called the Enclosure Acts, to consolidate land holdings and to specify that the enclosed land was surrounded by hedges and ditches.

The more than 5,000 separate Enclosure Acts enclosed more than 7 million acres of open fields and common land. Oliver Rackham estimated that more than 200,000 miles of new hedge were planted between 1750 and 1850. The percentage of open field land enclosed varied considerably across the country, which suited wealthy landowners with sporting interests. As examples, in Northants, 51% of open fields were enclosed, whereas in Shropshire this figure was approximately 7%.

The speed and scale of hedge establishment in this period led to the creation of large commercial nurseries specialising in the supply of hedging material, mostly Hawthorn, to create the new network of hedges.

The spread of Enclosure, particularly in the open field systems of the Midlands, led to the creation of many new, often very straight, Hawthorn hedges which divided up the countryside geometrically. These hedges were 'quickset', a word which indicates both the hedge itself and the act of planting it. They cut across the open field ridge and furrow system, and many of the fields that had previously been ploughed were converted to arable pasture. As the landscape changed, foxhunting began to spread, especially in the Midland areas of Rutland, Leics, Northants, and Buckinghamshire. This had a knock-on effect on hedge management, as dense low growing 'walls' of vegetation were desirable as jumps for riders. These tidy, well-managed hedge 'jumps' are still a feature of many hedges today.

Exotic importations

The 18th and 19th centuries also saw great collaborations of plant hunters, botanists, nurseries and private collectors. The huge array of trees on the west coast of North America still lay untapped until the remarkable efforts of David Douglas brought two of Britain's prime timber trees – the Douglas Fir and the Sitka Spruce – as well as several other maples, firs and pines.

In Ireland, during the 18th century, large houses were built, and their grounds were often planted with these new tree species, and as in Britain, Douglas Fir and Sitka Spruce were used, as were Lodgepole Pine and Western Red-cedar.

Archibald Menzies had brought the first Monkey-puzzles from Chile in 1795, but it was William Lobb who, in 1841, reintroduced the tree, and Veitch's nursery of Exeter popularised it. The Monkey-puzzle and the Giant Redwood (introduced by Lobb in 1853) became two of the most sought-after specimen trees for impressive centrepieces of many formal gardens.

The author with one of the first Giant Redwoods planted in Great Britain at Errol, Perthshire.

In the midst of all these exotic importations, the great landscapers, such as Charles Bridgeman, William Kent and Capability Brown, were busily transforming parks and gardens into vast rolling landscapes employing man-made 'natural' hills and vales, meandering rivers and tastefully distributed spreading specimen trees or picturesque clumps. Naturally occurring native species were the order of the day, and the new trees were virtually ignored for this manufactured countryside. A theory suggests that many of the larger specimen trees of parkland were hedgerow trees which were retained for the new schemes, when all signs of hedged farmland were obliterated. The linear alignments of many of these trees today may bear out this theory.

The 20th century

From 1870 until 1940, agriculture hit a period of recession, as America began to flood European markets with a glut of cheap grain. Farming neglected the poorer land, and the countryside reverted to a wilder state than it had been for many centuries. Hedges were managed less frequently, and there was a significant emergence of hedge trees as hedge management declined.

In Ireland by the end of the 19th century, more woodland clearance occurred, and by 1918, woodland cover was as low as 1% of Ireland's land mass. This started to be reversed following the Irish 1946 Forestry Act.

In Britain, from 1940 to the end of the 1990s, agriculture fortunes changed again with a period of expansion, enhanced by entry into the EEC in 1973. Arable farming and its intensification led to the development of larger machinery, making much bigger fields an economic necessity. As a result, hedges suffered, and estimates suggest that between 1946 and 1970 some 4,500 miles of hedge were removed every year.

During the 1960s and 1970s, there was a resurgence of Dutch elm disease, and, due to the importation of a more virulent strain from North America, most of the countryside's significant elm trees in hedges were lost.

In Ireland by 1951, woodland cover in the Republic of Ireland had grown to 1·8% and consisted mostly of exotic conifers. To improve the tree cover, in 1960 a national planting target was introduced which had pushed the percentage woodland cover up to 4·8% by 1983.

Trees in the 21st century

Climate change is bringing global temperature increases resulting in extreme weather events including drought, flooding and storms which plants will have to endure on a more regular basis in coming years. Climate modelling research by the Royal Botanic Gardens, Kew, suggests that London's climate by 2050 may be comparable to present-day Barcelona.

The first instinct is to see how our native trees will fare. Beech and birch both suffer in droughts, as does the naturalized Sycamore. Beech, being shallow-rooted and often growing on light soils, will probably become stressed in the southern counties. Sycamore, an important tree of northern Britain, might also struggle in the south. If the cover of some of our native species shifts to more northern climes it will be interesting to see what happens in the spaces they vacate. A mixture of natural succession and intervention planting to adapt to a changing climate may well create whole new treescapes. This will be at its most marked in towns and cities, where treescapes are needed that provide valuable cooling and shading benefits to cope with the changing climate.

There are numerous other concerns for future trees which affect commercial foresters and the conservators of ancient woodland, drawing together potential problems such as plant competition and an increased range of insect pests and pathogens. We are already aware of the predations of longhorn beetles, numerous fungi and *Phytophthora* water moulds which may be poised to tip some of our tree species into similar declines reminiscent of that seen from Dutch elm disease, and as is happening with Ash dieback.

However, there will be some positive aspects to the changes, and one, concerning our native Small-leaved Lime, has already been observed. Hotter summers are causing the species to set viable seed more often, which might mean limes will extend their range after several hundred years of standstill.

Current state of our treescape

Currently Britain & Ireland's treescape is expanding. In Great Britain in 2024, 13% of the total land area in the UK is woodland, 19% in Scotland, 15% in Wales, 10% in England, and 9% in Northern Ireland. In the Republic of Ireland, the current woodland cover is 11%. Recent data also show that in Britain, non-woodland trees make up another 3·2% of our landmass, although there is no equivalent number for Ireland.

Despite being significantly lower than many European countries, in Britain & Ireland these are the best tree cover numbers for many decades, and the current targets are to expand our tree cover further. Along with our wide range of native trees, over the past 500 years, trees from all over the globe have been brought into our islands, giving us a wealth of commercial timber trees and a startling collection of beautiful amenity trees.

Now there is little excuse not to plant trees. However, we also must remember the old maxim that 'We need to plant the right tree, in the right place' – and now add the extra thought – 'for the right reason'. If we do this, using a suitable palette of native and introduced species, the opportunities for tree planting have never been better.

Our treescape consists of a wide variety of woody habitats, from orchards to hedges, copses, scrub and woodland plus many more– some of which are on view here in Northern Ireland.

Trees and wildlife

Introduction

The biological value of trees and tree habitats to our wildlife is enormous. However, one of the most widely asked questions is, which of the native species is the most valuable for wildlife?

Answering this question is surprisingly difficult. In 1961, Professor Southwood compared the numbers of insect species feeding on the foliage of various trees and shrubs and in 1984 updated the paper to include newly available data.

These papers showed that oaks, willows and birches supported the greatest number of plant-eating insects and mites: five native willows supporting 450 other species, two native oaks supporting 423 species, and two native birches supporting 334 species.

At the other end of the spectrum, he showed that the lowest diversity of insects was found on Juniper (32 species), Holly (10), and Yew (6).

Southwood showed that the factors which impacted the number of insects feeding on a tree were:
- the abundance and distribution of the tree or shrub
- the length of time the species had been in Britain & Ireland
- whether the leaves were evergreen.

However, leaf-feeding insects are only one part of the species mix that uses our native trees, as there are many other organisms associated with trees, including:
- birds and animals that feed on the trees (including fruits and berries) or use them for breeding
- insects that feed on dead wood, pollen, nectar, fruits and seeds
- invertebrates which live on decomposing fallen leaves and organisms like nematodes and bacteria that live in the soil around the roots
- fungi that live on tree roots (mycorrhizal communities) or within the structure of the tree, like those that decompose wood
- lichens, mosses and liverworts that live on the bark and wood of the tree.

When the full range of species above is studied, the increase in species can be significant. For example, with the two native oaks, it was found that in fact more than 2,300 species of mammals, birds, invertebrates, fungi and mosses used oaks (as opposed to the 423 insects that feed on the leaves).

Of these, 326 species have come to rely solely on oak, including moths such as the Dark Crimson Underwing.

Unfortunately, the full range of species using our other tree and shrub species has not been studied to the same extent, so it is therefore very difficult to decide which tree species are 'best for wildlife'.

Dark Crimson Underwing

Oak Eggar

although it is obvious that some tree species support a wider range than others. However, it is the diversity of tree species and tree habitats that is often more important than the tree species alone.

The importance of dead and rotting wood

Another important factor which influences the number of species using any individual tree is the age of the tree, and especially the presence or absence of dead wood. Rotting wood in living and standing dead trees is especially important for providing habitats for species, such as the Lesser Stag-beetle. These specialist species which depend on dead and decaying wood are known as saproxylic invertebrates, and more than 650 British beetle species use decaying wood, especially in large old native broadleaved trees in wood pasture and parkland (see *p. 30*).

Violet Click Beetle

So specialised are some of these species, they even have preferences on the specific type of dead wood they need; from fallen branches to standing dead trees, the location and aspect of dead wood matters. The species of tree is also important, as for example the Violet Click Beetle lives only in ancient Ash or Beech, whilst the Variable Chafer survives only in dead wood in ancient oaks, Beech and Sweet Chestnut.

Variable Chafer

Holes in rotting trees provide nesting and sheltering opportunities for birds, including some rare and declining species such as the Tree Sparrow. Certain bats also use holes in trees, and two rare species (Barbastelle and Bechstein's Bat) are associated with ancient trees.

Unfortunately, our treescapes now have a serious lack of dead wood, as it is often 'tidied up' by overzealous tree managers, which means

Tree Sparrow

that many of the species associated with this invaluable habitat have become rare. This makes unique places like the New Forest in Hampshire vital, as here dead wood is still relatively abundant due to policies which leave dead and fallen trees in situ, wherever possible.

Treescapes and biodiversity

Within Britain & Ireland, as well as the species of tree and the presence of dead wood, the environment created by the treescape also significantly impacts species diversity. One of the most important factors which increases the number of species is the availability of light, with the edges of woodlands usually supporting a wider and more diverse species mix than the woodland interior.

The impact of light is perhaps best illustrated with hedges, which are essentially managed, thin woodlands, with plenty of light on either side. Here the 'edge effect' (as it is known) is most obvious, with large numbers of moths, including Small Eggar, plus nearly half of the 46 butterfly species found in lowland areas of Britain & Ireland (*e.g.* Gatekeeper), breeding in hedges.

Over large stretches of intensively managed farmland, hedges and their trees are an essential refuge for many plants and animals and can also act as corridors, allowing wildlife to move between habitats. Certain uncommon trees are also found growing in hedges as they like the light open situations created in this treescape, for example, the Plymouth Pear, which grows only in a few hedges in south Devon and Cornwall, the endemic Devon Whitebeam which also grows in hedges in the south-west of England and southern Ireland, and rare woodland-edge plant species such as Crested Cow-wheat which can be found in hedges in eastern England.

The beautiful Crested Cow-wheat, in a Cambridgeshire hedgerow.

The growth form of the tree (see *p. 18*) also impacts the range of species that live on any individual tree, with open-grown ancient trees providing the widest range of opportunities for any colonising species. Mature and ancient trees growing in any habitat, but particularly where there is light, have been described as 'keystone structures' for local biodiversity. This is because they have a range of features which support other species, including hollow trunks, partially dead canopies (stag-headed trees – see *p. 49*), holes in the rotting wood and dead wood itself. These features have been shown to increase if the tree grows in the open, rather than in a wood, presumably because of the impact of the weather – particularly the wind – which can cause non-fatal damage to the tree, opening up these new micro habitats.

Red-horned Cardinal Click-beetle

One interesting finding concerning the deadwood saproxylic beetles is that they are sensitive to the width of the tree's trunk and the exposure of the trunk to sunlight. Researchers have shown that individual trees in sunny habitats were selected by these insects in preference to shaded trees in darker woodlands, including many of the rarer click beetles. Species richness of rare lichens on ancient oaks in closed woodland is also half that compared with oaks growing in open conditions.

The importance for biodiversity of open grown, free-standing old and 'ancient' trees which benefit from sunshine cannot therefore be overstated. It is vital therefore that we make significant efforts to protect our current population of these special trees, whilst also striving to create the next generations of these ancient trees.

Trees as characters – the fascination of old trees

History of Ancient Trees

Our islands have a remarkable collection of ancient trees growing on their shores. Indeed, it has been said that there are more ancient oaks in Britain than in the rest of Europe combined. This is a by-product of our land and tree management history, and in particular the role of Royal Forests which were set up during The Middle Ages by, amongst others, King William I in the 11th century.

These medieval 'Forests' were a designation used for an area where the King or other eminent people had the right to hunt deer. 'Forest' was a legal term and did not mean that the whole area was wooded; indeed, there was no direct reference to trees or woodland. It was rather that the King had taken it upon himself to use the land to protect deer and, by default, other wildlife and trees for the royal hunt.

More than 130 different forests have been recorded through history, some large and famous like the New Forest in Hampshire, Sherwood Forest in Nottinghamshire and the Forest of Dean in Gloucestershire, the boundaries of which are still known and where much of the original landscape still remains. Others, like the Forests of Buckholt, Sapley and Chute (Hampshire), have been lost and remain only as small woodlands. At one point, it is thought that nearly 25% of England was covered by a 'Forest' which gave our trees a form of protection not found elsewhere in Europe.

Another route to the development of ancient trees was our national network of hedges. One of the main management techniques for hedge trees throughout history has been 'pollarding' – cutting the crown off a young tree at a height of 1·8–4·5 m (6–15 feet) from the ground, leaving a permanent trunk called a 'bolling' (see *p. 22*). This trunk then sprouts a range of shoots, at a height that keeps them away from grazing animals.

A beautiful 'stag headed' old tree in Hampshire (see *p. 49*).

The effect of pollarding on a tree is curious, as it often allows the tree to reach a much greater age than if it were left to grow into its normal mature shape. Pollarding also appears to retain the tree in a state of greater vitality, by interrupting the normal aging process and, since the crown of branches and leaves is smaller, also reduces the likelihood of storm damage. Pollarding trees has therefore allowed many to grow for several hundred years and some for much longer.

During the second half of the 18th century, these ancient, pollarded hedge trees were also used by the great landscape designers of the time. To allow grand new landscapes to be created, which carried with them a feeling of age and continuity, the designers – Humphry Repton, Lancelot 'Capability' Brown and others – used these old trees to their advantage. To create an 'instant park' which had the feeling of antiquity, these designers carefully removed the hedges from around the ancient trees, leaving them free-standing in the landscape, many of which still exist.

What is an 'ancient' tree?

The term 'ancient tree' is one that is not capable of precise definition, but it encompasses trees in three categories:
- trees of interest biologically, aesthetically, or culturally because of their age
- trees in the last stage of their life
- trees that are old relative to others of the same species.

One of the complications of ancient trees is that some tree species simply live longer than others. Since one definition of 'ancient' is that a tree must be old for its species, this means that an 'ancient' Yew may be 2,000 years old; an 'ancient' oak may be 1,000 years old, whilst an 'ancient' birch can be only 150 years old.

An ancient pollarded Hornbeam in Kent.

Birch trees have a shorter lifespan than many species – so become 'ancient' earlier.

There is also often confusion between the terms 'ancient' and 'veteran' trees. The term 'veteran' has been coined to describe the wildlife or habitat quality of trees which are not yet ancient (old) but have a range of 'ancient' features such as dead wood in the crown or signs associated with wood decay in ancient trees, like holes in the bark, root, trunk and branches.

In practice, the terms are often used together, with 'ancient and veteran' trees being used as a catch-all, to describe some of the most amazing trees in the country. To return to the claim that Britain has the largest collection of ancient oaks in Europe, research by the Royal Botanical Garden at Kew and the University of Oxford has shown that more than 115 oak trees with a girth of more than 9 m are known in Britain, compared with only 96 trees of similar girth across the remainder of Europe. Britain therefore does have an internationally important collection of ancient trees that need our care and protection.

Why do ancient and veteran trees matter?

As a tree ages, it creates valuable habitats for a wide range of species. An ancient oak, for example, can support hundreds or possibly thousands of other species which take advantage of the cracks, rot holes and cavities that can exist in a single ancient tree.

One of the characteristics of ancient trees is that they begin to lose the heartwood at the centre of the trunk, which is decomposed by a range of fungal species and eaten by a range of beetle and other insect larvae. This leads to many ancient trees becoming hollow, which we have come to realise over the last 30 or 40 years is an important part of healthy aging in many trees. This hollow tree trunk reduces the tree's mass, and recent exceptional storms have shown that hollow trees are blown down less frequently than 'solid' ones.

As a tree gets older, it can also begin to create dead wood in the upper canopy, which is described as 'stag-headed' (like a deer's antlers). These stag-headed trees can be produced when an old tree 'cuts off' water and nutrients to certain branches which die, giving the characteristic 'stag-headed' form. This does not mean that the whole tree is about to die, it is a condition that can persist for many decades or even centuries.

The ecological value of old trees is that they retain large quantities of decaying wood within the structure of the tree which provides valuable habitats for rare and endangered fungi, lichens and invertebrates plus roosting and nest sites for bats, birds and other small mammals. Even a single ancient tree can host rare and endangered species, providing a huge range of micro-habitats, as can be seen in the picture.

There are more than 2,000 invertebrate species in Britain and 650 in Ireland which are dependent on decaying wood, about 7% of the entire British invertebrate fauna.

For these creatures, a large standing living tree with columns of decay in the heartwood is a crucial resource. In the early stages of decomposition, rotting heartwood provides food for the larvae of species such as the Lesser Stag and Rhinoceros Beetles, whilst rare colourful insects like the red Cosnard's Net-winged Beetle can feed on the rotting heartwood of Beech and Ash.

Beautiful stag-headed oak in Suffolk.

As the tree further decomposes a different fauna develops, including the Hairy Fungus Beetle and the larvae of the rare Noble Chafer, which can occasionally be found developing in hollowing fruit trees, oaks and willows.

In rotting trees, the decomposing wood accumulates in the bottom of the hollow trunk. Some of our rarest insects develop in this environment of relatively constant temperature and humidity, protected from the outside world by the surrounding living trunk tissues. The Darkling Beetle is one of the most widespread specialists, while the rare Violet Click Beetle is one of Britain's very few legally protected beetles.

Rhinoceros Beetle

The adults of many insects that develop in decaying wood need nearby blossom on which to feed before they can start to reproduce. Blossom provides nectar – an energy-rich food – and pollen, which provides the energy needed for egg production. Blossom can be important throughout the spring, and species such as sallows, Holly, Wild Privet, Rowan, Crab Apple, Wild Pear, Guelder-rose and brambles are all beneficial.

Flowering trees and shrubs are by far the most important sources of nectar and pollen to these creatures. Insects in ancient trees have easy access to a whole range of species, especially Hawthorn, which provides the ideal insect blossom, due partly to its flowering in late spring when so many wood-decay insects are in the adult stage.

Unfortunately, as an old tree becomes 'ancient' or 'senile', it continues to decline and decay. Many ancient trees grow along roadsides and highways, where falling dead wood or limbs may become a hazard to passing traffic or pedestrians. These trees need careful management to ensure public safety whilst retaining them and their unique ecological value.

Is it good to leave a dead ancient tree standing?

From the point of view of the environment, yes. Standing dead trees are home and feeding station for many animals and birds and numerous insects. Some of the great historic landscape designers such as Humphry Repton recognized the importance of trees with decay, and William Kent (in the early 18th century) was known to import dead trees to create an immediate 'air of antiquity' in his landscape gardens. From a health and safety perspective, leaving a dead tree standing may not be such a good idea. Much depends upon the situation and amount of public access. A compromise may be to leave a safe stump and a pile of dead wood on the ground. A totally dead tree is unlikely to spread disease to live trees unless the adjacent trees are under stress already from some other cause.

Dead standing oak at Windsor Great Park.

TREE IDENTIFICATION

Tree identification

Getting started
It is quite common to find tree identification difficult to begin with. Why this is the case remains a little unclear, although it may be simply because it is hard to figure out which part of the tree to use when starting an identification, and that the available features change with the seasons. An identification might use flowers in the spring, leaves during the summer, fruit in the autumn, and twigs and buds throughout the winter. This shifting set of features can make tree identification tricky at some times of the year, particularly in winter until twig and bud features have been mastered.

Fortunately, many of our native trees have a single useful feature, or simple combination, which will help you to identify that species through the seasons. Learning these should enable a quick and confident identification of some of Britain & Ireland's commonest trees, and will also provide a good knowledge base from which to compare other tree species.

Getting to know ones trees is about practice and perseverance, and this book should enable identification to be reached for most trees you will encounter either through the 'one-step' method or by using the more detailed illustrated keys and species accounts. However, the trickiest species, such as willows and rare whitebeams, and those groups that readily hybridize, may prove impossible to identify and may require accessing specialist information that is beyond the scope of this book. Don't be put off as even with lots of practice, it is still possible to get it wrong!

One-step identification
On the next five pages, we will show you a few easy examples of this 'one-step' method of tree identification, as a way of confidently learning a base set of species which can be used as a starting point from which to compare other species.

Park and street trees and varieties
The nature of the British & Irish landscape is that there are many different species of exotic planted trees: those deliberately planted in parks as examples of their type and those amenity-planted en masse *e.g.* in business parks. In addition, cultivated varieties of even well-known species can look very different in form, and the colour and shape of the leaves and flowers. Only those that are relatively common, or may be found naturalized or self-seeding, are included in this book. Given the hundreds of species and varieties that have been planted, it is likely that trees will be found that are not in this book. Nonetheless, those that are most likely to be encountered are included in their own section, **Park, street & garden trees and shrubs** starting on *p. 338*.

Author's note – identification using bark | Speaking from personal experience, frankly, some species are next to impossible to identify from their bark alone without years and years of practice, and even then, it is easy to get it wrong. However, there are a few species where it is possible to use bark and others where a combination of bark, twigs and buds allow a quick and accurate identification. Consequently, although the vast majority of the species accounts include images of the bark, these are included for comprehensiveness and rarely referred to as identification features.

Bark varies in many species with age and environment – these are all examples of Wild Service-tree (p. 142)

TREES EASILY IDENTIFIABLE ALL YEAR

'One-step' ID – using different features at different seasons

Species may have one season during which they are easier to identify than at other times. Others are readily identifiable throughout the year using differing features – Horse Chestnut (*below*) is a good example of such a species, although there are two other much rarer similar species to be aware of.

In **winter**, Horse Chestnut twigs (*p. 290*), with their large leaf-scars and **sticky buds**, are obvious if a close look is taken. Horse Chestnut can be readily identified during the remainder of the year: in **spring to autumn** by the large **'hand'-like leaves** with **unstalked** leaflets; in **spring** by the white **'candelabra' flowers** (which are highly distinctive, even at distance); and, in **autumn** by the characteristic **spiky fruits**.

Similar species | The leaf-shape rules out any other British & Irish tree except **Indian** and **Red Horse Chestnuts** (*p. 291*) which both have stalked leaflets and almost smooth fruit; **Red Horse Chestnut** also has red flowers. Confusion with **Sweet Chestnut** (*p. 216*) is possible, but that has narrowly oval single leaves.

Trees easily identified all year

For a few species it is possible to make a quick and easy identification all year-round using bark and/or twigs and buds alone or in combination.

 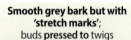

Smooth grey bark like an elephant's skin; buds **away from** twigs

Beech (*p. 214*)

Smooth grey bark but with 'stretch marks'; buds **pressed to** twigs

Hornbeam (*p. 278*)

Whitish bark with black vertical cracks and 'diamonds'

Silver Birch (*p. 268*)

Reddish brown to grey bark with distinctive 'rings'

Cherries (*pp. 176, 177*)

Conifer with salmon-pink upper trunk contrasting with a darker lower portion

Scots Pine (*p. 96*)

Downy Birch (*p. 268*) has horizontal cherry-like 'rings'; planted **ornamental birches** have bark that peels in strips

Most likely a **Wild Cherry** in woodland and an **ornamental cherry** in parks and streets

TREES EASILY IDENTIFIABLE IN WINTER ID

Four evergreen species easily identified all year

Prickly leaves

Holly (p. 302)

A few **Evergreen Oak** (p. 223) can have some holly-like leaves

Sharply pointed unbranched needles; arranged in **threes**

Juniper (p. 118)

Sharply pointed branched spines yellow flowers

Gorses (p. 318)

Identify to species by flower details

Unmistakeable form; large triangular needles

Monkey-puzzle (p. 129)

Trees easily identified in winter

It is often thought that winter is a tricky time for identifying trees, but there are a number of species, including all the ones covered above that can be done quite easily in winter.

Twigs and buds can be used, with some attention to detail, to identify most species in winter.

▶ see p. 78 for a key to winter twigs

Long, bright yellow catkins present all winter

Hazel (p. 276)

MATURE ♀ CONES present all winter

Purplish at a distance; with small woody 'cones'

Alders (pp. 272, 274)

In a river or wetland, it will most likely be **Common Alder** – but non-native varieties are planted in other places

Purplish orange-brown twigs are distinctive even at a distance

Bog Myrtle (p. 314)

End-bud black and sharply pointed; side-buds on twigs arranged oppositely

Ash (p. 298)

53

ID — TREES EASILY IDENTIFIABLE IN SPRING

Trees easily identified in spring

Flowering starts as early as February and other than Hazel, the first flowers of the year are the massed white blossom of Cherry Plum usually seen in early to late March, followed by Blackthorn.

In April many other species will appear in flower, most of which need a closer look and/or other features to identify.

▶ see *p. 68* for a key to flowers

Masses of white blossom before the leaves appear

large flowers; early–late March	smaller flowers; late March–early April
Cherry Plum (*p. 172*)	**Blackthorn** (*p. 171*)

Later in the season check for plum, apple and pear (see *p. 70*)

Yellow 'pussy willow' ♂ catkins; grey-green ♀ catkins

Goat and Grey Willows (*pp. 232–235*)

Long pendent catkins (Aspen and Grey Poplar grey hairy); ♂ red; ♀ yellow-green

Aspen and other poplars (*pp. 256–264*)

There are many **willow** and **poplar** species with similar flowers and these require examination of the leaves later in the year for confident identification (see *p. 228*). The long, pendent yellow ♂ catkins of **alders** and **birches** (*p. 68*) are somewhat similar but both have much smaller buds and different leaves to poplars.

Clusters of winged seeds in late spring are diagnostic of elms as a group	The earliest maple to flower (before the leaves) standing out against other bare trees	Flat clusters of small white flowers with a honey-like aroma; pinnate leaves
Elms (*pp. 197–212*)	**Norway Maple** (*p. 288*)	**Elders** (*p. 306*)
Taxonomy is complex – see *p. 202*	Other maples (*pp. 284–289*) flower with their leaves	Other white clustered species have different leaf shapes

54

TREES EASILY IDENTIFIABLE IN SUMMER ID

Trees easily identified in summer

Summer is the prime time for trees. Many species have finished flowering and are developing fruits but there are some distinctive later-flowering species. Leaves are the main identification features, and although many are similar there are some that are distinctive.

▶ see *p. 58* for a key to leaves

Domed clusters of small white 4-petalled flowers; oval leaves with 2–5 pairs of veins

Dogwoods (*p. 304*)

Other **planted dogwoods** have leaves with 6–7 pairs of veins

Flat clusters with flowers of two very different sizes; lobed leaves

Guelder-roses (*p. 308*)

The **planted American Guelder-rose** has minor leaf differences

Pinnate leaves with oval, untoothed leaflets; 'furniture polish' aroma if crushed

Walnuts (*p. 196*)

Black Walnut has pointed leaflets

Pinnate leaves with narrowly oval, untoothed leaflets; arranged alternately

Rowan (*p. 140*)

Ash (*p. 298*) has opposite leaves

Leaves with an uneven base
NOTE: a huge range of shapes and sizes between taxa

Elms (*p. 197*)

Toothed, narrowly oval leaves; distinctive clusters of very long, yellow ♂ catkins

Sweet Chestnut (*p. 216*)

Evergreen shrub; leaves oval, shiny; white flowers in cylindrical spikes

Cherry Laurel (*p. 178*)

Semi-evergreen shrub; leaves narrowly oval; white flowers in pyramidal clusters

Wild Privet (*p. 312*)

Garden Privet has broadly oval leaves

ID TREES EASILY IDENTIFIABLE IN AUTUMN

Trees easily identified in autumn

The fruits and seeds of some trees and shrubs are highly distinctive. Very few woody plants are in flower, but those that are can be readily identified.

▶ see *p. 74* for a key to fruit

Conical spikes of purple flowers can remain into late autumn

Butterfly-bush (*p. 328*)

Softly spiny fruit

Sweet Chestnut (*p. 216*)

Not to be confused with **Horse Chestnut's** (*p. 290*) spiked capsule

The long papery bracts are diagnostic of limes

Limes (*p. 292*)

Identify to species by the number of fruit and whether they are pendent or erect

Pink seed pods and orange seeds are unmistakable

Spindle (*p. 280*)

Red 'berries' that are actually soft flesh partially surrounding a hard seed are diagnostic

Yew (*p. 116*)

SESSILE OAK

PEDUNCULATE OAK

ACORNS on stalk from twig

ACORNS directly on twig

Acorns in scaly, hairless cups and leaves with rounded lobes

Native deciduous oaks (*pp. 218–221*)

Identify to species by seeing whether the acorn sits directly on the twig (**Sessile Oak**); or on a thin stalk (**Pedunculate** (English) **Oak**). Non-native oaks are evergreen, have pointed lobes, or 'hairy' acorn cups.

HAWTHORN
1 seed

MIDLAND HAWTHORN
2 seeds

Red berries, spiny twigs and/or lobed leaves are distinctive

Hawthorns (*p. 186*)

Identify to species by squishing the fruit and counting the stony seeds inside – one = **Common Hawthorn**; two = **Midland Hawthorn**.

56

TREE IDENTIFICATION

Tree identification

The following pages contain visual keys to the naturally occurring (rather than street or park planted) trees and shrubs in this book. The keys cover the leaves, flowers, fruit and winter twigs as separate sections. In the field it is highly likely that more than one feature will be used in an identification.

The keys indicate the pages where the relevant species account can be found or, in some cases, to additional keys.

The first distinction is to determine whether a tree is a broadleaf or a conifer

Broadleaf trees ▶ p. 130

FLOWERS diverse range of both conspicuous and tiny unisexual or bisexual flowers (solitary or ± clustered in various arrangements)

FRUIT diverse range of hard and soft fruits and cases that enclose and protect the seed

LEAVES typically broad (plus needle-like gorses) in a wide range of structures, outlines and shapes

HABIT most species lose their leaves in winter (deciduous), a few evergreen

MAR–NOV
NOV–APR

Key to leaves	▶ p. 58
Key to flowers	▶ p. 68
Key to fruit	▶ p. 74
Key to winter twigs	▶ p. 78

Conifers ▶ p. 88

'FLOWERS' ♂ pollen cones (sporangia) and ♀ seed cones, both very small in some species

'FRUITS' mature ♀ seed cones, typically scaled and woody (soft in Juniper); Yew with berry-like aril with exposed seed

LEAVES typically needle-like and thin (<8 mm across), or scale-like

HABIT most species evergreen, a few deciduous

ALL YEAR
MAR–NOV
NOV–APR

| Key to conifer leaves | ▶ p. 91 |
| Key to conifer cones | ▶ p. 94 |

57

Key to leaves

Using leaves for identification | The leaf feature terms used for identification are illustrated on these two pages. This book tries to avoid detailed botanical terms – technical terms for the descriptions used are given in *italic text*.

Leaf structure, shape and margin are of prime importance in identification: whether the leaf is a single continuous leaf-blade or comprises a number of separated smaller leaflets; the outline of the leaf blade and the shape of any lobes; and whether the edge of the leaf is smooth, with teeth or spines. The categories with page numbers below relate to the groupings used in the key.

Leaf or leaflet shape | see *Distinctively shaped leaves* (p. 60) for those that do not fit these categories.

simple — leaf-blade that, although it may be lobed or deeply divided, is an entire, single piece

UNLOBED

- **triangular** ▶ p. 64
- **heart-shaped** (*cordate*) **to ± circular** ▶ p. 64
- **broadly oval** (*ovate*) ▶ pp. 65–67
- **oval** (*ovate*)
- **narrowly oval** (*ovate, lanceolate*)
- **needles** ▶ **Conifers** p. 91
- **scale-like**

NOTE: oval leaves with tips that are obviously pointed are referred to as 'pointed-oval'

NOTE: the deciduous **Tamarisk** (p. 326) has scale-like leaves – see p. 60

LOBED

- **spiny** — broad or narrow spines on leaf-margin ▶ p. 63
- **lobed** — lobes irregular (includes spiny leaves) ▶ p. 63
- **pinnately lobed** — lobes arranged ± regularly and oppositely ▶ p. 63
- **palmately lobed** — lobes spreading radially from a single point ▶ p. 62

Leaf-margin

- **untoothed** (*entire*) smooth-edged (not divided, lobed or toothed)
- **toothed** (*entire*) serrated leaf-margin; the shape of the teeth is useful for identification (see *facing page*)

compound

LEAFLETS arising from a single point

NOTE: closely spaced simple leaves (especially those arranged oppositely) can look compound

▶ p. 61

LEAFLETS arising from multiple points

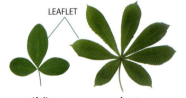

- **trifoliate** 3 LEAFLETS
- **palmate** >3 LEAFLETS

▶ p. 60

ID

1/5

NOTE leaf shape is the primary feature to observe; leaves are not scaled accurately

Choosing the right leaf | For most species any leaf will do, but for some it is important to look at the 'right' leaves for identification. Factors such as leaf age, whether it is on a vegetative (long shoot) or flowering twig (short shoot), and whether it is in sun or shade can be very important. Generally it is best to avoid those leaves on suckers. Those cases for which it is crucial to examine particular leaves for identification are highlighted.

suckers – new shoots produced from the roots of the parent tree

Leaf arrangement and growth

Arrangement is used as an identification differentiator and also applies to buds

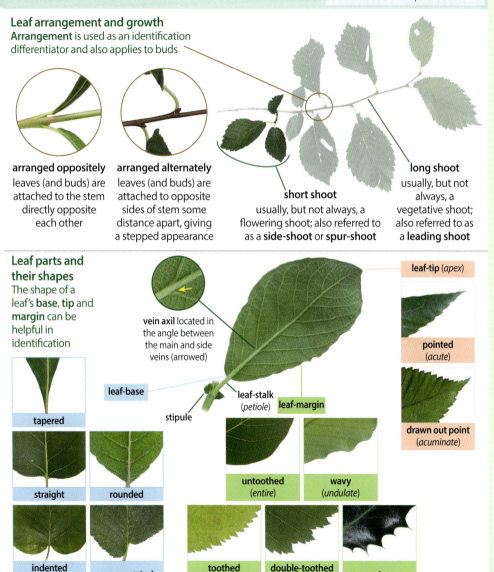

arranged oppositely
leaves (and buds) are attached to the stem directly opposite each other

arranged alternately
leaves (and buds) are attached to opposite sides of stem some distance apart, giving a stepped appearance

short shoot
usually, but not always, a flowering shoot; also referred to as a **side-shoot** or **spur-shoot**

long shoot
usually, but not always, a vegetative shoot; also referred to as a **leading shoot**

Leaf parts and their shapes

The shape of a leaf's **base**, **tip** and **margin** can be helpful in identification

vein axil located in the angle between the main and side veins (arrowed)

leaf-base

stipule

leaf-stalk (*petiole*)

leaf-margin

leaf-tip (*apex*)

tapered

straight

rounded

indented (*cordate*)

asymmetrical

untoothed (*entire*)

wavy (*undulate*)

toothed (*serrate*)

double-toothed (*biserrate*)

spiny

pointed (*acute*)

drawn out point (*acuminate*)

ID LEAF IDENTIFICATION

Key to leaves
PALE GREEN NAMES INDICATE PARK, STREET & GARDEN SPECIES

Distinctively shaped leaves which do not fit elsewhere

tiny; scale-like

DECIDUOUS SHRUB

Tamarisk *p. 326*

alternate

CONIFEROUS TREES

Cypresses *p. 93*

opposite

alternate

Giant Redwood *p. 123*

needle-like – see Conifers *p. 91*

spiny; leaves apparently absent / not obvious

SHRUBS **Broom** (*p. 320*) is not spiny but the trifoliate leaves are not always present

branched spines

Gorses [3 spp.] *p. 318*

Spp. ID – flower + fruit

needles in whorls of 3
Common Juniper *p. 118*

'leaves' are modified branches

Butcher's-broom *p. 315*

Spanish-dagger *p. 339*

TREE

Monkey-puzzle *p. 129*

Cabbage-palm *p. 339*

Compound leaves: leaflets from a single point

trifoliate (3 leaflets)

SHRUB OR SMALL TREE

Laburnums [2 spp.] *p. 330*

Spp. ID – flower-cluster, flower and leaf underside details

SHRUBS

Broom *p. 320*

NOTE leaves not always present

palmate (>3 leaflets)

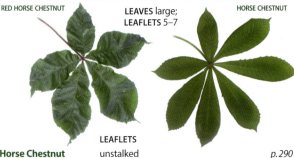

RED HORSE CHESTNUT

LEAVES large; LEAFLETS 5–7

HORSE CHESTNUT

LEAFLETS
Horse Chestnut — unstalked — *p. 290*
Red Horse Chestnut — very shortly stalked; red flowers — *p. 291*
Indian Horse Chestnut — stalked; narrowest of the three species — *p. 291*

LEAF IDENTIFICATION 2/5

NOTE leaf shape is the primary feature to observe; leaves are not scaled accurately

Compound leaves: leaflets from multiple points — Leaves opposite

LEAFLETS 3 or 5

Ash-leaf Maple *p. 289*

Elders [2 spp.] *p. 306*
Spp. ID – leaf-stalk + fruit

LEAFLETS 5 or 7

LEAFLETS 7–13

Ashes [3 spp.] *pp. 298, 340*
Spp. ID – bud + leaflet

MOCK-ORANGE

BOX

Mock-orange (*p. 334*) and **Box** (*p. 313*) have closely spaced opposite simple leaves which could be mistaken for leaflets

Compound leaves: leaflets from multiple points — Leaves alternate

SPINY TREES

FALSE-ACACIA

LFLTS oval; tip rounded

False Acacia	LEAF L 15–30 cm; LFLTS 5–25	*p. 300*
Honey-locust	LEAF L to 20 cm; LFLTS 14–32	*p. 343*
Pagoda Tree	LEAFLETS tip pointed	*p. 343*

SHRUBS

LEAFLETS spiny

TWIGS very hairy

Oregon-grape *p. 331*

Stag's-horn Sumach *p. 337*

TREES

LEAFLETS toothed

Rowan *p. 140*
Rare, localized *Sorbus* *p. 138*
Spp. ID – leaflets + fruit

LEAFLETS ±oval

Walnut *p. 196*

LEAFLETS pointed-oval

Black Walnut *p. 196*

LEAFLETS with large basal teeth

Tree-of-heaven *p. 340*

ID — LEAF IDENTIFICATION

Key to leaves
PALE GREEN NAMES INDICATE PARK, STREET & GARDEN SPECIES

Simple leaves: LOBED 1/2

palmately lobed – lobes radiating from a single point; NOTE also check lobed leaves (*opposite*)

SHRUBS

GOOSEBERRY   REDCURRANT
LOBES 3–5

Currants [6 spp.] *p. 322*
Spp. ID – leaf, flower, fruit

LOBES usually 3 (a few 5); **LEAVES** arranged oppositely

Guelder-roses [2 spp.] *p. 308*
Spp. ID – leaf, flower, fruit

shiny; deeply lobed

Fig *p. 337*

TREES; LEAVES ARRANGED ALTERNATELY

Tulip-tree *p. 339*

distinctive 'truncated' tip

Ginkgo *p. 338*

distinctive 'fan' shape

UNDERSIDE pale grey; at least slightly hairy

White Poplar *p. 260*
Grey Poplar *p. 258*

maple-like
Planes [3 spp.] *p. 282*
Spp. ID – leaf, bark, fruit

TREES; LEAVES ARRANGED OPPOSITELY (Maples)

LOBES variable; typically 3 main + 2 smaller basal

Field Maple *p. 284*

LOBES 5; no bristles; irregularly toothed

Sycamore *p. 286*

LOBES 3 large (± 2 tiny basal); margins toothed

Red Maple *p. 289*

LOBES 5–7; untoothed

Cappadocian Maple *p. 289*

LOBES 5–7; bristle-pointed; a few teeth

Norway Maple *p. 288*

LOBES deep; UNDERSIDE silvery white

Silver Maple *p. 289*

62

LEAF IDENTIFICATION

NOTE leaf shape is the primary feature to observe; leaves are not scaled accurately

3/5

Simple leaves: LOBED and/or spiny 1/2

lobed and pinnately lobed

Wild Service-tree *p. 142*

Swedish Whitebeam *p. 146*

GERMAN SERVICE-TREE

Some rare, localized *Sorbus* (*p. 136*) have similar leaves

Spp. ID – leaflets + fruit

MIDLAND HAWTHORN

HAWTHORN

Hawthorns *p. 186*

Spp. ID – leaves, flowers + fruit

DECIDUOUS OAKS WITH ROUNDED LOBES

LEAF-STALK long

LEAF-STALK short at most

LEAF-BASE with hairy stipules

TURKEY OAK

SESSILE OAK

PEDUNCULATE OAK

Oaks (widespread) [6 spp.] *pp. 218–222*

Spp. ID – leaf, fruit

Hybrid Oak (*p. 219*), the common hybrid between **Sessile** and **Pedunculate Oaks** has variably shaped leaves, even on the same tree in some cases

HYBRID OAK

OAKS WITH POINTED LOBES

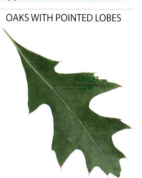

SCARLET OAK

American oaks [3 spp.] *p. 224*

Spp. ID – leaf shape

evergreen; with spines

All can have leaves that lack spines

BARK corky

BARK cracked

YOUNG LEAF

Cork Oak *p. 225*

Evergreen Oak *p. 223*

BARK ± smooth

Highclere Holly *p. 303*

Holly *p. 302*

TWIGS with sharp spines in groups

GAGNEPAIN'S BARBERRY

DARWIN'S BARBERRY

Barberries (*pp. 316–317*) Native **Barberry** (oval leaves) has inconspicuous fine spines; some other **planted barberries** have ± spiny leaves

63

ID LEAF IDENTIFICATION

Key to leaves

PALE GREEN NAMES INDICATE PARK, STREET & GARDEN SPECIES

Simple leaves: ±CIRCULAR, HEART-SHAPED TO TRIANGULAR;
also check **BROADLY OVAL** (*facing page*) + **OVAL (POINTED)** leaves (*p.66*)

arranged oppositely

TREES

Katsura *p.342*

distinctive
toffee aroma
in autumn

Indian Bean-tree *p.342*

very large
(L to 25 cm)

SHRUBS

Lilac *p.329*

FLOWER-CLUSTER
conical spike

Wayfaring-tree *p.310*

FLOWER-CLUSTER
flat-topped

'Snowberries'
[3 spp.] *p.332*

Spp. ID – leaves

arranged alternately

very large (L to 23 cm)

TREE with catkins
and/or cone-like fruit

wider
than long

Limes [4 spp.] *p.292*

Spp. ID – flower-cluster,
leaf underside details

Hazels
[3 spp.] *p.276*

Spp. ID –
bract shape, leaves

Black Mulberry *p.342*

Italian Alder *p.274* **Judas-tree** *p.342*

LEAVES variable in size and
leaf + leaf-base shape

BALM-OF-GILEAD

Silver Birch *p.268*

LF typically with longer
point than in Downy Birch

Downy Birch *p.268*

LF typically with shorter
point than in Silver Birch

Himalayan and **Paper Birches** (*facing page*) have more
pointed-oval leaves

BLACK POPLAR

Aspen *p.256*

very
thin

UNDERSIDE
slightly
paler green

Grey Poplar *p.258*

UNDERSIDE
typically
grey hairy

WESTERN
BALSAM-POPLAR

UNDERSIDE
paler than
upperside

HYBRID BLACK
POPLAR

UNDERSIDE
±same as
upperside

Black poplars and **Balsam-poplars** [8 spp.] *pp.261–264*

Spp. ID – leaf details, twig + sucker form

64

LEAF IDENTIFICATION

ID 4/5

NOTE leaf shape is the primary feature to observe; leaves are not scaled accurately

Simple leaves: BROADLY OVAL TO OVAL
also check HEART-SHAPED (*facing page*) + OVAL (POINTED) leaves (*p. 66*)

arranged alternately

SHRUBS AND SMALL TREES
TWIGS with sharp spines in groups

Barberries
[6 spp.] *p. 316*
Spp. ID – leaves

TWIGS dark; with sharp spines

Blackthorn *p. 171*
Plums (*below*) have larger leaves and **TWIGS** sparsely spiny at most

LEAVES pointed-oval; variable in size by species

FRANCHET'S COTONEASTER

BULLATE COTONEASTER

Cotoneasters
[7 spp.] *pp. 189–192*
Spp. ID – leaves + flowers

Dwarf Birch *p. 271*

CREEPING WILLOW

Willows range in size from low shrubs to tall trees. In leaf they will have characteristic catkins (*p. 68*) or '1-scaled' buds (*p. 81*) present. Leaves have a wide variety of leaf shapes, from broadly oval to very narrow.

TREES

Alder Buckthorn *p. 192*

Buckthorn (*p. 193*) has leaves arranged ± oppositely

Wild Plum *p. 173*

TREES with catkins and/or cone-like fruit

Common Alder *p. 272*
TIP rounded

TIP pointed **Grey Alder** *p. 274*

GOAT WILLOW

▶ **Willows** *p. 226*

Himalayan + Paper Birches *p. 270*

Silver and **Downy Birches** (*facing page*) have more triangular leaves

Apples more coarsely toothed than **Pears**

Apples [2 spp.] *p. 182*
Pears [3 spp.] *pp. 183–185*
Spp. ID – leaves, flowers, fruit

UPPERSIDE 'flat'; MARGIN untoothed

Beech *p. 214*

MARGIN finely toothed

Cherry Plum *p. 172*
Bird Cherry *p. 174*

Hornbeam *p. 278*

UPPERSIDE 'corrugated'; MARGIN toothed

UNDERSIDE usually whitish hairy

Whitebeams [44 spp.] *p. 136*
Spp. ID – leaf details

MARGIN finely to coarsely toothed

Cherries [4 spp.] *pp. 176, 181*
Spp. ID – leaves

Elms are a complex group taxonomically; leaves can be found in a wide range of sizes and shapes but, as a group, can be recognized by their asymmetrical leaf-bases, albeit small in some. **Limes** (*p. 292*) could be confused for elms.

FIELD ELM (ENGLISH ELM)

WYCH ELM (SOUTHERN)

▶ **Elms** *p. 197*

65

ID LEAF IDENTIFICATION

Key to leaves

PALE GREEN NAMES INDICATE PARK, STREET & GARDEN SPECIES

Simple leaves: BROADLY OVAL TO NARROWLY OVAL

arranged oppositely

FLOWERS in conspicuous round-topped clusters; **TWIGS** red

Butterfly-bush (Buddleia)
p. 328

Purple Willow
p. 238

FLOWERS clustered around twig; **TWIGS** grey to dark brown

Buckthorn
p. 193

leaves can be arranged slightly alternately

Dogwoods
[3 spp.] *p. 304*
Spp. ID – leaf vein count

TWIGS greenish brown; unridged

glossy

WILD PRIVET

GARDEN PRIVET

Privets [2 spp.] *p. 312*
Spp. ID – leaf + twig details

TWIGS green; usually ridged

dull

Spindles
p. 280

Fuchsia
p. 333

thick, evergreen leaves in close arrangement

Box
p. 313

± thick and leathery

'HEDGE' HEBE

LEWIS'S HEBE

Hebes [3 spp.] *p. 328*
Spp. ID – leaf details

distinctively mottled

Spotted Laurel
p. 330

UNDERSIDE lacking glands *cf.* Escallonia (*facing page*)

FLOWERHEAD flat-topped cluster

Mock-oranges
p. 334
Spp. ID – leaf + flower details

Weigelia
p. 341

Wrinkled Viburnum
p. 335

Laurustinus
p. 335

66

FLOWER IDENTIFICATION   ID
1/3

NOTE images are not scaled accurately

Catkins, catkin-like flower-clusters, inconspicuous and/or yellowish green flowers

Sweet Chestnut p. 216

Planes [3 spp.] p. 282
Spp. ID – leaves, bark

Oaks [8 spp.] pp. 218–221
Spp. ID – leaves

Beech p. 214

Maples [7 spp.] p. 284
Spp. ID – leaves

Ashes [3 spp.] p. 298
Spp. ID – leaves

Manna Ash (p. 340) has ± pendent clusters of white flowers

Limes [4 spp.] p. 292
Spp. ID – leaves, flowers

Sea-buckthorn p. 194

♂ and ♀ (L > 4 mm); on separate shrubs

Elms pp. 197–212
Spp. ID – leaves

Box p. 313

Currants [4/6 spp.] p. 322
Spp. ID – leaves, flowers

69

ID FLOWER IDENTIFICATION

Key to flowers
PALE GREEN NAMES INDICATE PARK, STREET & GARDEN SPECIES

5-petalled white flowers – petals separated (Rose Family – p. 131)

Blackthorn + Plums
[3 spp.] *p. 170*

Spp. ID – flower details, flowering time, leaves

generally flowering earlier (Feb–Apr) than other roses

BLACKTHORN

FLOWERS IN DOMED CLUSTERS
(*cf.* Wayfaring-tree (*facing page*))

FLOWERS (D > 6 mm); PETALS separate

Whitebeams, Rowans, Service-trees [44 spp.] *p. 133*
Spp. ID – leaves

PETAL-TIPS SLIGHTLY NOTCHED

Cherries I
[3 spp.]
pp. 176, 179

Spp. ID – flower details, flowering time, leaves

flowers in clusters

WILD CHERRY

TWIGS SPINY; LEAVES LOBED

Hawthorns [2 spp.] *p. 186*
Spp. ID – leaves, flowers

PETAL-TIPS ROUNDED

Cherries II
[3 spp.]
pp. 174, 178

Spp. ID – flower details, flowering time, leaves

flowers in spikes

BIRD CHERRY

CHERRY LAUREL

Medlar
p. 336

PETAL-TIPS ROUNDED

Apples and Pears
[5 spp.]
pp. 182–183

Spp. ID – flower details, leaves

flowers in loose clusters

Apples ANTHERS yellow

Pears ANTHERS purple

Juneberry
p. 336

FLOWER IDENTIFICATION

NOTE images are not scaled accurately

2/3

5-petalled white flowers – petals fused

FLOWERS 5-PETALLED; FUSED; SAME SIZE

FLOWERS 5-PETALLED; FUSED; 2 SIZES

FLOWERS (D < 6 mm); PETALS fused
LEAVES oval to heart-shaped

Wayfaring-tree
p. 310

LEAVES pinnate

Elders [2 spp.] *p. 306*
Spp. ID – pith, fruit

LEAVES lobed

Guelder-roses [2 spp.] *p. 308*
Spp. ID – leaves

4-petalled white flowers

Holly (*p. 302*) has 4-petalled white flowers but is instantly recognizable by its spiny evergreen leaves

LEAVES oval
PETALS separated

Dogwoods [3 spp.] *p. 304*
Spp. ID – leaves

LEAVES oval
PETALS fused

Privets [2 spp.] *p. 312*
Spp. ID – pith, fruit

LEAVES oval
PETALS separated

Mock-oranges
p. 334

Flowers with bilateral symmetry

Horse Chestnuts
[5 spp.] *p. 290*
Spp. ID – flower details, leaves

False Acacia
p. 300

FLOWERS pea-like; TWIGS with broad-based spines

Honey Locust (*p. 343*) has twigs with narrow-based spines

71

ID FLOWER IDENTIFICATION

Key to flowers

PALE GREEN NAMES INDICATE PARK, STREET & GARDEN SPECIES

Coloured flowers

FLOWER IDENTIFICATION **ID**

NOTE images are not scaled accurately 3/3

ID FRUIT IDENTIFICATION

Key to fruit

PALE GREEN NAMES INDICATE PARK, STREET & GARDEN SPECIES

TYPICALLY GREEN OR BROWN | catkins, 'cone'-like structures and cases containing seeds

In catkins; seeds with hairs

NOTE: it is not practical to identify willow or poplars by their fruit alone

Willow + poplars [29 spp.] *p. 226*
Spp. ID – leaves

In catkins; seeds other

YOUNG

DOWNY BIRCH   SILVER BIRCH

Birches [4 spp.] *p. 266*
Spp. ID – leaves, seeds, fruiting bracts

NOTE: immature ♂ catkins can look like unripe fruiting catkins

MATURE

LEAVES 'Eucalyptus' aroma

YOUNG   MATURE

NUT

Bog-myrtle *p. 314*

Alders [3 spp.] *p. 272–275*
Spp. ID – catkin size, leaves

The cones of a few conifer species could possibly be confused with alder fruits but conifers lack the broad leaves of alders; conifer cones are relatively easy to identify to species ▶ *p. 94*

Cases containing nuts

YOUNG   MATURE

HORSE CHESTNUT

INDIAN + RED HORSE CHESTNUTS

LEAVES pinnate

LEAVES oval

LEAVES narrowly oval

LEAVES palmate

Beech *p. 214*

Sweet Chestnut *p. 216*

Horse Chestnuts [3 spp.] *p. 290*
Spp. ID – fruiting cases, leaves

Walnuts [2 spp.] *p. 196*
Spp. ID – leaves

Exposed nuts

in scaly cups   in bracts

PEDUNCULATE OAK

TURKEY OAK

Oaks [8 spp.] *pp. 218–225*
Spp. ID – cupules, leaves

Hazels [3 spp.] *p. 276*
Spp. ID – bract shape, leaves

Cases containing seeds

distinctive long bracts

Limes [4 spp.] *p. 292*
Spp. ID – fruit form + quantity, leaves

Planes [3 spp.] *p. 282*
Spp. ID – fruit quantity, bark, leaves

PODS long, narrow (cf. Indian Bean-tree *p. 342*)

PEAS (False Acacia, gorses, brooms, greenweeds, laburnums etc.) [15 spp.] *pp. 300, 318–321, 330*
Spp. ID – leaves, flowers, pods

Mock-oranges *p. 334*

distinctive 3 'horns'

Box *p. 313*

FRUIT IDENTIFICATION 1/2

NOTE images are not scaled accurately

TYPICALLY GREEN OR BROWN | exposed winged seeds

	2 divergent wings	in clusters	in clusters	in clusters	in clusters
	FIELD MAPLE			seed ± halfway to tip	seed towards tip
	NORWAY MAPLE				
	SYCAMORE	1-winged	distinctive; 3-bracts		
	Maples, Sycamore [7 spp.] *p. 284–289*	**Ashes** [3 spp.] *p. 298, p. 340*	**Hornbeam** *p. 278*	**White Elm** *p. 199,* **'Wych' Elms** *p. 200*	**'Field' Elms** *p. 201*
	Spp. ID – wing angle, leaves	Spp. ID – buds, leaves		Spp. ID – fruit hairiness	Spp. ID – buds, leaves

LARGER SOFT, FLESHY FRUITS (usually L > 10 mm)   NOTE some Blackthorn fruits can be < 10 mm

LEAVES	L > 40 mm; ± shiny	L < 40 mm; dull	L > 40 mm; dull	
FRUIT	D 15–22 mm	D ≤ 15 mm	D 15–25 mm	D 25–80 mm

can be reddish

Cherry Plum *p. 172*

Blackthorn *p. 171*

Wild Plum *p. 173*

'Domestic' Plum *p. 173*

LEAVES pinnate

Fig *p. 337*

LEAVES oval

Plymouth Pear *p. 183* — globular; D < 2 cm

Wild Pear *p. 184* — ± globular; D < 5 cm

Cultivated Pear *p. 184* — pear-shaped; D > 5 cm

Medlar *p. 336*

True Service-tree *p. 148*

SKIN uniquely textured

SEPALS conspicuous

SKIN ± smooth

SKIN rough

AROMA rancid butter or vomit

Strawberry-tree *p. 301*

Domestic Apple *p. 181* — D > 40 mm

Wild Apple *p. 180* — D < 35 mm

Flowering quinces *p. 341*

Ginkgo *p. 338*

75

ID FRUIT IDENTIFICATION

Key to fruit

PALE GREEN NAMES INDICATE PARK, STREET & GARDEN SPECIES

SMALLER SOFT, FLESHY FRUITS (usually L < 10 mm) NOTE Cherries usually larger

Using soft fruits for identification
Although a few trees and shrubs can be identified by their individual fruit, for most species this is difficult. The structure of the fruiting cluster, together with the leaves and twigs are typically the best combination for a confident identification.

many; tightly clustered around twigs; black
Buckthorn p.193

Alder Buckthorn p.192

a few; loosely clustered; red and black often at the same time

pink; distinctively angled
Spindles [2 spp.] p.280
Spp. ID – leaves

greenish; **TWIGS** spiny
Gooseberry p.324

clustered; **LEAVES** aromatic
Bay p.331

DOGWOOD
WHITE DOGWOOD
RED-OSIER DOGWOOD
±domed clusters
Dogwoods [3 spp.] p.304
Spp. ID – fruit, leaves, twigs

white or pink; singly or in tight clusters
Snowberries p.332
Coralberries p.332
Spp. ID – leaves + fruit

ELDER
RED-BERRIED ELDER

±globular **LEAVES** lobed
Guelder-roses [2 spp.] p.308
Spp. ID – leaves

±conical clusters
Privets [2 spp.] p.312
Spp. ID – leaves, twigs

Cherry Laurel p.178
pendent branched clusters; **LEAVES** shiny, not aromatic

Elders [2 spp.] p.306
Spp. ID – fruit, pith

WRINKLED VIBURNUM   LAURUSTINUS
WAYFARING-TREE

ROWAN
red or brownish in ±tight ±slightly domed clusters

pendent branched clusters; **LEAVES** dull, not aromatic
Bird Cherry p.174   **Rum Cherry** p.179

Cherries [3 spp.] pp.176, 179
Spp. ID – leaves

egg-shaped **LEAVES** oval
Wayfaring-tree Other viburnums [4 spp.] p.310, p.335
Spp. ID – fruit, leaves

Rowans, Whitebeams [44 spp.] p.133
Spp. ID – leaves, twigs

76

FRUIT IDENTIFICATION 2/2

NOTE images are not scaled accurately

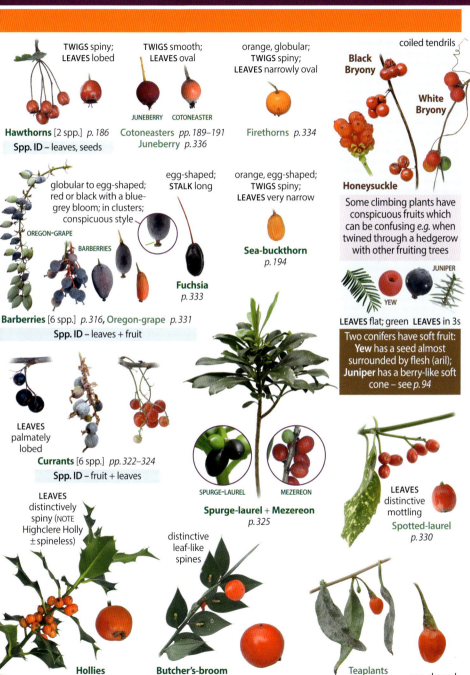

ID WINTER TWIG IDENTIFICATION

Key to winter twigs
PALE GREEN NAMES INDICATE PARK, STREET & GARDEN SPECIES

Using winter twigs for identification | This may seem daunting, but is actually reasonably straightforward for many species if the important features are understood, looked at in detail and measurements taken for some. The identification process needs some or all of these assessments:

- whether the side- (*lateral*) buds are arranged oppositely or alternately (same as leaf arrangement)
- whether the twig is spiny; its colour; the nature of any hairs if present and the texture of the bark
- the shape and size of the buds and how many scales they have. NOTE bud scales protect the new young flowers or leaves; in a few species there are no protective scales and next season's leaves or folded flowers can be seen as tiny versions of what they will become.

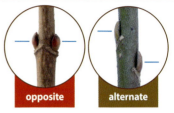

opposite | alternate

BUD SCALES 0 (*naked*)	BUD SCALES 1 – actually 2 fused	BUD SCALES 2–3 (4)		BUD SCALES ≥3		
		SCALES very different sizes 'boxing glove'	SCALES ± same size	BUDS pointed	BUDS conical	BUDS pointed egg-shape; egg-shaped to globular

- **details of other features**: *e.g.* distance between buds; whether buds are pressed to the stem, or angled away from it; pith colour; and other features used for identification as illustrated in the following pages. This book tries to avoid complex detailed botanical terms – any technical terms for descriptions used are given in *italic text*.

leaf-scar – the area to which the previous year's leaf-stalk was attached
scar ridges
end- (*terminal*) bud
side- (*lateral*) bud
bud scales (see *above*); scales can be ridged or flat (not ridged)
lenticels – openings that allow the passage of air
RIDGED | FLAT

BUDS ARRANGED OPPOSITELY

TWIGS hairy
Wayfaring-tree p.310
Other viburnums (p.335) are evergreen
FLOWER-BUD
LEAF-BUD

Dogwoods [3 spp.] p.304
LEAF-BUDS with FLOWER-BUD
DOGWOOD
TWIGS green; turning purplish red
RED-OSIER DOGWOOD dark red
WHITE DOGWOOD bright red

Bud scales absent (or apparently so)

green, maturing grey
new leaf-tips can be visible
Elders [2 spp.] p.306
ELDER PITH white
RED-BERRIED ELDER PITH orange

Butterfly-bush (Buddleia) p.328
BUDS 'naked'; a pair of tiny hairy leaves

78

WINTER TWIGS | BUDS OPPOSITE

NOTE images are not scaled accurately

BUDS ARRANGED OPPOSITELY

End-bud scales usually weakly ridged at least

Bud scales present

LEAF-SCARS RAISED

Privets *p. 312*

TWIGS pale greenish brown

WILD PRIVET

Wild Privet hairy
Garden Privet hairless

GARDEN PRIVET

END-BUDS large; green to purplish; usually in pairs

Lilac *p. 329*

LEAF-SCARS NOT RAISED; MATURE BUDS HAIRLESS

TWIGS grey- to greenish brown
Norway Maple *p. 288*

L 6–10 mm; reddish brown; green base

TWIGS grey- to greenish brown
Sycamore *p. 286*

L 5–10 mm; green with dark brown margin

TWIGS green to purplish red
Cappadocian Maple *p. 289*

L 3–7 mm; green to purplish green

TWIGS grey- to greenish brown
Field Maple *p. 284*

L 3–5 mm; reddish brown; pale margin

TWIGS green to purplish red
Red + Silver Maples *p. 289*

L 3–5 mm; red

INNER BARK Silver Maple with weak fetid aroma; Red odourless

LEAF-SCARS NOT RAISED; MATURE BUDS HAIRY

Ash-leaf Maple *p. 289*

Bud scales not keeled – 2 scales fused into 1

LEAF-BUDS like small leaves

FLOWER-BUDS globular

Guelder-roses *p. 308*

SIDE-BUDS very narrowly egg-shaped; pale, brown or black; pressed to twig

Purple Willow *p. 238*

twigs with at least some opposite or near-opposite buds unique amongst British & Irish willows

Bud scales not keeled – ≥4 scales

END-BUD 'BISHOP'S MITRE'; 4–6 SCALES

black
Ash *p. 298*

brown
Narrow-leaved Ash *p. 340*

pink- to grey brown
Manna Ash *p. 340*

END-BUD LARGE + BROAD (L > 8MM); 6–10 SCALES

Horse Chestnuts *p. 290*

HORSE CHESTNUT

Horse Chestnut
dark red-brown; 8–12 scales; usually very sticky; L 15–30 mm

Red H Chestnut
greenish brown; 8–10 scales; slightly sticky at most; L 15–20 mm

Indian H Chestnut
greenish red; 6–8 scales; sticky; L 8–15 mm

END-BUD SMALLER + NARROW (L > 10 MM)

TWIGS green; ridged

Spindle *p. 280*

egg-shaped green; margins dark

Dawn Redwood *p. 120*

cf. **Swamp Cypress** + **larches** (*p. 87*); both with alternate buds

TWIGS pale brown

egg-shaped; light red to yellow brown

Buckthorn *p. 193*

end-bud can be replaced by a thorn

NOTE: twigs can be spiny (see *p. 80*); buds can be alternate (see *p. 84*)

TWIGS dark brown or grey

conical; blackish

Key to winter twigs

WINTER TWIGS | BUDS ALTERNATE – SPINY TWIGS

NOTE images are not scaled accurately 2/5

TWIGS WITH SPINES OR THORNS

Buds rounded to egg-shaped

TWIGS reddish brown with broad-based spines
Honey Locust (*p. 343*) has similar buds but differs in its narrow based spines
BUDS minute
False Acacia *p. 300*

Barberries [6 spp.] *pp. 316–317*
BARBERRY — egg-shaped
BARBERRY   THUNBERG'S
TWIGS greyish; grooved; most species evergreen with 3–7-parted spines; a few 1-parted

TWIGS grey-brown
Hawthorn *p. 186*
± spherical (L 3–5 mm); shiny; dark red-brown

Midland Hawthorn *p. 186*
TWIGS grey-brown and often shiny, with fewer, shorter spines than Hawthorn

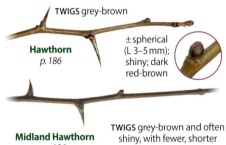

TWIGS blackish grey, can have spine at tip of branch
Blackthorn *p. 171*
± egg-shaped (L 1–2 mm); shiny; dark red-brown

TWIGS dark brown
Sea-buckthorn *p. 194*
egg-shaped;
♀ 2–4 scales;
♂ 6–8

bud arrangement alternate (Buckthorn ± opposite)

Buds pointed egg-shaped to conical

TWIGS dark brown or grey
end-bud can be replaced by a thorn
BUDS opposite or nearly so
conical; blackish
Buckthorn *p. 193*

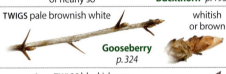

TWIGS pale brownish white
whitish or brown
Gooseberry *p. 324*

TWIGS blackish grey
Wild Pear *p. 184*
± conical; red-brown to blackish

Other fruit trees (Rosaceae) can have thorns present and separated by bud characters as follows:

Wild Apple *p. 180*
1ST-YEAR TWIGS grey-brown; BUDS pointed egg-shape; red- to purple brown

Plymouth Pear *p. 183*
1ST-YEAR TWIGS purplish; BUDS conical; pink- to purplish brown

Wild Plum *p. 173*
1ST-YEAR TWIGS grey-brown to green; BUDS conical (L 2–6 mm); dark brown

Cherry Plum *p. 172*
1ST-YEAR TWIGS green to purplish; BUDS conical (L 1–2 mm); pale to mid-brown

BUDS ARRANGED ALTERNATELY – no spines

Buds scales absent | twigs densely hairy

Stag's-horn Sumach *p. 337*
TWIGS fuzzy brown hairs
± conical; sunk in leaf scar

Wild Cotoneaster *p. 189*
STIPULES prominent

PLANTED COTONEASTERS *p. 190* are recognizable as a group by the densely hairy buds that have prominent stipules and the twigs which generally have yellowish hairs. There are many species, which require specialist identification beyond the scope of this book.

80

WINTER TWIGS | BUDS ALTERNATE – BUD SCALES 0–3   ID

BUDS ARRANGED ALTERNATELY | Bud scales absent | twigs slightly hairy

LEAVES IN BUD covered in dense hairs

Alder Buckthorn *p. 192*

BUDS ARRANGED ALTERNATELY | Bud scales apparently 1

WILLOWS, as a group, are easily recognized by their apparently **single-scaled buds** (actually two fused modified leaves) which detach as the new leaf emerges.

Willow twig identification ▶ *p. 230*

BUDS willow twigs and buds come in a range of shapes, sizes and colours which can help in identification

Purple Willow is unique among British & Irish willows in having opposite buds (*p. 238*)

TWIGS usually grey-brown; hairless

shiny; reddish; conical; characteristically curved

Planes [3 spp.] *p. 282*

Spp. ID – fruit quantity, bark

TWIGS usually olive-brown; tiny hairs at most

green; conical with long tapering tip; usually curved

Fig *p. 337*

BUDS ARRANGED ALTERNATELY | Bud scales 2–3

Bud scales ± the same size

Sweet Chestnut (*p. 216*) can have 2 bud scales but usually has more (up to 6)

TWIGS hairless; BUDS L 6–10mm; on **hairless** stalk (L 2–4mm)

Alder *p. 272*

TWIGS hairy; BUDS L 6–10mm; on **hairy** stalk (L 2–8mm)

Grey Alder *p. 274*

TWIGS hairless; BUDS L 4–8mm; on hairless stalk (**L 3–10mm**)

Italian Alder *p. 274*

LEAF-SCARS large

LEAF-SCARS large; BUDS L 8–20mm; on hairless stalk (L 2–4mm)

Tulip-tree *p. 339*

Buds obviously different sizes 'boxing glove'

Hazel (*p. 276*) has similar 'boxing glove' buds but these are green, with 6–8 scales

BUDS **Small-leaved Lime** smaller, rounder and more closely spaced than those of the other limes

Small-leaved Lime *p. 294*

BUD GAP >5 cm

Common Lime *p. 295*

BUD GAP <4 cm

Large-leaved Lime *p. 296*

TWIGS greyish; hairy

hairy

Silver Lime *p. 297*

81

ID WINTER TWIGS | BUDS ALTERNATE – BUD SCALES >3

Key to winter twigs
PALE GREEN NAMES INDICATE PARK, STREET & GARDEN SPECIES

Bud shapes on twigs with alternate buds The following twigs rely on an assessment of bud shape as the primary separator. Bud shape and size can vary between trees of the same species, and even on an individual tree, so it is recommended that these categories are regarded as broad and it may be necessary to look at one or more categories to find a match.

BUDS ARRANGED ALTERNATELY Scales ≥3 | end-buds larger, mostly L >8 mm; pointed

End-bud sticky; balsam aroma; L >15 mm | BALSAM-POPLARS

Balsam-poplars *p. 264*

HYBRID BALSAM-POPLAR
BALM-OF-GILEAD

usually obviously ridged   WESTERN BALSAM-POPLAR

TWIGS usually **ridged**; green- to yellow-brown	
Western	BUDS green- to red-brown
TWIGS usually **not ridged**; dark green- to red-brown	
Hybrid	BUDS green with brown margins
Balm-of-Gilead	BUDS reddish brown
Eastern	Spp. ID – suckers + leaf

End-bud ± sticky at most; no strong aroma | POPLARS, EXCEPT WHITE POPLAR

TWIGS YELLOWISH BROWN

TWIGS usually unridged

Black Poplar *p. 261*
END-BUD L <12 mm

TWIG TIPS more upswept in Black Poplar than in Hybrid Black Poplar

TWIGS usually ridged

Hybrid Black Poplar *p. 262*
END-BUD L usually >12 mm

TWIGS REDDISH BROWN

TWIGS hairless with orange lenticels; **SIDE-BUDS** ± appressed

Aspen *p. 256*
END-BUD L <12 mm; pale margins

TWIGS whitish flat hairs (some approaching White Poplar in hairiness) or hairless; **SIDE-BUDS** ± spreading

GREY POPLAR

Grey Poplar *p. 258*
END-BUD L <10 mm; dark margins

LEAF-SCARS pale brown (White Poplar (*p. 260*) has darker leaf-scars and smaller buds – see *p. 85*)

WHITE POPLAR

WINTER TWIGS | BUDS ALTERNATE – BUD SCALES >3

NOTE images are not scaled accurately

BUDS ARRANGED ALTERNATELY | Scales ≥3 | buds L < 10 mm; very narrow

TWIGS ±'ZIG-ZAGGING'

TWIGS dull purple-brown to grey

Beech *p. 214*
SIDE-BUDS
L 11–25 mm;
reddish brown;
held away from the twig

TWIGS green-brown to dark brown

Hornbeam *p. 278*
SIDE-BUDS L 6–10 mm;
greenish brown
pressed against the twig

BUDS ARRANGED ALTERNATELY | Scales ≥3 | buds L < 15 mm, ± narrow cone to 'pointed egg'

Small tree (H < 3 m)

TWIGS reddish brown with grey skin

Juneberry *p. 336*
BUDS pale yellowish green tinged purplish

Low shrubs (H < 3 m)

BUDS PALE; NOT AROMATIC

TWIGS grey-brown

Mountain Currant *p. 324*
BUDS very pale green or pink; scales 5–9

BUDS PALE; AROMATIC

TWIGS pale to purplish brown

Black Currant *p. 322*
BUDS with yellow glands

TWIGS reddish brown

Flowering Currant *p. 322*
BUDS with clear or reddish glands

BUDS DARK

TWIGS pale grey-brown

Red Currant *p. 323*

Downy and Red Currants indistinguishable in winter

Mezereon *p. 325*
TWIGS grey-brown; dotted black

ID — WINTER TWIGS | BUDS ALTERNATE – BUD SCALES >3

Key to winter twigs
PALE GREEN NAMES INDICATE PARK, STREET & GARDEN SPECIES

BUDS ARRANGED ALTERNATELY | Scales ≥3 | buds L <15 mm, ± narrow cone to 'pointed egg'

Trees (H >3 m); bark white or pale brown with horizontal fissures | **BIRCHES**

END BUDS L >8 MM – much longer than those of native birches | **PLANTED BIRCHES**

Paper Birch *p. 270*
BUDS L 9–15 mm

2ND-YEAR TWIGS hairless; white-dotted

Himalayan Birch *p. 270*
BUDS L 8–12 mm

2ND-YEAR TWIGS hairy; white-dotted

END BUDS L <8 MM | **NATIVE BIRCHES**

TWIGS HAIRLESS

YOUNG TWIGS hairless; white-dotted

Silver Birch *p. 268*
BUDS L 5–7 mm; typically more pointed and greener than in Downy Birch

TWIGS DOWNY

YOUNG TWIGS softly hairy

Downy Birch *p. 268*
BUDS L 5–7 mm; typically more rounded and reddish than in Silver Birch

Dwarf Birch (*p. 179*) is a small shrub with globular buds (see *p. 87*)

BUDS ARRANGED ALTERNATELY | Scales ≥3 | buds L <15 mm, ± broad cone to 'pointed egg'

BUDS usually some on twig ± opposite

Buckthorn can have alternate buds, but is usually thorny (*p. 80*) with buds opposite (*p. 79*)

Buckthorn *p. 193*

Rowan *p. 140*
BUDS dark brown; covered in grey hairs

Whitebeams *p. 136*
Whitebeams are not safely identifiable to species by their winter twigs

BUDS green-brown to brown; typically hairless
Wild Service-tree (*p. 142*) similar but end-buds rounded (see *p. 87*)

True Service-tree *p. 148*
BUDS green to reddish brown; scales with dark margins

Bird Cherry *p. 174*

TWIGS shiny, dark brown | BUDS dark brown with paler tip

Other cherries (*p. 86*) buds less pointed; usually clustered

WINTER TWIGS | BUDS ALTERNATE – BUD SCALES >3   ID
4/5

NOTE images are not scaled accurately

BUDS ARRANGED ALTERNATELY — Scales ≥3 | buds L < 10 mm, ± broad cone to 'pointed egg'

TWIGS smooth; red-brown to grey; ± 'zig-zagging'

Ginkgo *p. 338*
BUDS conical; brown; LEAF-SCARS spiral

TWIGS densely hairy, especially near tip

Medlar *p. 336*
BUDS conical

TWIGS greenish; can be hairy

Laburnum *p. 330*
BUDS green; usually covered in white hairs

WHITE POPLAR

LEAF-SCARS dark (*cf.* Grey Poplar – see *p. 82*)

GREY POPLAR – hairy twig example

White Poplar *p. 260*
TWIGS + BUDS covered in whitish grey downy hairs; END-BUDS with 3–5 scales (Grey Poplar 7–12)

'FRUIT' TREES (ROSACEAE)

* twigs can be spiny – see *p. 80*

1ST-YEAR TWIGS green to purplish

* **Cherry Plum** *p. 172*
BUDS (L 1–2 mm); pale to mid-brown

Wild Plum twigs can be green and best separated from Cherry Plum by bud size and colour.

1ST-YEAR TWIGS grey-brown to green

* **Wild Plum** *p. 173*
BUDS (L 2–6 mm); dark brown

TWIGS purplish; hairless

* **Plymouth Pear** *p. 183*
BUDS pink- to purplish brown

1ST-YEAR TWIGS grey- to reddish brown

Cultivated Pear *p. 185*
BUDS L 4–8 mm; hairless; scales can have grey skin

Wild Pear (*p. 184*)
TWIGS usually spiny; END-BUDS L 3–5 mm

WILD PEAR

1ST-YEAR TWIGS grey-brown

* **Wild Apple** *p. 180*
BUDS L 3–8 mm; red- to purple brown; scales usually with a few hairs at least

Domestic Apple (*p. 181*) END-BUDS usually more rounded

DOMESTIC APPLE

85

ID
WINTER TWIGS | BUDS ALTERNATE – BUD SCALES >3

Key to winter twigs
PALE GREEN NAMES INDICATE PARK, STREET & GARDEN SPECIES

BUDS ARRANGED ALTERNATELY | Scales ≥3 | buds L <10 mm, ±'pointed egg'

BUDS SOLITARY AT TIP | ELMS

TWIGS can be corky

'Field' Elms *p. 201*
BUDS L 3–6 mm; hairless

TWIGS never corky

'Wych' Elms *p. 200*
BUDS L 5–8 mm; hairy

TWIGS greyish brown; usually with tiny hairs

European White-elm *p. 199*
BUDS reddish brown with darker margins

BUDS IN CLUSTERS ONLY AT TIP | OAKS

TWIG greenish brown with minute greyish hairs

Turkey Oak *p. 222*
BUDS with hair-like stipules

TWIG red- to green- or grey-brown

Pedunculate Oak *p. 218*
LEAF-BUD L 3–8 mm; reddish brown; usually hairless; BUD-SCALES <20

TWIG reddish brown to greenish brown

Sessile Oak *p. 221*
LEAF-BUD L 5–15 mm; pale brown; usually hairy; BUD-SCALES >20

Other planted deciduous oaks, such as Red Oak, are similar but can be separated as a group by their lower bud scales (*insets right*) which are notched or split at the tip – see *p. 224* for differences between the species within the group.

PEDUNCULATE OAK   RED OAK

BUDS IN CLUSTERS AT TIP AND/OR FURTHER ALONG GREY TO REDDISH BROWN TWIGS | CHERRIES

END-BUD HAIRLESS; MAY HAVE GREY SKIN

SIDE-BUDS L 5–7 mm

Wild Cherry *p. 176*
END-BUD hairless; shiny red-brown L 2–8 mm; BUD-SCALES ± 10

End-buds can be solitary and lacking adjacent side-buds

SIDE-BUDS L 2–4 mm

Dwarf Cherry *p. 179*
END-BUD hairless; shiny red-brown; L 2–8 mm; BUD-SCALES 4–7

END-BUD WITH HAIRY TIP

Japanese Cherry *p. 179*
END-BUD hairy at tip; shiny red-brown; L 7–12 mm; BUD-SCALES 8–10

WINTER TWIGS | BUDS ALTERNATE – BUD SCALES >3   ID

NOTE images are not scaled accurately

BUDS ARRANGED ALTERNATELY | Scales ≥3 | buds L < 10 mm, ± 'pointed egg' to globular

CONIFEROUS TREES
NOTE: there are usually cones in the vicinity which can confirm identification – see *p. 94*

European Larch *p. 106*

TWIGS dull brown

Hybrid Larch *p. 107*

YOUNG TWIGS can be slightly hairy with a slight bloom

Japanese Larch *p. 107*

YOUNG TWIGS typically reddish brown; slightly hairy with a grey waxy bloom

Swamp Cypress *p. 121*

TWIGS thin; usually purplish brown

BROAD-LEAVED TREES OR LARGE SHRUBS

Walnut *p. 196*

LEAF-SCARS large, distinctive 'Y'-shape

BUDS egg-shaped to globular; end-bud larger than side-buds

TWIGS quite stout; grey to purple-brown; PITH CHAMBERS (*inset*) 8–12 per cm

Black Walnut (*p. 196*) PITH CHAMBERS 14–18 per cm

Sweet Chestnut *p. 216*

TWIGS olive to red-brown with rough, whitish granules

BUDS orange-red to purplish with 2–6 scales

Hazel *p. 276*

TWIGS flexible; pale brown to olive-brown

BUDS 'boxing-glove'; ±8 scales

Limes (*p. 81*) BUDS similar but brown with 2–3 scales

Wild Service-tree *p. 142*

TWIGS hairless; with numerous lenticels

BUDS globular; scales green with narrow brown margin

Other Whitebeams (*p. 84*) END-BUDS more pointed

Tamarisk *p. 326*

TWIGS very thin

BUDS globular; usually part hidden behind papery scale

SMALL SHRUBS (H < 2 m)

Bog Myrtle *p. 314*

TWIGS dark reddish brown

♂ 20–30 scales
♀ 6–8 scales

Creeping Willow (*inset, right* and *p. 230*) is similar in form and found in similar habitat, but appears to have only 1 bud scale

Dwarf Birch *p. 271*

TWIGS dull brown

BUDS smaller (L 1–2 mm) than those of other birches

Downy Birch (*p. 268*) can be found in similar habitats as a very short, low-growing plant but has much larger buds (L 4–7 mm)

87

Introduction to conifers

Conifers are usually thought of as trees that bear cones; more precisely, they are gymnosperms – plants that have 'naked' seeds that can be seen in the cone or the fruit without having to cut it open. There are approximately 600–630 species of conifer globally, including kauris, junipers, pines, redwoods and yews. There are also many hundreds of subspecies and varieties in cultivation.

Despite a relatively low number (approx. 600) of conifer species worldwide, conifer forests make up a large percentage of the world's forests, with the taiga (or boreal forest) forming a transcontinental belt of coniferous forest that encircles the Northern Hemisphere at higher latitudes, with the Scots Pine forests of the Scottish Highlands forming part of the belt. Conifer forest is the largest terrestrial habitat on Earth, making up one-third of the world's forested land (dominated by spruce, fir, pine and larch depending on the region) and covering some 15 million square kilometres (equivalent to 42× the size of Germany). About 60% of this area is in Russia, 30% in Canada, with 10% shared between Alaska, the Baltic states, Scotland and Scandinavia. Conifers make up 51% of all woodland in Britain and Northern Ireland, with a total area of 1·65 million hectares of trees.

In Britain & Ireland, there are more conifers growing than broadleaved trees, although the majority of these are introduced conifers in plantations. Conifers can be found growing almost everywhere from dry chalk downland to salt laden coasts, from lowland heathlands to upland mountains as well as in churchyards and gardens throughout the country.

The majority of the species are evergreen conifers, retaining their leaves and needles throughout the year, making them more obvious in the winter when the few deciduous conifers have lost all their leaves. In the following conifer species accounts, all are evergreen except where stated. Many conifers have adapted to survive the winter whilst retaining their leaves, by developing a narrow conical shape, with limbs which droop downward to help reduce the impacts of wind and snow. Many of the species also have small needle-like leaves and an ability to produce chemicals which help the tree to resist freezing. These adaptations have all added to their ability to survive harsh winter conditions.

These adaptations to survive many of the rigors of life have allowed conifers to become the world's tallest, largest and oldest trees. Coast Redwoods (*p. 122*) are the tallest at more than 115 m (taller than a 30-storey building); Giant Redwoods, historically known as Wellingtonia (*p. 123*), are the largest, with a tree called General Sherman in California holding the record for being the largest volume tree; whilst the oldest are Bristlecone Pines *Pinus longaeva* (N/I), some having been found to be more than 4,800 years old, such as Methuselah, a tree which grows in the White Mountains of eastern California; however, there is now a new challenger to the title of oldest tree alive – another conifer – the Patagonian Cypress *Fitzroya cupressoides* (N/I) having recently been suggested to be 5,000 years old.

What is the difference between evergreen and deciduous trees?

An evergreen tree retains green leaves all year-round. A deciduous tree loses its leaves completely for part of the year, becoming bare and leafless. The terms 'evergreen' and 'deciduous' are sometimes taken as shorthand for conifers (evergreen) and broadleaves (deciduous). However, this does not work in every case as there are deciduous conifers, for example, Dawn Redwood (*p. 120*) and Larches (*p. 106*), and evergreen broadleaf trees, including Holly (*p. 302*) and Evergreen Oak (*p. 223*).

Native conifers

Numerically, most of the conifer trees planted in Britain & Ireland are not native, with the introduced Sitka Spruce making up more than half of all the conifers growing in woodlands in Britain, due to large plantations in Scotland. However, we do have three native species – Scots Pine (*p. 96*), Common Juniper (*p. 118*) and Yew (*p. 116*) – each of which form distinctive woodlands and habitats.

All three native species are highly distinctive. The most abundant species is Scots Pine, which used to occur over large areas of Scotland although now covers only 1% of the area where it was once found. However, there are still sizeable forests of Scots Pine in the Grampians and the Highlands of Scotland.

The next most-abundant species is the Yew, which has an unusual distribution, preferring lime-rich soils on chalk or limestone. Yew is most frequently seen in parks, gardens and churchyards, where individuals have become some of our oldest trees, such as the Fortingall Yew in Perthshire, which is several thousand years old. However it also occurs uncommonly across the country as pure Yew woodland.

The remaining native species is Juniper, which also has an unusual distribution – growing on chalk and limestone soils in the south of England as well as in high mountains and on beaches in Scotland. It also occurs on coastal sites in Cornwall and Pembrokeshire, where it grows as a small hemispherical shrub.

All three of our native conifers are important to our ecology and landscape. They should all be celebrated for their contribution to our treescape and recognized for what they are – some of the greatest survivors and oldest living things on our islands.

Non-native conifers

Conifers have been planted across Britain & Ireland from all over the globe. Some of the most widely planted are Sitka Spruce (*p. 104*), Corsican Pine (*p. 98*), Norway Spruce (*p. 105*), European Larch (*p. 106*), Douglas Fir (*p. 112*) and Lodgepole Pine (*p. 102*), all of which have been used for commercial plantations. Within the different countries the proportion of conifers varies, with approximately one-quarter (26%) of England's woodland made up of conifers compared to almost three-quarters (73%) of Scotland's.

Along with the commercial conifers, a wide range of other conifers have been introduced especially for ornamental purposes. This has led to a wide range of conifer forms and varieties – particularly columnar, pyramidal, drooping, and globular forms.

This book covers a range of the most widely encountered introduced conifers, including some like the Dawn Redwood known as 'fossil trees' – so called because fossil records of these species date back more than 150 million years.

Conifers have been planted extensively in some upland areas, as here in Cumbria.

CONIFERS | IDENTIFICATION

Key to conifers

PALE GREEN NAMES INDICATE PARK, STREET & GARDEN SPECIES

Identification tips | Identify to family by: ● tree shape/form ● leaf shape and arrangement ● mature female cone details ● male cone details

PINE [Pinaceae]

PINES [9 spp.]
FORM typically tall, evergreen trees (1 small and shrubby). LEAVES typically long (L > 7 cm); needle-like in **groups of 2, 3 or 5**. ♀ CONES typically reddish purple when young; turning green; woody when mature (L > 5 cm). ♂ CONES small; typically yellow; clustered at shoot base.

FIRS + SPRUCES [6 spp.]
FORM tall evergreen trees. LEAVES short (L < 6 cm); needle-like or very narrow and flat, **not paired or grouped**. ♀ CONES typically reddish purple when young; turning green and then brown (mature L > 5 cm). ♂ CONES typically small and reddish/yellowish.

LARCHES [3 spp.]
FORM typically tall, deciduous trees. LEAVES short (L < 3 cm); needle-like in rosette-like clusters or singly. ♀ CONES green-purple; woody when mature (L > 5 cm); cones fall from tree. ♂ CONES along twig underside.

CEDARS [3 spp.]
FORM tall, evergreen or deciduous trees. LEAVES short (L < 4 cm); needle-like in rosette-like clusters or singly. ♀ CONES along twig upperside; woody when mature (L > 5 cm); cone spikes remain on tree. ♂ CONES yellowish.

JUNIPER (incl. Cypresses) [Cupressaceae]

JUNIPER [1 spp.]
FORM evergreen shrub. LEAVES needle-like in whorls of 3. ♀ CONES tiny (L ± 2 mm); yellowish; soft-scaled, berry-like (D to 13 mm); green then black when mature. ♂ CONES small (L ± 8 mm); on twigs.

CYPRESSES [13 spp.]
FORM evergreen or deciduous trees or evergreen shrubs. LEAVES very narrow or scale-like. ♀ CONES small, woody with fused scales that open when mature (L > 5 cm). ♂ CONES tiny, at tips of shoots.

YEW [Taxaceae]

YEW [1 spp.]
FORM evergreen tree. LEAVES very narrow and flat (L < 4 cm). ♀ CONES berry-like; red when mature (L 8–15 mm). ♂ CONES tiny, yellowish.

MONKEY-PUZZLE [Araucariaceae]

MONKEY-PUZZLE [1 spp.]
FORM tall, evergreen tree. LEAVES distinctive broad triangles. ♀ CONES ± globular (L to 18 cm). ♂ CONES cylindrical.

CONIFERS | IDENTIFICATION | ID

NOTE images are not scaled accurately

PINES | NEEDLES long; in pairs, 3s or 5s

Species ID | also use mature ♀ cone shapes (p. 94)

NEEDLES in pairs

NEEDLES in 3s
  Monterey Pine p. 101

NEEDLES in 5s

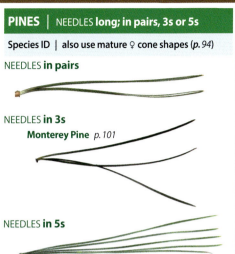

NEEDLES in pairs

Scots Pine	L 2–8 cm; bluish green; usually twisted	p. 96
Lodgepole Pine	L 3–8 cm; dark or yellow-green; twisted	p. 102
Dwarf Mountain Pine	L 3–8 cm; bright green; straight	p. 103
Austrian Black Pine	L 8–12 cm; dark green; straight	p. 98
Corsican Pine	L 10–18 cm; dark green; slightly twisted	p. 98
Maritime Pine	L 15–20 cm; dark green; straight	p. 100

NEEDLES in 5s

Weymouth Pine	L 5–13 cm; bluish green; twisted	p. 103
Macedonian Pine	L 7–12 cm; ±stiff; point towards tip	p. 103
Bhutan Pine	L 8–20 cm; flexible; hang down	p. 103

NEEDLES single, in groups; on pegs or broadly triangular

MONKEY-PUZZLE | NEEDLES triangular

Monkey-puzzle
p. 129

JUNIPER | NEEDLES in whorls of 3

Common Juniper
p. 118

LARCHES + CEDARS | NEEDLES in rosettes

SPRUCES | NEEDLES on pegs

Norway Spruce
p. 105

UPPER

UNDER

NEEDLE CROSS-SECTION
4-angled

NEEDLES single; in rosettes on short shoots
(NOTE singly on long shoots)

LARCHES	deciduous; FORM narrower; NEEDLE CROSS-SECTION ±flat; CONES smaller (L<4 cm)	p. 106
	Species ID	mature ♀ cone details	
CEDARS	evergreen; FORM broader; NEEDLE CROSS-SECTION angled; CONES larger (L<5 cm)	p. 108
	Species ID	needle details	

Sitka Spruce
p. 104

UPPER

UNDER

NEEDLE CROSS-SECTION
±flattened

91

ID

Key to conifers

PALE GREEN NAMES INDICATE PARK, STREET & GARDEN SPECIES

NEEDLES single; flattened

NEEDLES attached to branchlet by 'sucker'

Grand Fir *p.114*

European Silver-fir *p.115*

NEEDLES all similarly sized; more angled and shorter than those of Grand Fir

NEEDLES of two different sizes; flatter arrangement and longer than in European Silver-fir

UNDERSIDE 2 white (usually) bands

UNDERSIDE 2 greenish white bands

NEEDLES attached to branchlet 'normally'

1/2

NEEDLES WITH DISTINCTIVE ARRANGEMENTS

Noble Fir *p.113*

NEEDLES distinctively curving up from the base of the branchlets

Douglas Fir *p.112*

NEEDLES arranged all around twig

CONES 3-pointed bracts diagnostic (see *p.94*)

NEEDLES ARRANGED SPIRALLY

Western Hemlock-spruce *p.111*

Yew needles are spirally arranged but appear to be opposite (see *facing page*)

Japanese Red-cedar *p.125*

UPPERSIDE grooved

NEEDLES arranged in two spiral ranks; UPPERSIDE grooved; BASE held away from twig; TIP rounded

NEEDLES spirally arranged; BASE pressed to twig; TIP pointed

92

NOTE images are not scaled accurately

ID

NEEDLES attached to branchlet 'normally' 2/2

NEEDLES OPPOSITE (OR APPEARING TO BE)

UPPERSIDE bright green; **UNDERSIDE** with two faint pale green lines

Dawn Redwood p. 120

Yew p. 116

UPPERSIDE dark glossy green; **UNDERSIDE** paler yellowish green

NEEDLES ALTERNATE

Swamp Cypress p. 121

UPPERSIDE green; **TIPS** ± rounded

UPPERSIDE dark green; **TIPS** ± pointed

Coast Redwood p. 122

NEEDLES scale-like 1/2

BRANCHLETS rounded; 'rope'-like

Giant Redwood p. 123

NEEDLES tightly arranged on branchlets

NEEDLES scale-like 2/2

BRANCHLETS flattened

NOTE: the deciduous **Tamarisk** (p. 326) has scale-like leaves – see p. 60

AROMA WHEN RUBBED lemon; **UNDERSIDE** glands typically **absent**

Monterey Cypress p. 125

AROMA WHEN RUBBED fruity resinous; **UNDERSIDE** resin glands **absent**

Leyland Cypress p. 126

AROMA WHEN RUBBED parsley; **UNDERSIDE** stomatal patches x-shaped

Lawson's Cypress p. 127

AROMA WHEN RUBBED pear-drops or pineapple; **UNDERSIDE** stomatal patches white, 'butterfly'-shaped

Western Red-cedar p. 124

Other less widespread planted species are somewhat similar and differentiated by details of the needles

Nootka Cypress	dark green; L 3–5 mm; **UNDERSIDE** wholly green, resin glands present but no visible white stomata	p. 128
Northern White-cedar	yellowish green; L 1–4 mm; **UNDERSIDE** paler yellowish stomatal patches; **AROMA** reminiscent of apples	p. 128
Sawara Cypress	light to dark green; L 1·5–2·0 mm; **TIP** long, slender point; **UNDERSIDE** small gland and white 'butterfly'-shaped stomatal patches	p. 128

93

ID CONIFERS | IDENTIFICATION

PALE GREEN NAMES INDICATE PARK, STREET & GARDEN SPECIES

Large to medium-sized cones; generally pendent to erect

Small cones of flat- and scale-leaved species | soft 'fruit' of Juniper and Yew

In most cases **small-coned species** are best identified by leaf details (see p. 90)

94

NOTE images are not scaled accurately

CONIFERS | IDENTIFICATION **ID**

Large cones; generally spreading to erect — disintegrating; leaving central spike on tree

NEEDLES in rosettes

NEEDLES flat; more openly arranged

European Silver-fir *p. 115*

NEEDLES very closely arranged

Noble Fir *p. 113*

Cedrus **Cedars** [3 spp.] *p. 108*

Spp. ID – form + needle details

Cone size differs slightly between species but is not reliable for identification

BRACTS hidden

Grand Fir (*p. 114*) similar but with bracts visible

BRACTS visible

Large cones; generally spreading to erect — fallen cones entire; no central spike left on tree

SCALES no sharp point

Dwarf Mountain Pine *p. 103*

SCALES slender prickle

Lodgepole Pine *p. 102*

SCALES point or bump

SCALES central transverse ridge

Maritime Pine *p. 100*

Monterey Pine *p. 101*

SCALES top angled; central bump

Scots Pine *p. 96*

SCALES top curved; central transverse ridge

Black Pines *p. 98*

95

PINACEAE (PINES) | LONG, THIN NEEDLES   Conifers compared *pp. 90–95*

LC LC Scots Pine   *Pinus sylvestris*

The region's only native pine is tall (H ≤ 37 m in B&I, to 50 m around the Baltic). The oldest trees in Britain & Ireland are approximately 300 years old although a tree in Finnish Lapland was recorded as 780 years old in 2007.

It only occurs naturally in Scotland where it is widespread, especially in the Spey Valley and Upper Dee areas. It is also widely planted and generally naturalized throughout Britain & Ireland and is a feature of many heathlands in southern England. It can tolerate a wide range of soils but has a preference for light, dry, sandy soil and gravels. **Features:** BARK thick, scaly or flaky; **distinctively two-toned**. WOOD strong, moderately hard; used historically for charcoal and telegraph poles and is still used for doors, floors and furniture. TWIGS young shoots green at first and grey-buff by the end of the first summer. BUDS ± conical, typically thinly covered in white resin. LEAVES **blue-green** needles (L 2–10 cm; W 1–2 mm) in **twisted pairs** that hang down in clusters. CONES ♂s clustered at a shoot base; ♀s singly at a shoot tip; ripening over three years – purple once pollinated, turning green, and then woody in their second year (L ≤ 7·5 cm when mature).

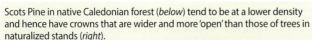

J F M A M J J A S O N D

> **Did you know?** Native Scots Pine forests in Scotland are known as the Caledonian Forest. Pollen records from peat bogs in north-west Scotland show that Scots Pine has been present in Scotland for at least 8,500 years, and some modern trees may be the direct descendants of those that first colonized the area after the last ice age. The habit, bark, leaf colour, leaf length and cone form of Scottish trees have led to the suggestion they are different enough from other populations to be classified as a subspecies or variety (*scotica*). There is also a unique population around Loch Maree, Wester Ross, which has adaptations to climate that are unlike any other Scottish trees, including a distinctive resin chemistry. One theory is that this population may have survived the last ice age in a refuge, centred off western Scotland, and then spread to Loch Maree.
>
> Scots Pine forests once covered an estimated 70% of Scotland but only 1% now remain due to centuries of felling and intensive sheep grazing. Growing in these forests are impressive, solitary, ancient individuals, known as 'Granny Pines' which produce large quantities of seed and, consequently, are favoured by a wide range of wildlife as a food source.

Scots Pine in native Caledonian forest (*below*) tend to be at a lower density and hence have crowns that are wider and more 'open' than those of trees in naturalized stands (*right*).

96

PINACEAE (PINES) | LONG, THIN NEEDLES

NEEDLES bluish green twisted pairs; much shorter than in Black Pine

♀ **NEW CONE** small; globular; red; at tip

♂ **CONE** yellow or pinkish; clustered

BUDS orange-brown; less resinous than in Black Pine

♀ **YOUNG CONE**

BARK the lower trunk has thick, grey or reddish dark brown bark; the upper third is usually flaky or scaly and shows a distinctive salmon pink colour

♀ **MATURE CONE** cone scales with a long raised bump at their centre; cones open to release seeds when ripe

Similar species | Black Pines (*p. 98*); Dwarf Mountain Pine (*p. 103*)

RED SQUIRREL

OSPREY

Associated species | Scots Pine has more than 172 invertebrate species, including **Pine Hawkmoth**, that feed on it. It also provides nesting sites for **Osprey** and **Scottish Crossbill**. The cones also provide food for **Red Squirrel** which, in turn, are eaten by the rare **Pine Marten**.

One of the more unusual aspects of Scottish pinewoods is a group of rare 'tooth fungi', named after their odd tooth-like spore-producing structures. The rare species are concentrated in the pine forests of the Central and Eastern Highlands and include species such as **Orange Tooth** and **Greenfoot Tooth**.

Plants that are strongly associated with the pinewoods include **Lesser Twayblade**, **Creeping Lady's-Tresses**, **Intermediate Wintergreen**, **One-flowered Wintergreen** and **Twinflower**.

There are also many rare insects associated with old or dead pine trees such as the specialist hoverfly *Callicera rufa*, which lays its eggs in rot holes.

PINACEAE (PINES) | LONG, THIN NEEDLES   Conifers compared *pp. 90–95*

Black Pines   *Pinus nigra*

The historical naming of Black Pines has been complicated, but it is now accepted that *Pinus nigra* has two main subspecies: ssp. *nigra* (the eastern European subspecies, known as **Austrian Pine**) and ssp. *laricio* (the western European subspecies, known as **Corsican Pine**).

LC ■ Austrian Black Pine *Pinus nigra* ssp. *nigra*

A tall conifer (H to 45 m) with a straight trunk (girth to 6 m), and a **broad crown of stout, densely needled branches** that are level with upswept tips. Found naturally in Austria and the Balkans and planted/naturalized in Britain & Ireland. **Features:** BARK entire trunk **usually grey-brown**; becoming thick and scaly or plated in older trees. TWIGS young shoots green at first, becoming orange-buff by the end of the first summer. BUDS red-brownish with whitish fringes to the scales and usually with some patchy grey-white resin. LEAVES pairs of long, stout, dark green needles (**L 8–12 cm**, a few up to 18 cm; W 1·5–2·0 mm) that are usually curved and slightly twisted, with finely toothed margins. CONES ♂ (similar to Scots Pine) yellow; 1·0–1·5 cm long in clusters at the base of a shoot. ♀ (L 3–9 cm when mature) are red when young, purple once pollinated, turning green and ripening grey-buff to shiny yellow-buff. Cone scales are rounded with a transverse ridge, lacking the raised bump found in Scots Pine.

J F M A M J J A S O N D

CROWN broad | BRANCHES tips upswept; densely needled

> **Did you know?** Corsican Pine was introduced into Britain & Ireland in 1759 as *Pinus sylvestris v maritima* and widely planted for its timber and for shelter in parks, gardens and on railway banks. It became *Pinus maritima* (1810), then *Pinus laricio* (1822) and most recently *Pinus nigra* ssp. *laricio*. Austrian pines were introduced in 1835 by Charles Lawson, a Scottish nurseryman after whom the Lawson Cypress is also named. The timber has been widely used for buildings, fencing, telegraph poles and railway sleepers. Unfortunately, a widespread disease called Red Band Needle Blight (caused by the fungus *Dothistroma septosporum*) has impacted Corsican Pine. Symptoms include the older needles developing yellow and brown spots and bands, with the ends of the needles turning reddish brown whilst the needle base remains green. The infected needles are most visible in June and July.

BARK entire trunk grey-brown

LC ■ Corsican Pine *Pinus nigra* ssp. *laricio*

A tall conifer (H to 50 m) with a straight trunk and a **narrow crown of slender, sparsely needled branches** that are level along their length. Found naturally in Corsica, Sicily and southern Italy and used in Britain & Ireland in plantings – particularly on acidic, freely draining, sandy loams. It can cope with exposure to wind and pollution, but is susceptible to frost damage, and grows best in southern lowland areas, including sand dunes as long as not directly exposed to salt spray. **Features:** BARK entire trunk usually a **dark greyish brown to black** (hence the scientific name *nigra*); becoming deeply furrowed in older trees. TWIGS as Austrian Black Pine. BUDS red-brown with grey fringes to the scales and usually thickly covered in greyish resin. LEAVES pairs of long, slender dark green needles (**L 10–18 cm**; W 0·8–1·5 mm) that are usually slightly twisted. CONES ♂ and ♀ cones are similar to those of Austrian Black Pine. Mature ♀ cones are 5–8 cm long.

CROWN narrow | BRANCHES level; sparsely needled

BARK dark greyish brown to black

PINACEAE (PINES) | LONG, THIN NEEDLES

AUSTRIAN

♀ **MATURE CONE** cone scales rounded with a transverse ridge

NEEDLES paired; **thicker and usually shorter** than in Corsican Pine; longer than in Scots Pine

BUDS reddish brown; scales white-fringed with patchy greyish resin

CORSICAN

NEEDLES paired; **slender; longer** than in Corsican or Scots Pines

♀ **MATURE CONE** as Austrian Black Pine

Red Band Needle Blight

Similar species |
Scots Pine (*p. 96*)
Dwarf Mountain Pine (*p. 103*)

Corsican Pines on sand dunes (*left*); and a parkland Austrian Black Pine (*right*).

PINACEAE (PINES) | LONG, THIN NEEDLES

LC ■ Maritime Pine — *Pinus pinaster*

A tall conifer (H to 40 m), usually with an open crown of upswept branches. Found naturally around the Mediterranean from Spain and Portugal to Italy, from sea level up to 600 m, and also in North Africa where it can be found at an elevation of 2,000 m in Morocco. Records suggest that it was introduced to Britain sometime before 1597– the year John Gerard published his *Herball, or Generall Historie of Plantes*. It is planted in small plantations and shelterbelts and has become naturalized in warm, sandy, southern heathland areas, especially near the coast. Some of the most famous Maritime Pine areas in Britain are around Bournemouth and the Dorset coastal heathlands. **Features:** BARK thick; dark brown; with fissures between scaly plates revealing the reddish brown inner bark. TWIGS buff to yellow-brown. BUDS non-resinous; large (L 25–32 mm); bud scales red-brown with long free tips fringed with white hairs. LEAVES green to yellow-green, stout, **straight needles (L 15–20 cm**; W 2–3 mm) with a persistent basal sheath (L to 2 cm). CONES ♂ cylindrical (L 1·0–1·5 cm), yellow. ♀ in whorls of 4–8; stalkless or on very short, stout stalks; very large (**L 10–18 cm when mature**); **rich chestnut-brown** with thick, stout, woody cone scales.

J F M A M J J A S O N D

♀ MATURE CONE ± stalkless; in clusters of 4–8; cone scales with raised horizontal ridge

BARK dark brown; thick; splits into scaly plates

BUDS lack resin and have scales with long, fringed tips

NEEDLES ± straight and stiff; pairs (can be 3)

Maritime Pine on the coast near Wareham, Dorset.

Did you know? Historically resin was collected from Maritime Pine to produce oils, varnishes, turpentine, soap and waxes, and the species was the most important source of resin in Europe; however, over the last 80 years this practice has largely ceased. To harvest the resin, the tree is 'tapped' by cutting 'grooves' into the bark (usually 12 cm wide × 3 cm long) on the south side of the trunk and the resin that oozes out is collected in a pot. Resin exports from Portugal and Spain are sent mainly to Germany, Italy, Spain, Switzerland, France and the United Kingdom.

PINACEAE (PINES) | LONG, THIN NEEDLES

EN Monterey Pine

Pinus radiata

A tall conifer (H to 40 m), although more likely found as a contorted, stunted tree (trunk girth to 6 m). Taller (H to 50 m) in its native range along the coast of central California, USA and Baja California, Mexico. Some native populations are legally protected. It was first described by the plant hunter David Douglas in 1831 and brought to Britain as seed in 1833. It is salt- and wind-tolerant and grows well in areas with a mild climate and is probably under-recorded in Britain & Ireland. **Features:** BARK reddish brown to grey and deeply furrowed between scaly rectangular patches. TWIGS red-brown, becoming grey with age. BUDS resinous; red-brown (L 10–15 mm). LEAVES light to dark green, **straight or slightly twisted needles** (L 10–15 cm; W to 1·8 mm) in **clusters of 3**. There is a 2-leaved form (var. *binate*) known as 'Guadalupe Island Pine'. CONES ♂ cylindrical (L 1·0–1·5 cm), orange-brown. ♀ solitary, or in clusters, on stalks (L to 1 cm); egg-shaped; large (**L 7–15 cm when mature**); yellow-brown with rigid cone scales that can have a sharp point. Cones can persist for 20–40 years.

J F M A M J J A S O N D

NEEDLES slender; typically in 3s

♀ MATURE CONE large; cone scales rigid, usually with sharp point

BUDS resinous; unfringed

BARK reddish brown turning grey with age; rectangular plates

Did you know? Pitch Canker, a fungal disease introduced into California c. 1986 is pushing some native populations towards extinction. Monterey Pine is the world's most widely planted forest tree.

A typically stunted coastal Monterey Pine (*left*); and the distinctive, persistent, large clustered cones (*right*).

PINACEAE (PINES) | LONG, THIN NEEDLES   Conifers compared *pp. 90–95*

LC ■ Lodgepole Pine   *Pinus contorta*

Lodgepole Pine occurs in two forms: In planted forestry it is slim and conical, but in the open it can have a flattened crown. In the wild there are three (possibly four) subspecies ranging in heights from approximately 3 m to over 50 m. It occurs naturally along the western coast of the USA and Canada, particularly along rocky shorelines and in dry conifer woodlands. Some subspecies are adapted to wildfires with their seed cones needing the high temperatures of a fire to open. First found and described by David Douglas in the 1820s near Cape Disappointment in Washington State, it was successfully introduced to Britain & Ireland by *c.*1855 (appearing in a catalogue as *Pinus macintoshiana*) and has been widely planted in northern and western regions, often in wetter habitats and upland areas (>500 m elevation). **Features:** BARK thin, greyish brown, developing a furrowed appearance as it ages, when it can appear scaly or flaky. TWIGS orange- to red-brown; darkening with age. BUDS resinous; egg-shaped (L to 12 mm); bud scales dark red-brown. LEAVES dark to yellowish green needles (L 2–8 cm) **in pairs** and usually twisted with fine stomatal lines and a finely serrated edge. CONES ♂ cylindrical (L 0·5–1·5 cm); reddish orange. ♀ egg-shaped, often asymmetric (L to 7 cm when mature); stalkless or very short-stalked (L to 3 mm); pale red-brown usually with **distinctive, sharp-tipped cone scales**.

J F M A M J J A S O N D

♀ MATURE CONE scales with a slender prickle

BARK grey-brown; thin; with plates or furrows

NEEDLES in pairs; can be distinctly twisted

BUDS resinous

Lodgepole Pine can be found in forestry stands, especially in Scotland.

> **Did you know?** The common name 'Lodgepole', appears to have arisen as the tree was used by Native Americans to construct lodges, particularly in the Rocky Mountains. Lodgepole Pine spreads well from seed, which has allowed the species to become naturalized in Britain, and caused it to become an invasive species in New Zealand. It has also been widely planted in Iceland. One of the main timber producing trees, the species has been impacted by Red Band Needle Blight fungus.

PINACEAE (PINES) | LONG, THIN NEEDLES

Other long-needled pines

LC ■ Weymouth Pine *Pinus strobus*

Tall (H 30–75 m). TWIGS tiny hairs at base of short shoots. BUDS slightly resinous. LEAVES deep green to blue-green **straight or slightly twisted needles in 5s** (L 5–13 cm; W ± 1 mm). CONES ♀ pendent on stalks (L 2–3 cm); narrow (L 8–20 cm); cone scales splayed. Weymouth Pine is native to the eastern USA and Canada, occurring to 1,500 m elevation with a preference for cool, humid conditions and well-drained soils. It was introduced to Britain probably in 1705 but has not been planted recently due to its susceptibility to White Pine Blister Rust that originates from Asia.

♀ MATURE CONE narrow; pendent; scales splayed

NEEDLES in 5s; straight or slightly twisted

VU ■ Macedonian Pine *Pinus peuce*

Tall (H to 50 m). TWIGS hairless. LEAVES more or less stiff needles (L 7–12 cm) in 5s and angled towards the shoot tip, especially when young with cones. CONES ♀ pendent, straight to slightly curved cylindrical (L 9–18 cm); green when young, maturing to orange-brown. Cone scale tips are slightly incurved. Native to the Balkans and introduced to Britain in 1863, but never widely planted.

♀ MATURE CONE cylindrical; pendent; scale tips **slightly incurved**

NEEDLES in 5s; ± stiff

LC ■ Bhutan Pine *Pinus wallichiana*

Tall (H to 40 m). TWIGS hairy. LEAVES flexible needles (L 8–20 cm) in 5s. CONES ♀s long (**L 20–30 cm**) and narrowly cylindrical; bluish green when young, maturing to light brown; in groups of 1–6. They are erect at first, becoming pendent with age. Cone scale tips are straight or slightly out-curving at the base of the cone. Native to the Himalayas and introduced to Britain in 1823, although never widely planted.

♀ MATURE CONE long; pendent; scale tips straight or slightly out-curved

NEEDLES in 5s; ± flexible

LC ■ Dwarf Mountain Pine *Pinus mugo*

Small (**H 1–3 m**) and bushy. LEAVES straight, dark green, smooth needles (L 3–8 cm) in pairs; needle margins finely toothed. CONES ♀ egg-shaped; small (**L 2·5–5·0 cm when mature**); without sharp points on the scales (*cf.* Scots Pine). Native to central and south-east Europe and introduced to Britain in 1774.

Similar species | Scots Pine (*p. 96*); Black Pines (*p. 98*)

♀ MATURE CONE small; rounded; scales lacking points

NEEDLES **in pairs**; margins finely toothed

PINACEAE (PINES) | SHORT NEEDLES ON PEGS   Conifers compared *pp. 90–95*

LC ■ Sitka Spruce   *Picea sitchensis*

A tall conical tree (H to 80 m) with a straight trunk (girth to 16 m in North America; to 9 m in Britain & Ireland) that can have a buttressed base. Native to the west coast of the USA from Alaska to California, Sitka Spruce has been planted as an important timber species in Britain & Ireland. It is nowadays by far the most commonly planted tree in the UK, with a coverage of 66,500 km² (mostly in Scotland). **Features:** BARK smooth and grey, **ageing dark purplish brown with red-brown scales**. TWIGS pinkish brown. BUDS reddish brown (L 5–10 mm). LEAVES **sharp-tipped** rigid needles (L 15–25 mm) on short pegs, with a blue-green upperside lacking obvious stomata and a noticeably paler underside with conspicuous stomata in two white bands. CONES ♂ small; orange to red. ♀ pendent, cylindrical (L 5–9 cm when mature) with rounded to diamond-shaped scales.

J F M A M J J A S O N D

♀ CONE (prev. year)

BRANCHLET foliage 'thinner' than in Norway Spruce

♀ MATURE CONE L < 10 cm; shorter than in Norway Spruce

♀ CONES (fresh)

UPPER

UNDER

TIP pointed

NEEDLES upperside dark green; underside glaucous with two white bands

NEEDLE X-SECTION ± flattened

BARK dark, purplish brown

The distinctive rigid blue-green needles (*left*); an open-grown tree (*centre*); and one in woodland (*right*).

Did you know? Sitka Spruce is one of the species found by Archibald Menzies (surgeon and naturalist to Captain Vancouver's expedition) in Puget Sound in May 1792. However, it was only introduced into cultivation in 1831 from collected seed sent to Britain by David Douglas.
A versatile tree, its timber was used by the US Army at the end of the First World War to make military aircraft. In contrast, the soft spring shoots, which have a sweet citrus and strawberry scent, are used in a delicately flavoured beer.

PINACEAE (PINES) | SHORT NEEDLES ON PEGS

LC Norway Spruce — *Picea abies*

A tall conical tree (H to 55 m), with a straight trunk (girth to 5 m). Native to mountains from Norway through central Europe to Greece, Norway Spruce is now widely planted for timber and as an ornamental tree in parks and gardens. It has a preference for moist soils and can grow in cold, wet and shallow soils. Prior to the last ice age, it was probably native to Britain, but was one of the species that never naturally recolonized after the ice's retreat.

Features: BARK finely flaking, orange-brown, ageing grey-brown. TWIGS orange-brown. BUDS reddish brown (L 5–7 mm). LEAVES **blunt-tipped** rigid needles (L 10–25 mm) on short pegs, with both the upper and underside light to dark green with stomata. CONES ♂ small; reddish. ♀ pendent, cylindrical (L 12–16 cm when mature) with diamond-shaped scales. Var. *acuminata*, has cones to 18 cm in length, the longest of any spruce.

J F M A M J J A S O N D

BRANCHLET foliage 'thicker' than in Sitka Spruce

♀ MATURE CONE L > 10 cm; longer than in Sitka Spruce

UPPER
'peg' attachment
UNDER
TIP blunt

NEEDLES upperside and underside the same colour

NEEDLE X-SECTION 4-angled

BARK grey-brown and scaly on old trees

Young green ♀ cones with old brown ♂ cones (*left*); the young and distinctive shape of a young Norway Spruce, best known as a Christmas tree (*centre*); older tree (*right*).

Did you know? Introduced into Britain & Ireland *c*.1500, Norway Spruce is mentioned by Turner in his *Names of Herbes* in 1548. In Europe, spruce fronds were often strewn about the floors of houses (sometimes with Juniper), giving off a refreshing aroma when crushed under foot. It was originally planted for its ornamental value, but 19th-century writers realised the tree had commercial uses, as scaffolding poles, ladders, and spars and masts for ships. As a result, the tree began to be widely planted, for example, at Kielder Forest in Northumberland, and its use as the main type of Christmas tree has ensured that it can be found growing in large areas of Britain & Ireland.

PINACEAE (PINES) | NEEDLES IN ROSETTES (LARCHES) **Conifers compared** *pp. 90–95*

Larches *Larix* species

Larches are largely native in the colder Northern Hemisphere, and have become abundant trees in Siberia, Europe and Canada. Larches, unlike most conifers, are deciduous, shedding their needles by November, with fresh green needles present from early April. There are 10 species worldwide, the tallest of which can reach 60 metres. In Britain & Ireland two species (both introduced) and one hybrid occur. Larches have dense rosette-like clusters of 20–50 flattened needles on short shoots and single needles on new leading shoots. Needle arrangement is similar to that of the cedars (*p. 108*) but these are evergreen, broader in form and with much larger cones. Other deciduous conifers include Dawn Redwood (*p. 120*), Swamp Cypress (*p. 121*) and bald cypresses (N/I).

LC ■ European Larch   *Larix decidua*

A large deciduous conifer (H to 35 m; trunk girth to 8·5 m) with pyramidal crown of thick, downswept branches which ascend near their tip. The native range includes upland areas in the Alps, Russia and Poland. It is now planted widely in Britain & Ireland in plantations, shelterbelts and parkland, and regenerates freely from seed. **Features:** BARK grey to red-brown; flaking; deeply fissured; inner bark reddish. TWIGS leading shoots slender, flexible; pale yellowish grey to yellowish brown; usually hairless. LEAVES needles (L 2–4 cm; W ± 1 mm): light green, darkening through the summer before turning yellow and falling in the autumn; stomata mainly on underside in two inconspicuous pale green bands separated by a keel. Needles on short shoots in rosette-like bunches of 30–40; those on long shoots sparse and single. CONES ♀ narrowly egg-shaped (L 2–4 cm; length 1·25–1·50 × width) with convex scales that have the upper margins turned in or straight (slightly wavy at most).

J F M A M J J A S O N D

BARK grey to red-brown, flaking and deeply fissured

> **Did you know?** European Larch was introduced to Britain & Ireland in the early 17th century and grown as an ornamental tree, before being taken to Scotland in 1738 by a Mr Menzies. Five trees were given to Duke James of Atholl. The potential of the tree for timber was noted and by 1830 the fourth Duke of Atholl (nicknamed the 'Planting Duke') is reputed to have planted '8,604 Scotch acres or 10,324 imperial acres' of Larch (Loudon 1838). The wood has been used in yacht building on account of its toughness, flexibility and durability.

Vibrant green Larch in spring (*left*); autumnal Hybrid Larch (*centre*); and winter Japanese Larch (*right*).

PINACEAE (PINES) | NEEDLES IN ROSETTES (LARCHES)

LC ■ Japanese Larch *Larix kaempferi*

Native to the main Honshu Island of Japan, this species has been used for forestry in Britain & Ireland since it arrived here in 1861. However, like European and Hybrid Larches it is susceptible to infection by *Phytophthora ramorum*, a fungus-like water mould which has reached Britain & Ireland, so it is likely that Japanese Larch will be less widely planted in the future.

■ Hybrid Larch *Larix ×marschlinsii*

A hybrid between European Larch and Japanese Larch was found growing in the grounds of the Dunkeld Estate around 1904 and was named Dunkeld Larch *Larix ×eurolepis*. However, this same pairing had occurred in Switzerland a year or two earlier and named *Larix ×marschlinsii*, and so this is the recognized name. Hybrid Larches are often more vigorous than their parents, so this cross is often used in plantations to maximise the potential timber that can be grown.

Associated species | Larch seeds are important for birds such as **Redpoll** and **Siskin**. In Scotland the buds and cones are food for the **Capercaillie**.

Similar species | The three larches are all similar; the easiest way to identify them is by the length/width of the mature ♀ cone and the shape of its cone scales, though there are other less obvious differences (see *below*).

JAPANESE LARCH

HYBRID LARCH

ALL IMAGES HYBRID LARCH

NEEDLES single on new leading shoots

NEEDLES in 'rosettes' on older short shoots

♀ CONE

♂ CONES

CONES ♂ + ♀ on same tree

EUROPEAN	HYBRID	JAPANESE
♀ MATURE CONE L 1¼–1½ × W; scales wavy at most; **tips never turned outwards**	♀ MATURE CONE **shape as European; scale tips turned outwards (like Japanese)**	♀ MATURE CONE L 1–1¼ × W; **scale tips turned outwards**
NEEDLES L ≤ 4 cm on new shoots; underside bands inconspicuous	NEEDLES L ≤ 5 cm on new shoots; underside bands greyish	NEEDLES L 4–6 cm on new shoots; underside bands obvious
TWIGS greyish yellow-brown; usually **hairless**	**YOUNG TWIGS** can be slightly hairy with a slight bloom	**YOUNG TWIGS** typically **reddish brown**; slightly hairy with a grey waxy bloom

PINACEAE (PINES) | NEEDLES IN ROSETTES (CEDARS) — Conifers compared *pp. 90–95*

'True' Cedars *Cedrus* species

The *Cedrus* cedars are evergreen conifers native to the Mediterranean and western Himalayas. They occur naturally at relatively high elevations (1,000–2,200 m in the Mediterranean; 1,500–3,200 m in the Himalayas. Three species (all introduced and planted) occur in Britain & Ireland. Cedars have needles that are angled in cross-section. The needles are in a similar rosette-like arrangement as found in the larches (*p. 106*) which differ in being deciduous, much narrower in form and with much smaller cones.

Similar species | The three *Cedrus* cedars can be very hard to separate. The profile of a tree, the size and number of needles and details of the needle tip can help but many trees can show features that are intermediate and may not be identifiable (see *p. 110*).

Typical *Cedrus* cedar branch (*left*); ♂ and young ♀ cones (*centre*); and maturing ♀ cone (*right*).

ATLAS CEDAR

♀ MATURE *CEDRUS* CEDAR CONES
Green when young; maturing brown and breaking up on the tree, releasing the seeds; leaving the woody central spike on the branch

BARK red-brown to dark grey, fissured with age

Branch profiles: **Deodar** – drooping (*left*); **Cedar of Lebanon** – level (*centre*) and **Atlas Cedar** – ascending (*right*).

PINACEAE (PINES) | NEEDLES IN ROSETTES (CEDARS)

Deodar *Cedrus deodara*

Deodar can be tall (H to 50 m) and has a rounded profile with a thick trunk (girth to 10 m); a drooping leader; and branches which typically **droop at their tips**. Found naturally in the wild in India and Pakistan, on deep, well-drained soils at elevations of 1,100–3,000 m. In Britain & Ireland it is frequently found planted in parks, gardens and churchyards. It appears to have been introduced as seed to Britain in 1831 by Leslie Melville and was sown at Melville in Fifeshire. However, it was not until 1841 that large quantities of seed arrived from the Himalayas. Deodar produces a dark brown oil with a strong unpleasant smell that has been used as an antiseptic and insect repellent. The wood also has antifungal properties and this, together with its insect-repelling qualities, meant that Deodar wood was used in the construction of food storerooms.

J F M A M J J A S O N D

Cedar of Lebanon *Cedrus libani*

Mature Cedar of Lebanon (H to 35 m, trunk girth to 9 m) has a distinctive flat-topped profile with **level branches**. A native of the mountains of the Mediterranean region in Turkey, Lebanon and western Syria. It was planted in Britain & Ireland, particularly during the 18th century, and can often be found in the grounds of Georgian manor houses. It grows best on warm, deep, well-drained soils although, once established, it is tolerant of drought as well as of chalk and dry sites. The species was probably first brought to Britain & Ireland as seed by Dr Edward Pococke, a scholar of Arabic at Oxford University, who made several journeys to Syria in 1638–39. Although it is the national emblem and on the flag of Lebanon, few old trees remain in the country due to a long history of felling. However, there are active conservation efforts in place. An oil distilled from the wood was used in ancient Egypt to embalm the dead.

J F M A M J J A S O N D

Atlas Cedar *Cedrus atlantica*

Mature Atlas Cedars (H to 40 m), can be a very similar shape to Cedar of Lebanon, although the **branches are typically ascending**. Atlas Cedar is native to the mountains of Algeria and Morocco, where it grows in forests on mountainsides at elevations of 1,370–2,200 m. It was introduced into Britain & Ireland in 1841. The two species appear to have diverged about 23 million years ago and their close relationship means there are few obvious differences between them (Atlas Cedar was historically often considered a subspecies of Cedar of Lebanon). Although Atlas Cedar cones are slightly smaller on average than those of Cedar of Lebanon, significant variation between trees make this feature unreliable. A blue form (Glauca) is also widely planted as it is tolerant of drier, warmer conditions.

J F M A M J J A S O N D

PINACEAE (PINES) | NEEDLES IN ROSETTES (CEDARS) Conifers compared *pp. 90–95*

Cedrus cedar identification

| DEODAR | CEDAR OF LEBANON | ATLAS CEDAR |

BRANCHES (RULE OF THUMB)

DEODAR
Drooping

LEBANON
Level

ATLAS
Ascending

NEEDLES

DEODAR average needle length 30–40 mm

tapered to translucent tip
(L ± 0·4 mm)

LEBANON average needle length 15–25 mm

abruptly tapered to ± translucent tip
(L ± 0·2 mm)

ATLAS average needle length 10–25 mm

tapered to translucent tip
(L ± 0·5 mm)

♀ MATURE CONES

DEODAR
L 7–10 cm;
W 5–6 cm

LEBANON
L 8–12 cm;
W 4–6 cm

ATLAS
L 5–9 cm;
W 3–5 cm

PINACEAE (PINES) | FLAT NEEDLES

Western Hemlock-spruce — *Tsuga heterophylla*

LC

A tall, conical tree (H to 75 m) with a trunk girth up to 6 m; branchlets have drooping tips. Native to western North America in cool moist forests from sea level to 1,800 m elevation. It is distributed continuously from Alaska south to northern California. Introduced to Britain & Ireland in 1852, it is widely grown, found in plantations (most frequently in Scotland) and occasionally as an amenity tree. The name *Tsuga* appears to derive from the Japanese for **'Yew-like'; a species it superficially resembles.**
Features: BARK grey-brown; scaly and fissured. **TWIGS** yellowish brown; softly hairy. **BUDS** egg-shaped (L 2–5 mm). **LEAVES** flat, shiny, dark green, grooved needles (L 6–20 mm) with a rounded tip and minute teeth on the margin (use hand lens); **arranged in 2 spiral ranks** varying in size along a branch – becoming smaller towards the tip; underside blue-green with 2 broad white bands. **CONES** ♂ small (L 3–4 mm), yellow; ♀ egg-shaped (L 20–25 mm); scales rounded with rounded to pointed tips.

J F M A M J J A S O N D

TWIGS hairy

♂ **CONES** single or in small clusters

♀ **CONE**

UPPER

UNDER

♀ **MATURE CONE rounded** with rounded or pointed ends to the scales

NEEDLES arranged in two spiral ranks; grooved; tip **rounded**; two white bands on underside

BARK scaly, grey-brown and fissured

NEEDLES taper to tip

EASTERN HEMLOCK-SPRUCE

Similar species | Eastern Hemlock-spruce, *Tsuga canadensis* is planted in parks and gardens. Compared with Western Hemlock-spruce it has longer (L 10–20 mm) **needles that taper towards a narrow tip;** smaller (L 12–20 mm) cones; and buds that are a pointed egg-shape.

Did you know? A shade-tolerant species, this tree can grow slowly for long periods before finding a gap in the canopy to exploit, particularly in its native forests alongside Sitka Spruce or Douglas Fir. Once a tree finds light, it can grow at rates of up to a metre a year, sometimes more.
In its native habitats, the tallest trees (H to over 85 m) are in California, with some trees more than 1,200 years old.
The bark has a high tannin content, which has been used to make a variety of dyes to colour everything from human skin to goats' wool and even for camouflaging fishing nets.

PINACEAE (PINES) | FLAT NEEDLES   Conifers compared *pp. 90–95*

LC ■ Douglas Fir *Pseudotsuga menziesii*

A very tall, substantial tree (H to 100 m; trunk girth to 13 m) with a narrow conical crown that typically becomes irregular in shape with age. Naturally occurring over a wide area of the western USA from the Pacific coast to east of the Rocky Mountains, Douglas Fir is now widely planted in Britain & Ireland in forestry plantations and as specimen trees in parks and large gardens. **Features:** BARK variable; from black or red-brown to grey; corky with deep longitudinal fissures making scaly, often flaking, ridges. TWIGS greyish brown to reddish brown. BUDS pointed egg-shape (L 10 mm); red-brown; non-resinous. LEAVES flat, **straight** needles (usually L 25–35 mm; W 1·0–1·5 mm) with a **rounded tip**; yellow- to blue-green (colour can be variable by branch on a single tree); apparently in 2 rows but actually all around the twig; **underside with 2 bands of white or dull grey stomata**. CONES ♂ yellow or red (L 15–20 mm). ♀ green ripening brown; pendent; egg-shaped (L 4–10 cm; W 3·0–3·5 cm when mature) on stalks (L 5–10 mm); **bracts distinctive**.

J F M A M J J A S O N D

BUDS sharply pointed

♀ MATURE CONE
3-pointed bracts extend beyond the scale

BARK corky; black or red-brown to grey

NEEDLES flat, straight with rounded tip; 2 stomatal bands on underside

UPPER

UNDER

Did you know? The Douglas Fir was discovered in 1797 by Archibald Menzies at Nootka Sound during an expedition with Captain Vancouver to the north-east Pacific Ocean. Seed from these trees was brought back by the celebrated Scottish botanist and explorer David Douglas – hence its name. It is most likely that the seed was collected from the lower reaches of the Columbia River, near Fort Vancouver in 1825, arriving in Britain & Ireland in 1827. The resultant seedlings were raised in a nursery and planted out at Scone in Perthshire in 1834.

Douglas had accepted a post at the Glasgow Botanic Gardens in 1820, under Dr William Hooker, and he began his plant-hunting expeditions in 1823. Among his other discoveries are the Noble and Grand Firs and the Sitka Spruce, widely recognized as one of our most important timber trees. The scientific name celebrates its finder, Menzies; its English name, Douglas.

PINACEAE (PINES) | FLAT NEEDLES

Noble Fir   *Abies procera*

LC

A narrow, tall conifer (H 45–75 m) with a conical crown. A native of Washington and Oregon, Noble Fir was introduced into Britain & Ireland in 1830 by David Douglas. They are planted for timber in western and northern areas of Britain, with some of the finest trees at Diana's Grove at Blair Atholl, Scotland. **Features:** BARK smooth pale greyish brown; reddish purple and furrowed with age. WOOD hard, close-grained; often used for interior joinery. TWIGS reddish brown; with fine hairs when young. BUDS hidden by needles. LEAVES long (**L 25–35 mm**) **blunt-tipped** (or notched) needles arranged spirally around the branch with a pungent, somewhat turpentine-like, aroma; bluish green with 2–4 pale stomatal bands on the underside. CONES ♂ purple or reddish brown. ♀ large (L 10–15 cm), **cylindrical; upright** with purple scales and downturned, yellow-green, feathery bracts when young; **maturing brown; the central spike remaining on the tree after the cone breaks up.**

J F M A M J J A S O N D

NEEDLES distinctive arrangement; bluish green; pale stomatal bands on underside

♂ CONES

♀ YOUNG CONE purple scales; yellow-green downturned bracts

BARK smooth, pale grey to reddish purplish; deeply furrowed in older trees

♀ MATURE CONE bracts downturned

cone breaks up to leave central spike on tree

Did you know? Young Noble Firs are used as Christmas trees, as their striking blue-grey needles are retained for a longer period than many alternatives.

Similar species | Caucasian Fir *Abies nordmanniana* (N/I) is also used as a Christmas tree but differs in its leaves not being as tightly packed as in Noble Fir and the young twigs being hairless.

PINACEAE (PINES) | FLAT NEEDLES   Conifers compared *pp. 90–95*

LC ■ Grand (Giant) Fir   *Abies grandis*

A tall tree (H to 75 m; trunk girth to 5 m) with a conical crown. Found in western Canada and the USA (Montana, Idaho, Washington, Oregon and California) from sea level to 1,500 m in moist conifer forests, it is grown in Britain & Ireland largely for forestry or as a specimen tree in parks and gardens. **Features:** BARK grey, becoming reddish brown with age, and developing shallow furrows, with reddish wood visible between hard flat ridges with thick scales. TWIGS light brown, softly hairy. BUDS purplish brown or green; resinous. LEAVES **flexible, shiny,** needles (L **20–60 mm**; W 1·5–2·5 mm) with a **notched (rarely rounded) tip**; needles **arranged in 2 lateral sets** on either side of the twig with the upper set (rank) about half the length of those below, and those at the centre of the branch usually longer than those at both the base and tip; upperside light to dark green, grooved; underside with 2 white bands of stomata; **citrus aroma when crushed.** CONES ♂ blue-red, purple or orange-yellow. ♀ light green, purplish or grey; **cylindrical** (L **6–10 cm**; W 3 cm) with a rounded tip; ± 100–150 cone scales; bracts concealed; central spike remains on tree after cone breaks up.

J F M A M J J A S O N D

NEEDLES upper rank shorter than lower rank

'sucker' attachment

♀ YOUNG CONE typically **100–150 scales**; bracts **hidden**

BARK grey, ageing reddish brown

NEEDLES straight with notched tip; shiny upperside; typically **2 white stomatal bands on underside**

UPPER

UNDER typically **longer** than in European Silver-fir

Did you know? This is one of the many North American conifers first described by David Douglas, who discovered it in 1825 on the Columbia River, though he does not seem to have sent seeds to Britain & Ireland until *c.*1830. The next consignment of seed was sent by William Lobb in 1851 to Messrs. Veitch at Exeter.

This tree has been used in many ways, with the inner bark used to treat colds and fever, whilst the foliage, which has an attractive citrus-like scent, has been used as Christmas decorations. The wood has also been used for papermaking, packing crates and in construction.

PINACEAE (PINES) | FLAT NEEDLES

European Silver-fir *Abies alba*

LC

A tall tree (H to 60 m; trunk girth to 8 m) with a pyramidal to flat-topped crown. It occurs naturally in central Europe, the Alps and the Pyrenees. Although mountains are its natural habitat, the species thrives throughout Britain & Ireland with a preference for moist, but not waterlogged, soils and heavy clay. It is a shade-tolerant tree, hence it is used for under-planting in existing forests. Once widely planted throughout Britain & Ireland for forestry, it is now less used, but it can still be found in Cornwall, Wales, Scotland and south-east Ireland. **Features:** BARK distinctive, smooth, silver-grey often with cracks and spiralling. TWIGS grooved; pale brown or dull grey with soft dark hairs. BUDS reddish brown; softly hairy; can be resinous. LEAVES **flexible, glossy** needles (L to 30 mm; W 2 mm) with a **rounded or slightly notched tip arranged in 2 angled upwards to slightly flattened ranks**; upperside grooved, dark green; underside with 2 white or greenish white bands of stomata; **pine aroma when crushed**. CONES ♂ blue, purple or red. ♀ green, maturing red-brown; **cylindrical (L 7 cm; W 3–5 cm)** with a slightly pointed tip; approx. 150–200 cone scales; bracts visible; central spike remains on tree after cone breaks up.

J F M A M J J A S O N D

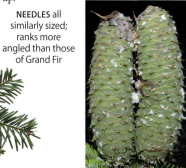

NEEDLES all similarly sized; ranks more angled than those of Grand Fir

♀ YOUNG CONE typically **150–200 scales**; bracts **visible**

BARK distinctive; smooth, silver-grey; usually with cracks

'sucker' attachment

NEEDLES glossy, dark green upperside, straight; **2 greenish white stomatal bands on underside**

UPPER

UNDER

typically **shorter** than in Grand Fir

Did you know? The species was introduced into Britain & Ireland before 1596, and the earliest trees recorded are two mentioned by John Evelyn in his book *Sylva* (1666) as '*being planted by Serjeant Newdigate in Harefield Park in 1603*'.
Once used as Christmas trees, European Silver-fir have been largely replaced by Norway Spruce; however, they are still good for this purpose. One Silver-fir growing at Ardkinglas in Scotland is thought to be more than 250 years old and was described by the Duke of Argyll as '*undeniably the mightiest conifer, if not the biggest bole of any kind in Europe*'.

TAXACEAE (YEW) | FLAT NEEDLES

Conifers compared *pp. 90–95*

Yew

Taxus baccata

A native evergreen (H to 25 m) with a preference for lime-rich soils on chalk or limestone. Although Yew and Beech can form dominant woodlands on these soils, Yew woods are uncommon in Britain & Ireland. It can also grow on acidic soils and has been widely planted in parks, gardens and churchyards. Trees usually mature into a rounded or pyramidal profile with multiple spreading trunks but, if left unmanaged, will develop many low horizontally spreading branches which touch the ground, root, and develop as clonal trees. Many new trees can develop around a single parent tree in this way – and led to the early British pagan belief that the Yew was immortal (see *p. 20*). **Features:** BARK thin; **reddish brown and usually flaky or scaly**. TWIGS arranged alternately on branches which can be very long. LEAVES flat, **parallel-sided** needles (L 1–4 cm; W 2–3 mm) appearing as two parallel ranks but actually arranged spirally. CONES trees normally single-sex, but a few have both ♂ and ♀ cones; ♂ small (D 3–6 mm), yellow, scaly; ♀ located on the underside of a shoot; starts small and scaly and ripens into a **fleshy red structure** (aril) which partially surrounds a **single, small, shiny brown-black seed** (aril and seed L 8–15 mm). The aril and seed fall as a unit.

J F M A M J J A S O N D

BARK grey to red-brown, flaking and deeply fissured

> **Did you know?** Britain's oldest tree is an approximately 2,000-year-old Yew growing in the churchyard at Fortingall, Perthshire. By 2023 the Ancient Yew Group had recorded almost 1,000 ancient or veteran Yews (more than 400 years old) in churchyards in England and more than 400 in Wales, some of which predate the churches.
>
> Yews cast a deep shade which limits the other plant species that can grow around them. The seeds and foliage are full of toxic alkaloids which can be rapidly absorbed and cause a fatal shock to the heart in humans. The only part of the Yew that is not toxic is the red aril, which is eaten by birds, especially thrushes. Yew is now a declining or rare species throughout the world, and it is thought that Britain may have more ancient Yews than anywhere else on the planet.
>
> The hard wood was historically much prized for the making of the finest longbows and also for furniture and bowls.

Not widely seen as a hedgerow tree, the distinctive shape of the Yew is visible in this hedge (*left*); Irish Yew (*right*).

TAXACEAE (YEW) | FLAT NEEDLES

NEEDLES upperside dark glossy green; underside paler yellowish green

UPPER

UNDER

UPPER

NEEDLES appear to be arranged in parallel ranks

♀ YOUNG CONE small, green

♀ CONE ripens green at first

♀ MATURE CONE distinctive fleshy soft red aril partially surrounding central black seed

♂ CONE small, scaly, yellow (left), expanding when releasing pollen (right)

Ripe male cones (left); and the distinctive red arils (right).

Similar species
The Andean **Plum-fruited Yew** *Podocarpus andinus* (N/I) is planted in a few parks and gardens and has distinctive yellow fleshy fruits like small plums. Other flat-leaved conifers (pp. 118–123).
A compact, more upright form of Yew – the Irish Yew *Taxus baccata* 'fastigiata' (*facing page, right*) is also a common feature of many churchyards and gardens. This naturally occurring ♀ variant of the Common Yew was found around 1770 in the Florence Court area of the Cuileagh Mountains of Fermanagh. Propagated cuttings from the tree became commercially available from around 1820, and it was widely planted throughout the 19th and 20th centuries.

GALL PRODUCED BY YEW GALL MIDGE

Associated species
The arils and seeds are widely eaten by birds, which then disperse the seeds.

In comparison with other native British trees, Yew is notable for the low number of insects and other invertebrates it supports, with 6 recorded in Britain & Ireland and fewer than 30 in western Europe. This is almost certainly due to chemical compounds (such as 10-deacetylbaccatin III and 10-deacetylbaccatin V) in the leaves and bark which deter feeding.

The **Yew Gall Midge** lays its egg into a leaf bud, which then develops into a gall comprising 60–80 terminal leaves – similar in appearance to the 'artichoke galls' found on oak. The midge larva feeds and then pupates within the gall. The midge is itself parasitised by two wasps: *Mesopolobus diffinis* and *Torymus nigritarsus*.

There are two uncommon Yew-dwelling spiders, *Hybocoptus decollatus* and the rare **Triangle Spider**.

One fungus particularly associated with the decomposition of the wood of Yew trees is **Sulphur Polypore**. This beautiful orange or sulphur yellow fungus is very easy to spot when it is in fruit. Also known as Chicken of the Woods, this species decomposes dead and dying heartwood in a variety of species, including Yew, oaks, Sweet Chestnut and Beech. As the heartwood is decomposed by the fungus, Yews also have the remarkable ability to produce aerial roots which grow into the centre of the tree and recycle all the decomposing material. This can produce amazing intricate patterns of roots within the centre of the hollowing tree.

CUPRESSACEAE (JUNIPER) | SHORT, THIN NEEDLES IN 3S | Conifers compared *pp. 90–95*

Common Juniper *Juniperus communis*

A usually multi-stemmed shrub (H to 11 m), Juniper is one of the most widely distributed native trees in Northern Europe. In Britain & Ireland it has a curious distribution, being found on chalk and limestone in southern England as well as on a range of soils in the uplands of Wales, northern England, Scotland, plus the north-west and west of Ireland. A very variable species, with the extremes regarded as distinct subspecies by some authorities (see *facing page*). Despite its small size, Juniper is slow-growing and can reach a considerable age: with some trees on the chalk more than 100 years old; an individual in Upper Teesdale more than 250 years old; and a few in Scandinavia thought to be more than 1,000 years old. Juniper occurs from sea level to ± 1,000 m in a wide range of habitats, from sea cliffs to moorland and mountains. **Features:** BARK reddish brown; fibrous; usually thin and often peeling in long strips. TWIGS slender, smooth; shiny in some. LEAVES stiff, **sharp, blue-green needles** (L 5–15 mm) **in whorls of 3**; upperside with stomatal band on the upperside. CONES ♂ (L ± 8 mm); ♀ smaller (L ± 2 mm); green when young, ripening over 2–3 years into a berry-like closed soft cone (D 6–13 mm) that is **black with a blue bloom**. The cone typically contains 1–3 (rarely up to 6) seeds.

J F M A M J J A S O N D

> **Did you know?** Juniper is well known for its medicinal and culinary uses but is perhaps best known for its blue-black fruit, which is used to flavour gin. It is thought that Juniper has been used as flavouring for hundreds, possibly thousands of years, but gin has had a mixed history. Historically, it was used as a 'cure' for indigestion and gout and, during the mid-18th century, its cheapness led to devastating impacts on British society. These were famously depicted in Hogarth's 1751 picture **'Gin Lane'**. In London alone, there were more than 7,000 shops selling gin, and about 10 million gallons were distilled annually in the capital. Gin Acts were passed to restrict its availability, and with a subsequent rise in enthusiasm for beer, the worst impacts of gin – 'Mother's Ruin' – faded into history.

Juniper (ssp. *communis*) develops into a small tree or shrub on the chalk and limestone (*below*); the ripening green fruit can be seen from mid-summer (*above right*).

CUPRESSACEAE (JUNIPER) | SHORT, THIN NEEDLES IN 3S

♂ BRANCH
♀ BRANCH
NEEDLES blue-green; in 3s; distinctive white band on upperside
♂ CONE L ±8mm
♀ YOUNG CONE L ±2mm; green when young
♀ MATURE CONE fleshy; green maturing black

TONGUES OF FIRE
JUNIPER CARPET

Juniper subspecies

ssp. *communis* – a shrub or small tree on chalk and limestone in southern England and on a range of soils in the Scottish Highlands and western Ireland.

ssp. *nana* (**Dwarf Juniper**) – a low-spreading shrub on well-drained bogs and rocky outcrops in Scotland, northern England, Wales and western Ireland.

ssp. *hemisphaerica* – a low clump-forming shrub of sea cliffs confined to a tiny area of The Lizard, Cornwall.

Associated species | More than 113 species of fungi and slime mould have been recorded on Juniper, plus the introduced water mould *Phytophthora austrocedri* which was first reported in Britain & Ireland in 2011 and is killing trees across Scotland and the north of England. One spectacular fungus found on Juniper is known as the **Tongues of Fire**, a bright orange fruiting body of the fungal rust *Gymnosporangium clavariiforme*. These can be up to 1 cm long and appear in damp spring weather, growing directly from the wood of Juniper branches.

There are also more than 50 species of insects and mites associated with Juniper, with at least 42 specific to the species. These include the **Juniper Carpet**, a scarce moth species, which feeds on Juniper. The moth naturally occurs in only a few localities throughout Britain, although it can occasionally be discovered feeding on Junipers in residential gardens.

Dwarf Juniper ssp. *nana*

ssp. *hemisphaerica* with Black Bog-rush and Bell Heather

CUPRESSACEAE (CYPRESSES) | FLAT NEEDLES Conifers compared *pp. 90–95*

EN Dawn Redwood — *Metasequoia glyptostroboides*

A **deciduous** conifer (H to 45 m) with a tapering trunk that can have a buttressed base. Young trees have a conical crown which becomes broad and rounded with age. Occurring naturally in China, Dawn Redwood can be found planted in parks, gardens and other public places.
Features: BARK red/brown when young; becoming greyish and cracked with age and shedding long narrow strips of bark. TWIGS reddish brown; hairless. BUDS non-resinous; egg-shaped (L 2·5–5·0 mm); light red to yellow-brown; typically **arranged oppositely**, although can be solitary between branchlets. LEAVES small (L **10–30 mm**; W 1·6 mm), needle-like; **straight and flattened in 2 ranks**; upperside bright green with a narrowly grooved midvein; underside with faint, lighter green lines of stomata; turn **reddish brown in autumn** before falling. CONES ♂ (L to 5 mm) numerous; clustered along branches; ♀ solitary; pendent; egg-shaped (L to 25 mm) with 20–30 broadly triangular woody scales each with a horizontal groove.

J F M A M J J A S O N D

♂ **CONES** clustered

♀ **CONE**

♂ **CONES**

NEEDLES flat, **opposite**; wider than those of Swamp Cypress

WINTER TWIG with **oppositely arranged egg-shaped buds**

♀ **CONE** green and closed when young, ripening brown and open

BARK red to brown when young; becoming greyish and cracked with age

Did you know? In 1941 palaeobotanist Shigeru Miki was looking at fossil sequoias in Japan and noticed a new genus which he named 'Metasequoia' (meaning '*near sequoia*'). Amazingly, in the same year a living tree was found in Szechuan province, China. Cuttings sent to Beijing were confirmed by Professor Cheng and Dr Hu as being the same species as Miki's 3–5 million year old fossil.
The Chinese botanists informed Professor Merrill at Harvard University's Arnold Arboretum. In September 1947 an expedition was sent to Szechuan to bring back seeds and by 1948 the Arnold Arboretum was distributing seed to botanical gardens and collectors around the world. Seed was sent to Cambridge by a Dr Silow who worked with the British Council in Beijing, enabling Cambridge University to be the first to plant out a Dawn Redwood on British soil.

CUPRESSACEAE (CYPRESSES) | FLAT NEEDLES

Swamp Cypress — *Taxodium distichum*

LC

A tall **deciduous or semi-deciduous** conifer (H to 40 m), usually with a pyramidal crown and a single straight (or gradually tapering) broad-based trunk (girth to 9.5 m). Occurring naturally in the USA from Delaware to Texas, southern Illinois and Indiana, it was introduced to Britain & Ireland in c. 1638 by John Tradescant and can be found planted in parks and gardens. It has a preference for deep, moist, loamy soils. In swampy areas, trees usually have upright woody growths (H to 3 m) that transport oxygen from the air to the submerged roots and also provide additional support and stability in the soft ground. **Features:** BARK dull red-brown, turning grey, fissured and splitting into long, thin strips. TWIGS green- to red-brown. BUDS non-resinous; egg-shaped. LEAVES flat, green, needle-like (**L 6–17 mm**) in 2 ranks, **arranged alternately**; turn **yellow or copper red in autumn** before falling. CONES ♂ (L to 5 mm) usually numerous and clustered along branches; ♀ solitary; pendent; globular (L 20–35 mm) with 20–30 broadly 4-sided scales.

J F M A M J J A S O N D

♂ CONE clustered

NEEDLES flat, **alternate**; thinner than those of Dawn Redwood; yellow or copper in autumn

WINTER TWIG egg-shaped buds alternate

♀ CONE green and closed when young, ripening brown and open

BARK dull reddish brown; grey and fissured with age

The air-roots (or pneumatophores) of Swamp Cypress – like woody vertical projections (*left*); the autumn colours can be striking (*right*).

CUPRESSACEAE (CYPRESSES) | FLAT NEEDLES   Conifers compared *pp. 90–95*

EN ■ Coast (Coastal) Redwood   *Sequoia sempervirens*

The tallest trees on the planet (H to 110 m), with a conical tapering trunk (girth 6–9 m at 1·3 m above the ground), that usually has buttresses as well as rounded burrs or growths. Found naturally along an often foggy, 60 km wide strip of coast from northern California to south-west Oregon, this species was introduced into Britain & Ireland in 1843 and has been widely planted. The first detailed description of a Sequoia was written by Father Juan Crespi, a friar of the Portola expedition in 1769. Father Pedro Font took the first measurements (girth and height) of a Coast Redwood in 1776.
Features: BARK **distinctively red-brown and spongy/fibrous;** furrowed into broad, scaly ridges up to 35 cm thick; inner bark cinnamon-brown. TWIGS slender; dark green. BUDS small; scaly. LEAVES **flat, dark green, single** needles (L to 3 cm); arranged alternately; underside with 2 prominent white bands. CONES ♂ (L 2–5 mm); singly on short stalks. ♀ located at the end of twigs; **egg-shaped; pendent;** woody (L 12–35 mm) with 15–30 flat, short-pointed scales.

J F M A M J J A S O N D

TWIGS green

♂ CONE single; on short stalks

♀ MATURE CONE red-brown, oval with 15–30 flat, short-pointed scales

BARK distinctive spongy red-brown

NEEDLES arranged alternately; flat; tip pointed; 2 white bands on underside

UNDER

Similar species | Giant Redwood (*facing page*) has 'rope'-like branches.

Did you know? During the summer of 2006, researchers from Humboldt State University discovered the tallest recorded tree in the world – which they named Hyperion. This tree was 115·7 m tall. Since 2006 it appears to have grown at about 3·9 cm per year, now standing at just over 115·9 m, and has an estimated age of 1,260 years. One tree at Humboldt Redwoods State Park has been recorded with 2,267 annual rings.

CUPRESSACEAE (CYPRESSES) | FLAT NEEDLES

EN Giant Redwood — *Sequoiadendron giganteum*

An enormous trees (H to 90 m; trunk girth to 34 m) with down-sweeping branches that have upturned ends. Native to the western foothills of the Sierra Nevada, California, where it occurs as groves of trees at elevations of 900–2,700 m. They are widely planted in Britain & Ireland, frequently as ornamental trees in parks. **Features:** BARK reddish brown; **spongy and fibrous** (to 60 cm thick) with strong ridges and furrows. LEAVES needle-like (L to 15 mm); triangular in cross section; arranged alternately. CONES ♂ yellowish brown (L 4–8 mm), at tips of outer branches; ♀ **egg-shaped** (L 4–9 cm), maturing from green to brown over 2 years.

J F M A M J J A S O N D

♀ CONE distinctive; large; egg-shaped; green when young, ripening brown

NEEDLES tightly, alternately arranged on branchlets giving a 'rope-like' appearance

♂ CONE at tips of outer branches

BARK thick; spongy; reddish brown

Similar species | Japanese Red-cedar (*p. 125*) has larger leaves and less spongy bark; **Coastal Redwood** (*facing page*) has flat needles.

Did you know? The introduction of Giant Redwood into Britain & Ireland is split between two individuals: William Lobb and John Mathews. During 1852 Lobb located a grove of trees in San Francisco and brought seeds and seedlings to England in December 1853. His discovery was published on 24th December 1853 in the *Gardeners' Chronicle* as the first account of the tree. However, John Mathews had earlier posted a letter (dated 10th July 1853) containing seeds to his father in Scotland, who subsequently planted them. Almost a year later, on 23rd June 1854, the letter was published in the *Gardeners' Chronicle*. Therefore, although Mathews' seeds were the first to arrive in Britain & Ireland, Lobb was credited with having secured the supply of seeds and seedlings. John Lindley of the Horticultural Society named the species, and as it was the '*world's most impressive tree*', he thought it should commemorate the 'Great' Duke of Wellington who had died in 1852. Lindley therefore gave the tree the name *Wellingtonia gigantean* in 1853, triggering a storm of protest from American botanists, who were outraged that the world's largest tree was named for an English war hero by a botanist who had never seen the tree. The name was subsequently changed to *Sequoiadendron* in 1939 by John Buchholz.

123

CUPRESSACEAE (CYPRESSES) | SCALE-LIKE NEEDLES   Conifers compared *pp. 90–95*

Western Red-cedar   *Thuja plicata*

A tall conical tree (H to 50 m; trunk girth to 16 m at 1·3 m above the ground) with trunks that can have flared buttresses and older trees can have multiple leading shoots. It is found along the Pacific coast from south-east Alaska to California and in the Rocky Mountains with a preference for wetter sites in mixed coniferous forest. The species was first described by a Mr Nee during a 1789–94 around-the-world journey and introduced into cultivation in 1853 by Mr W. Lobb. It was then distributed from Veitch's nursery at Exeter as *Thuya lobbi*. Widely used in forestry and as a screening or hedging tree, sometimes in gardens. **Features: BARK** fibrous red-brown or grey-brown (to 2·5 cm thick) with shallow fissures leaving narrow plates. **TWIGS** brown; ascending at the tips with leaf-sprays (L 15–50 cm; W 5–15 cm) that consist of 2-years' leaf growth arranged oppositely in flat branchlets (twigs over 3 years old normally bare). **LEAVES** scale-like, small (L 1–6 mm); green; **underside with whitish stomatal patches**; strong, sweet **aroma of pear-drops or pineapple**, particularly on warm days. **CONES** ♂ dark red; cylindrical (L 1–3 mm) in a 'cup' created by two leaf pairs at the tips of branchlets. ♀ oblong (L 10–12 mm); light brown with 8–12 cone scales arranged in 4 ranks.

J F M A M J J A S O N D

♀ **MATURE CONES** typically **conspicuously clustered near the ends of branches**; narrowly egg-shaped with 8–12 scales arranged in 4 ranks

BARK fibrous; red-brown or grey-brown

♂ **CONE** dark red; at tip of branchlets

NEEDLES oppositely arranged in flat branchlets; aroma of **pear-drops or pineapple**; white 'butterfly' stomatal patches on underside

UPPER   'butterfly' marks   UNDER

Did you know? In its native range, one tree has been recorded as more than 1,400 years old. Some of the most impressive stands are found on the west coast of Vancouver Island. Used extensively throughout history for a wide range of items from canoes and totem poles to roof shingles, wall cladding and sheds.

CUPRESSACEAE (CYPRESSES) | SCALE-LIKE NEEDLES

Monterey Cypress   *Cupressus macrocarpa*

An evergreen tree (H to 25 m) that typically has a broad spreading crown (especially in exposed areas), although trees can be more upright in sheltered locations. A rare native of California, only found in the wild in groves at Cypress Point and Point Lobos on the Pacific Coast near Monterey. Seeds were collected and sent to Europe some time before 1838, when A. B. Lambert gave some of unspecified provenance to the Horticultural Society. Monterey Cypress is found planted in parks, churchyards and gardens (sometimes as hedging), particularly in the south and west of Britain. Being salt- and wind-tolerant means it thrives in coastal locations where it is often planted in windbreaks. **Features:** BARK greyish brown; rough and fibrous. LEAVES scale-like (L 2–5 mm); bright green to dark-green; can have a faint white bloom; **never waxy or resin-coated**. CONES ♂ (L 4–6 mm) yellow. ♀ broadly egg-shaped (L 2–4 cm), grey-brown with 8–12 scales with irregular margins and a slight central ridge.

J F M A M J J A S O N D

BARK rough and fibrous

NEEDLES opposite; scale-like on flat branchlets; can have faint white bloom; **lemon** aroma

♀ MATURE CONE
8–12 scales; each with raised centre and irregular margin

♂ CONE yellow

Similar species | Other planted *Cupressus* such as Smooth Arizona Cypress *Cupressus glabra* (N/I) differ in cone shape and/or the cone scale margins being straighter.

Japanese Red-cedar   *Cryptomeria japonica*

A very large evergreen forest tree (H to 70 m) that is endemic to Japan and was introduced into Britain & Ireland in 1846. **Features:** BARK red-brown; peeling. LEAVES needle-like (L 5–10 mm); arranged spirally. CONES ♀ globular (D ± 10 – 15 mm) with 20–40 scales; solitary or in groups up to 6.

Similar species | Giant Redwood (*p. 123*) has smaller leaves and spongier bark.

NEEDLES spirally arranged; base appressed; tip pointed

♀ MATURE CONE
globular; 20–40 scales with recurved teeth

125

CUPRESSACEAE (CYPRESSES) | SCALE-LIKE NEEDLES   Conifers compared *pp. 90–95*

■ Leyland Cypress   *Cupressus ×leylandii*

A tall, fast-growing, evergreen tree (to at least 40 m); the ultimate height of this naturally occurring hybrid between Nootka Cypress (*p. 128*) and Monterey Cypress (*p. 125*) is as yet unknown as it has only been in existence for just over 130 years. It was found at Leighton Hall, Powys in 1888, the home of Mr C. J. Leyland. However, although there is no official record, there is a possibility that the hybridization also happened earlier (*c.*1870) in Rostrevor, Co. Down, Northern Ireland. This cross does not appear to have occurred in the wild as the parent species have a 400-mile gap between their natural ranges. Leyland Cypress is often grown as garden hedging and grows well in a range of soils, including acidic, alkaline, loamy, rich, sandy, well-drained or clay soils. **Features:** BARK red-grey, fibrous and with ridges and furrows. LEAVES scale-like (L ± 1 mm); **rich green to yellowish green; soft and overlapping** in long-stalked flat sprays which are relatively 'open' with relatively straight branch tips. CONES ♂ located at tips of twigs; small (L 2–4 mm), pollen sacs yellow. ♀ ball-shaped (L 15–20 mm); brown when mature with 8 scales that each have a central spine. Female cones are only found on a few cultivars *e.g.* 'Leighton Green'.

J F M A M J J A S O N D

BRANCHES relatively 'open'; tips straightish

NEEDLES scale-like; **soft** and overlapping in long-stalked flat sprays; **fruity resinous** aroma when rubbed; **resin glands absent**

UPPER   UNDER
no glands

♀ **MATURE CONE** ball-shaped; 8 scales, **each with a small central spine**

♂ **CONE** yellow

BARK red-grey with ridges and furrows

Did you know? In 1888 there were six seedlings at Leighton Hall. One of these has been bred as a clone called 'Haggerston Grey' after the Northumberland estate where it was first planted. It has very dark green, slightly grey foliage and rarely flowers. Another clone, 'Leighton Green', which flowers and fruits from an early age, has foliage that is thicker rich green and yellowish green beneath.

CUPRESSACEAE (CYPRESSES) | SCALE-LIKE NEEDLES

Lawson's Cypress *Chamaecyparis lawsoniana*

A tall, narrow, evergreen tree (H to 70 m; trunk girth 9.5 m). Naturally occurring in the USA from south-west Oregon to north-west California. In Britain & Ireland it is widely planted *e.g.* in parks and churchyards. William Murray first sent seeds from the valley of the upper Sacramento River, California, to Lawson's nursery in Edinburgh in 1854 from which four seedlings were grown. By 1855 a larger quantity of seed was sent by the same collector, and it soon became one of the most widely used ornamental conifers in Britain & Ireland. Today, there are more than 550 named cultivars, although only 30 or so are widely planted. **Features:** BARK red-brown, stringy, with a spongy outer layer (up to 20 cm thick on older trees) divided into wide, rounded ridges with scales. LEAVES scale-like (L ±2 mm); **blue-green with angled to sharp-pointed tips; resin glands present; underside with X-shaped white stomatal bands**; leaves in sprays that are relatively 'shaggy' in appearance and branch tips that tend to be drooping. CONES ♂ located at tips of twigs; small (L 2–4 mm); pollen sacs red. ♀ broad (L 7–12 mm when mature) with 6-10 scales that each have a blunt spike or knob; pale blue-green when young, ripening dark purplish brown.

J F M A M J J A S O N D

BRANCHES relatively 'shaggy'; tips drooping

NEEDLES scale-like; **sharp-tipped**; parsley aroma when rubbed; UPPERSIDE with **glands**; UNDERSIDE with **x-shaped stomatal patches**

roundly pointed tips

♀ **YOUNG CONE**

UPPER UNDER
resin glands **x-shaped white marks**

♀ **MATURE CONE** broad; 6–10 pairs of scales, **each with a small spike or knob**

♂ **CONE** red

BARK stringy, red-brown

Did you know? In its native forest in Oregon, where it grows with Douglas Fir (*p. 112*) and Incense Cedar (N/I), areas of old-growth trees are now rare as they have nearly all been logged or are dying from *Phytophthora lateralis* (a fungus-like water mould). This pathogen has now been reported in England in (at least) Devon, Sussex and Yorkshire as well as in Scotland, Wales and Northern Ireland. Signs include the foliage turning pale green initially, then rusty reddish brown – a process which can spread rapidly within a tree's crown.

Conifers compared *pp. 90–95*

Other infrequently seen cypresses and redwoods

LC Nootka Cypress
Xanthocyparis nootkatensis

A tall (H to 40 m) long-lived cypress (to more than 1,800 years old) native to coastal northwestern North America. First found on land owned by the Nootka on Vancouver Island, it was introduced to Britain *c.* 1854 and is one of the parents of Leyland Cypress (*p. 126*). **Features:** BARK grey, smooth and thin, turning flaky with age. LEAVES dark green, scale-like (L 3–5 mm); in flat sprays; underside **wholly green with resin glands but no visible stomata**. CONES ♀ cones (L 8–12 mm) have 4–6 scales and pointed triangular bracts (crescent-shaped without points in *Cupressus*).

♀ MATURE CONE 4–6 scales; each with central triangular projection

BARK grey, smooth, thin; flaking with age

♂ CONE

no white stomata

UPPER   UNDER
NEEDLES

LC Sawara Cypress
Chamaecyparis pisifera

A tall (H 35–50 m) cypress native to Japan and introduced into Europe in 1843. **Features:** BARK red-brown with a stringy texture and vertical cracks. LEAVES scale-like (L 1·5–2·0 mm); in flat sprays; light to dark green; tip slender and pointed; underside with a small gland and **white 'butterfly'-shaped stomatal bands**. CONES ♀ (L 5–6 mm) greenish; like a wrinkled pea when young, maturing to dark brown with 6–10 scales that each have a tiny point at their centre.

sharply pointed tips

♀ MATURE CONE 6–10 scales; each with tiny central point

BARK grey, smooth, thin; flaking with age

resin glands

white 'butterfly' stomatal patches

UPPER   UNDER
NEEDLES

LC Northern White-cedar
Thuja occidentalis

A medium-sized (H to 15 m) evergreen conifer native to eastern Canada and the northeastern USA and introduced into Britain & Ireland by 1596. **Features:** LEAVES scale-like (L 1–4 mm); in flat sprays; yellowish green; underside with **paler yellowish stomatal patches**; aroma reminiscent of apples. CONES ♀ slender oval (L 9–14 mm) with 4 scales; green when young, maturing to brown.

♀ MATURE CONE narrowly egg-shaped; 6–8 overlapping scales

yellowish stomatal patches

♀ YOUNG CONE

UPPER   UNDER
NEEDLES

ARAUCARIACEAE (MONKEY-PUZZLE) | BROAD, TRIANGULAR NEEDLES

Monkey-puzzle
Araucaria araucana

A tall evergreen tree (H to 35 m) with a cylindrical trunk (girth to 6 m). Found in the wild in southern Chile and Argentina growing both in pure stands as well as in mixed forests, Monkey-puzzle trees are highly adapted to withstand fire damage. Introduced and planted in Britain & Ireland, the tree prefers a deep well-drained soil, dislikes atmospheric pollution, although will tolerate coastal conditions. The name 'Monkey-puzzle' arose when a proud owner of a young specimen made the remark *"It would puzzle a monkey to climb that"* and the name stuck. **Features:** BARK grey-brown, smooth, resinous and with **ring-shaped scars** from where old branches had been. BRANCHES in whorls of five. LEAVES **broad, triangular and scale-like** (L 3–5 cm; W 8–25 mm); shiny green and persistent on branches and the trunk. CONES ♂ erect; ± cylindrical; either solitary or in groups. ♀ ± spherical (L to 18 cm; W to 15 cm); yellowish green when young, maturing dark brown before falling.

J F M A M J J A S O N D

OLDER BRANCH

NEW GROWTH

NEEDLES shiny green, large, broad, spiky and triangular; persisting, even on the trunk

♂ MATURE CONES ± cylindrical; yellowish green when young

♀ YOUNG CONE yellowish green; ± spherical; maturing dark brown before falling

BARK smooth, grey-brown with **ring-shaped scars**

Did you know? Monkey-puzzle is a native of Chile introduced into Britain by Archibald Menzies. While on an expedition which reached the coast of Chile in 1795, Menzies took seeds *"from the dessert table of the Governor"* which he later planted, bringing seedlings back to England.

When roasted or boiled, the seeds have a nutty flavour reminiscent of almonds. Unfortunately, a large percentage of the native forests have been destroyed as a result of logging, grazing and fire, including 74,000 acres of protected forests burnt during 2001–02, with some ancient trees (more than 2,000 years old) lost. To help protect the species, it was declared a 'Natural Monument' in Chile, and it is currently illegal to log and export the timber.

BROAD-LEAVED TREES | Leaves *pp. 58–67* | Flowers *pp. 68–73* | Fruit *pp. 74–77* | Twigs *pp. 78–87*

Introduction to broadleaves

The broadleaved trees are angiosperms – plants with their seeds hidden inside a fruit. Angiosperm trees can be of two types: one produces two leaves (or more) from the seeds (dicotyledons), for example, oaks and birches; the other produces a single leaf from the seed (monocotyledons) – for example, palms, aloes and yuccas. Most of the world's trees are broadleaved, with about 73,000 species worldwide, the majority of which are deciduous. In the following accounts the trees are deciduous unless specifically mentioned otherwise in the text.

The broadleaved trees of Britain & Ireland are a diverse range of species. The most commonly occurring broadleaved species by woodland area in Great Britain are birches (± 18% of broadleaf woodland), oaks (16%) and Ash (12%). Birch is more dominant in Scotland, accounting for 43% of the broadleaf woodland area there.

There are 113 native broadleaved tree and shrub species in Britain & Ireland, and they grow from sea level at the edges of our coasts (like Blackthorn and elms), to the highest points trees grow in Britain & Ireland, at just over 1,300 m on Ben Nevis (Dwarf Willow).

Some of our broadleaf species like oaks and Ash have featured widely in our culture and history; others like Spindle are named for the task for which they were used, but many of our native broadleaf trees and shrubs have been little studied and are still poorly understood – like Wild (Crab) Apple and Wild Pear. Our native broadleaved trees are vital to the landscape and wildlife.

What makes our broadleaf trees important is that they form the majority of our native treescapes and support a vast range of biodiversity. They represent the natural flora that will eventually establish in most areas (the so-called climax community), and the nature of that treescape will be shaped by the interplay between the soil, the elevation, the weather conditions and the genetics of the tree species, plus how they are managed by people.

> **Did you know?** In 2015, a 2,300-year-old Iron Age shield was located in an archaeological dig in Leicester near the M1 motorway. There are few organic finds recorded in Britain & Ireland of this age, and this shield gives a unique insight into the extensive and varied use of many tree species. Its main element is the bark of willow – probably White Willow, as it is formed of a single sheet of bark, which must have come from a large tree. The bark had been painted and backed by wooden spars of apple, pear, quince or Hawthorn. The shield's rim was made of split Hazel and it had a central boss, also made of woven willow, to protect the hand of the user.

Broadleaved trees can create many habitats, shaping and framing the landscape.

Introduction to the rose family (Rosaceae)

The Rosaceae family is a group of trees and shrubs which include hawthorns (*Crataegus*), cotoneasters (*Cotoneaster*), apples (*Malus*), plums and cherries (*Prunus*), pears (*Pyrus*) and whitebeams, rowans and service-trees (*Sorbus*) as well as two groups which are woody but not included in this book as they are not generally regarded as shrubs: blackberries and raspberries (*Rubus*) and roses (*Rosa*).

This family accounts for 51 of the native tree and shrub species in Britain & Ireland – nearly half of the total number of native species. The preponderance of this family is due in large part to the complexity of the *Sorbus* group, which accounts for 42 of the 51 species – see *p. 133* for more details on the *Sorbus* genus.

One of the features of the Rosaceae is that many of the species have flowers which are attractive to insects, and the fruits which develop are edible. Some of these fruits have been utilised, and varieties cultivated, to produce fruit commercially (*e.g.* apples, pear, plums and cherries). However, another feature of these fruits is that they can contain tiny amounts of amygdalin, a chemical compound which can release cyanide when the fruit is damaged.

The flowers are typically radially symmetrical (with 5 sepals and 5 petals) and bisexual (having ♂ and ♀ reproductive organs in the same flower). The majority of trees and shrubs in the rose family also have white flowers, some of which can be difficult to distinguish if used as a single identification feature.

Apple Wild Plum Whitebeam sp.

Rose (like this Northern Dog-rose) and brambles, even though woody, are generally not regarded as shrubs.

ROSACEAE (ROSES) | INTRODUCTION Leaves *pp. 61, 63, 65, 67* | Flowers *p. 70* | Fruit *pp. 75–77* | Twigs *pp. 80, 84–86*

Identification of rose trees/shrubs

All the trees and shrubs in the rose family have white to pinkish 5-petalled flowers. Differentiating the species groups within the rose family is relatively straightforward for most, although identification of species within a group typically requires a detailed examination of features, the nature of which varies between groups as outlined below.

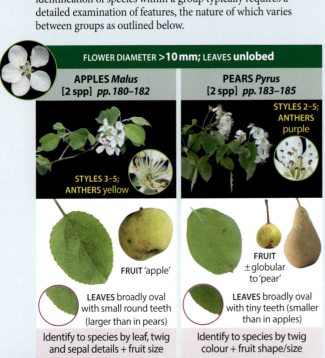

FLOWER DIAMETER >10 mm; LEAVES unlobed

APPLES *Malus* [2 spp] *pp. 180–182*
STYLES 3–5; ANTHERS yellow
FRUIT 'apple'
LEAVES broadly oval with small round teeth (larger than in pears)
Identify to species by leaf, twig and sepal details + fruit size

PEARS *Pyrus* [2 spp] *pp. 183–185*
STYLES 2–5; ANTHERS purple
FRUIT ± globular to 'pear'
LEAVES broadly oval with tiny teeth (smaller than in apples)
Identify to species by twig colour + fruit shape/size

PLUMS *Prunus* [3 spp] *pp. 170–173*
STYLES 1; SCALES AT FLOWER-CLUSTER BASE inconspicuous or absent
FRUIT 'sloe' or 'plum' on short stalks
LEAVES oval to broadly oval
Identify to species by twigs, flowering time + fruit

CHERRIES *Prunus* [6 spp] *pp. 176–179*
STYLES 1; SCALES AT FLOWER-CLUSTER BASE conspicuous
FRUIT 'cherry' on ± long stalks
LEAVES narrowly to broadly oval
Identify to species by leaf and flower-cluster details

FLOWER DIAMETER >10 mm; LEAVES lobed (in most spp.)

HAWTHORNS *Crataegus* [2 spp] *pp. 186–188*

TWIGS spiny | STYLES 1–5

FRUIT 'haw'

LEAVES lobed (oval with strong teeth in one planted species)
Identify to species by leaf shape + number of styles/seeds

Flower diameter <10 mm

WHITEBEAMS, ROWANS, SERVICE-TREES
Sorbus [47 spp] *pp. 133–169*

FLOWERS in flat-topped clusters

FRUIT clusters of red or brown berries
LEAVES variably shaped
Identify to species by fine details of leaf shape

COTONEASTERS [7 spp] *pp. 189–191*

FLOWERS 'closed' or open in branched clusters, unbranched groups or solitary

LEAVES oval | FRUIT 'haw' | Very difficult to identify; flowers, mature leaves and fruit all required

132

ROSACEAE (ROSES) | *SORBUS* (WHITEBEAMS, ROWANS AND SERVICE-TREES)

Introduction to the whitebeams (*Sorbus*)

Sorbus in Europe

There are 454 tree species considered native to Europe, of which 170 (nearly 40%) are from the genus *Sorbus* (the rowans, whitebeams and service-trees). The majority of these species are rare and occur only in a single country, many of which have highly restricted ranges, with concentrations of *Sorbus* species in Britain & Ireland, Hungary, Germany, the Czech Republic and the Balkans.

Of the *Sorbus* species featured in the European Red List of Trees published by the International Union for Conservation of Nature IUCN:

- 57 (14 B&I) are considered **Critically Endangered**
- 48 (13 B&I) are considered **Endangered**
- 24 (5 B&I) are considered **Vulnerable**

Indeed, *Sorbus* species account for 129 of the 168 most threatened tree species across Europe, and in Britain & Ireland we have 39 of these species, amounting to 19% of Europe's most threatened trees.

Observatory Whitebeam

Stirton's Whitebeam

Watersmeet Whitebeam

Devon Whitebeam

Welsh Whitebeam

133

ROSACEAE (ROSES) | *SORBUS* INTRODUCTION   **Leaves** *pp. 61, 63, 65* | **Flowers** *p. 70* | **Fruit** *p. 76* | **Twigs** *p. 84*

Sorbus diversity

The huge diversity of *Sorbus* occurs because of their unusual method of reproduction, which involves a mixture of hybridization, polyploidy (the ability to inherit more than two complete sets of chromosomes) and apomixis (the ability to reproduce without sexual reproduction with the resulting seeds germinating into a plant which is an identical clone).

In Britain & Ireland, our complex range of species have arisen starting with the three most widespread ancient diploid species – **Common Whitebeam** (*p. 144*), **Rowan** (*p. 140*), and **Wild Service-tree** (*p. 142*) – which each contain two complete sets of chromosomes (one from each parent). These have then crossed and increased their chromosomes through polyploidy (see *p. 137*) and then some have reproduced clonally to create stable species.

In Britain & Ireland the rarer species are grouped into three subgroups which outline their origin.

- **Group 1, subgroup** *aria*: contains species that originated from Common Whitebeam *Sorbus aria* but which have multiple sets of chromosomes.

The other two groups originate from hybridizations of these trees:
- **Group 2**: where subgroup *aria* has crossed with Wild Service-tree *Sorbus torminalis*
- **Group 3**: where subgroup *aria* has crossed with Rowan *Sorbus aucuparia*

Group 1: Common Whitebeam *Sorbus aria* subgroup

Common Whitebeam *Sorbus aria* group	
Common Whitebeam *S. aria*	ancient diploid (with 2 sets of chromosomes)
Herefordshire Whitebeam *S. herefordensis*	triploid probably derived from *S. aria*
Round-leaved Whitebeam *Sorbus eminens* group	
Round-leaved Whitebeam *S. eminens*	a tetraploid derived from *S. aria* and *S. porrigentiformis*
Doward Whitebeam *S. eminentiformis*	probably a hybrid between *S. aria* and *S. porrigentiformis*
Twin Cliffs Whitebeam *S. eminentoides*	probably originating from a hybrid between *S. porrigentiformis* and *S. aria*
Ship Rock Whitebeam *S. parviloba*	a distinct tetraploid of unknown parentage
Rich's Whitebeam *S. richii*	tetraploid related to *S. eminens*
Grey-leaved Whitebeam *Sorbus porrigentiformis* group	
Grey-leaved Whitebeam *S. porrigentiformis*	unclear origins but probably arising from *S. aria* × *S. rupicola*
Welsh Whitebeam *S. cambrensis*	origins unclear but *S. porrigentiformis* is probably a parent plus *S. rupicola*
Cheddar Whitebeam *S. cheddarensis*	probably *S. aria* × *S. porrigentiformis* hybrid
Evan's Whitebeam *S. evansii*	a triploid related to *S. porrigentiformis*
Green's Whitebeam *S. greenii*	a triploid related to *S. porrigentiformis*
Leigh Woods Whitebeam *S. leighensis*	triploid – probably *S. aria* × *S. porrigentiformis* hybrid
Symonds Yat Whitebeam *S. saxicola*	probably *S. porrigentiformis* × *S. aria* hybrid
Observatory Whitebeam *S. spectans*	triploid related to *S. porrigentiformis*
Llanthony Whitebeam *S. stenophylla*	probably derived from *S. porrigentiformis* × *S. rupicola* hybrid
White's Whitebeam *S. whiteana*	probably *S. porrigentiformis* × *S. aria*
Wilmott's Whitebeam *S. wilmottiana*	a triploid from *S. aria* × *porrigentiformis*
Irish Whitebeam *Sorbus hibernica* group	
Irish Whitebeam *S. hibernica*	origins uncertain
Menai Strait Whitebeam *S. arvonicola*	origins uncertain – appears to be a *S. porrigentiformis* × *S. hibernica* hybrid, but neither occurs naturally in North Wales

ROSACEAE (ROSES) | *SORBUS* INTRODUCTION

Thin-leaved Whitebeam *Sorbus leptophylla* group	
Thin-leaved Whitebeam *S. leptophylla*	probably derived from *S. porrigentiformis*
Stirton's Whitebeam *S. stirtoniana*	tetraploid but it is unclear how this species arose – but it appears to be related to *S. leptophylla*
Rock Whitebeam *Sorbus rupicola* group	
Rock Whitebeam *S. rupicola*	probably an ancient tetraploid
Margaret's Whitebeam *S. margaretae*	almost certainly derived from *S. rupicola and S. vexans*
Lancastrian Whitebeam *S. lancastriensis*	tetraploid derived from *S. rupicola*
Bloody Whitebeam *S. vexans*	almost certainly derived from *S. rupicola*
Gough's Rock Whitebeam *S. rupicoloides*	probably derived from *S. porrigentiformis* × *aria* although looks narrow-leaved like *S. rupicola*

Group 2: *Sorbus aria* subgroup (Group 1) × *S. torminalis* (Wild Service-tree)

Wild Service-tree *S. torminalis*	ancient diploid
Bristol Whitebeam *S. bristoliensis*	a hybrid between *S. torminalis* and *S. eminens*
Devon Whitebeam *Sorbus devoniensis* group	
Devon Whitebeam *S. devoniensis*	probably derived from *S. torminalis* and a species from *S. subcuneata*
Somerset Whitebeam *S. subcuneata*	uncertainty exists about the origin of this species but it is probably a cross between *S. torminalis* and *S. margaretae*
Watersmeet Whitebeam *S. admonitor*	derived from *S. subcuneata* and *S. torminalis*

Group 3: *Sorbus aria* subgroup (Group 1) × *S. aucuparia* (Rowan)

Rowan *S. aucuparia*	ancient diploid
English Whitebeam *Sorbus anglica* group	
English Whitebeam *S. anglica*	probably *S. aucuparia* × *S. porrigentiformis*
Llangollen Whitebeam *S. cuneifolia*	probably a mutation of *S. anglica*
Arran Whitebeam *Sorbus arranensis* group	
Arran Whitebeam *S. arranensis*	triploid – cross between *S. aucuparia* and *S. rupicola*
Ley's Whitebeam *S. leyana*	unclear origins, but one parent is *S. aucuparia* and the other either *S. rupicola* or *S. porrigentiformis*
Least Whitebeam *S. minima*	triploid, probably a cross between *S. aucuparia* and *S. rupicola* or *S. porrigentiformis*
Scannell's Whitebeam *S. scannelliana*	probably a cross between *S. aucuparia* and *S. rupicola*
Swedish Service-tree *Sorbus hybrida* group	
Arran Service-tree *S. pseudofennica*	half its genetics from *S. aucuparia*, half from *S. arranensis*
Hybrids and back crosses	
False Rowan *S. pseudomeinichii*	cross between *S. pseudofennica* and *S. aucuparia*
True Service-tree *S. domestica*	ancient diploid (with 2 sets of chromosomes)

Most British *Sorbus* species prefer open sunlit conditions, where they can flower and fruit easily. There are however exceptions, such as the Wild Service-tree and Herefordshire Whitebeam, which are more tolerant of shade and grow in woodland. Many of the species grow on cliff faces and rough slopes where they can avoid being eaten by deer or sheep.

The complex origins of the different species set out above mean they share a wide range of similar characteristics, and it can be very difficult to identify many of the species in the wild. Genetic testing can often be necessary, especially with young trees which have leaves that may not exhibit the full range of adult characteristics.

ID — SORBUS | IDENTIFICATION

Leaves *pp. 61, 63, 65* | Flowers *p. 70* | Fruit *p. 76* | Twigs *p. 84*

Sorbus identification

For the best chance of identifying whitebeams, **fully developed leaves on non-flowering side-shoots that are in a sunny position** (*right*) should be used. The features that will help identification include:

Leaf
- outline shape and extent/shape of any lobes that may be present
- vein pattern and angle to midrib
- total number of veins
- underside hair colour
- shape, size and location of any teeth on the leaf-margin

Fruit
- colour
- shape and size
- lenticel pattern (can be of help)

Sorbus branch showing the location of leaves suitable for identification

BRANCH FROM A SUNNY POSITION

Fully developed leaf on non-flowering side-shoot

Flowering shoot

ORANGE WHITEBEAM

Non-flowering side shoot

Sorbus keys and descriptions

Rather than producing a full key to all the species, with full descriptions and separating species based on the whole range of physical characteristics, the identification approach in the following pages is based on a pragmatic geographical approach on the basis that most of the endemic species have very restricted distributions. For example, although Arran and Ley's Whitebeams are very similar, Arran Whitebeam is found only on the Isle of Arran, Scotland, and Ley's Whitebeam only in the Bannau Brycheiniog (Brecon Beacons) in Wales. In addition, many botanists will visit precise known locations to see particular species and use published data to confirm identification.

Species are grouped by geographical distribution (as per the map and table – *pp. 138–139*) and are accompanied by notes that separate out the species from those that occur in the same local areas. Although this identification process is not infallible, as planted species and introductions do occur, and new species are occasionally located, it should provide a mechanism to allow the identification of a wide range of these rare trees for those who want to look for them.

If an individual is discovered in a new location or a specimen doesn't seem to fit the descriptions given, there are other resources that go into the subject in far greater detail – see https://bsbi.org/publications/ebooks/14-whitebeams-rowans-and-service-trees-of-britain-and-ireland.

The groupings used in this book are as follows:

Nationally widespread
Rowan (*p. 140*); Wild Service-tree (*p. 142*); Orange Whitebeam [introduced] (*p. 143*); Common Whitebeam (*p. 144*); Rock Whitebeam (*p. 145*); Swedish Whitebeam [introduced] (*p. 146*)

Localized and/or disparately distributed
English (*p. 147*); Lancastrian (*p. 149*); Grey-leaved (*p. 147*)

Broad regional 'hotspots' of rare and localized species (*see map and table overleaf for full details*)
Isle of Arran (*p. 152*); Ireland (*p. 150*); North Devon and Somerset (*pp. 158–161*); Cheddar Gorge, Somerset (*p. 162*); Avon Gorge, Avon (*p. 164*); Lancashire (*p. 149*); Wye Valley, Gloucestershire (*p. 167*); Bannau Brycheiniog (Brecon Beacons), Wales (*p. 156*); Mid- to North Wales (*p. 154*).

Rare species that are found in more than one region are ordered as best as possible and are cross-referenced from the relevant 'Comparison Species' box in the species accounts.

SORBUS | IDENTIFICATION

Identification features of *Sorbus* with ± entire leaves

LEAF-TIP whether rounded or pointed.

LEAF OUTLINE whether **lobed** or **unlobed** and, if lobed, and the distance they are **cut to the midrib**. (A ÷ B as a percentage)

CUT TO MIDRIB DISTANCE (A)

LOBED

MIDRIB DISTANCE (B)

UNLOBED

BISERRATE TEETH

UNISERRATE TEETH

TEETH ABSENT FROM BASAL PORTION OF LEAF-MARGIN

VEIN COUNT the total number of veins (best seen on underside).

VEIN COUNT 17

UNDERSIDE colour and hairiness differences, though subtle, can be useful evidence.

LEAF-BASE whether rounded or wedge-shaped, and the angle the leaf-base makes with the midrib.

TEETH shape; direction; and where on the margin they start. **uniserrate** = a single tooth; **biserrate** = a tooth with teeth.

LENTICELS are small pores which take in oxygen and release CO_2 and water.

BUDS generally show little variation between species; being either brown or green. However, where buds are useful for identification these are shown in the relevant species accounts.

BARK tends to be grey-brown in most species. Bark features are generally of little help in identification but are shown if useful or of interest.

FRUIT the colour (typically red, but some species orange or brown); the size and shape; and the presence, size and arrangement of lenticels can be useful in identification.

Common *S. aria* thought to be an ancient diploid with two sets of chromosomes (1 set inherited from each parent)

Rock *S. rupicola* origins uncertain but thought to be an ancient tetraploid with four sets of chromosomes

Leigh Woods *S. leighensis* [TRIPLOID] probably from hybridization between **Common** and **Grey-leaved**

Grey-leaved *S. porrigentiformis* [TETRAPLOID]

Thin-leaved *S. stenophylla* [PENTAPLOID] probably derived from hybridization between **Rock** and **Grey-leaved**

Polyploidy is when additional sets of chromosomes are inherited, resulting in some species having three (triploid), four (tetraploid) or rarely five (pentaploid) sets of chromosomes.

SORBUS | INTRODUCTION   Leaves pp. 61, 63, 65 | Flowers p. 70 | Fruit p. 76 | Twigs p. 84

Rare *Sorbus* by region

Across Britain & Ireland there are a few hotspots with high concentrations of rare species, especially within the Cheddar Gorge, south Wales and north Devon, making these areas some of the most important sites for rare trees in Europe. This composite shows the *Sorbus* hotspots (darker colour = more species).

Key | ■ ■ Very widespread species (Britain & Ireland); ■ Widespread species;
■ *S. hibernica* (Ireland); ■ ■ ■ Rare, localized species

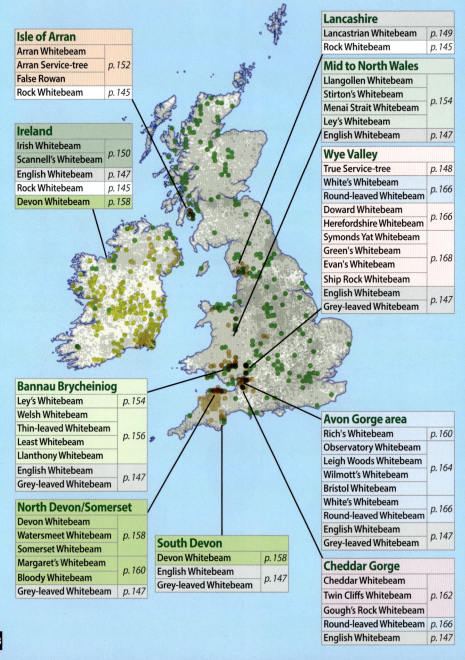

Isle of Arran
Arran Whitebeam	
Arran Service-tree	p. 152
False Rowan	
Rock Whitebeam	p. 145

Ireland
Irish Whitebeam	p. 150
Scannell's Whitebeam	
English Whitebeam	p. 147
Rock Whitebeam	p. 145
Devon Whitebeam	p. 158

Lancashire
Lancastrian Whitebeam	p. 149
Rock Whitebeam	p. 145

Mid to North Wales
Llangollen Whitebeam	
Stirton's Whitebeam	
Menai Strait Whitebeam	p. 154
Ley's Whitebeam	
English Whitebeam	p. 147

Wye Valley
True Service-tree	p. 148
White's Whitebeam	p. 166
Round-leaved Whitebeam	
Doward Whitebeam	p. 166
Herefordshire Whitebeam	
Symonds Yat Whitebeam	
Green's Whitebeam	p. 168
Evan's Whitebeam	
Ship Rock Whitebeam	
English Whitebeam	p. 147
Grey-leaved Whitebeam	

Bannau Brycheiniog
Ley's Whitebeam	p. 154
Welsh Whitebeam	
Thin-leaved Whitebeam	p. 156
Least Whitebeam	
Llanthony Whitebeam	
English Whitebeam	p. 147
Grey-leaved Whitebeam	

North Devon/Somerset
Devon Whitebeam	p. 158
Watersmeet Whitebeam	
Somerset Whitebeam	
Margaret's Whitebeam	p. 160
Bloody Whitebeam	
Grey-leaved Whitebeam	p. 147

South Devon
Devon Whitebeam	p. 158
English Whitebeam	p. 147
Grey-leaved Whitebeam	

Avon Gorge area
Rich's Whitebeam	p. 160
Observatory Whitebeam	p. 164
Leigh Woods Whitebeam	
Wilmott's Whitebeam	
Bristol Whitebeam	
White's Whitebeam	p. 166
Round-leaved Whitebeam	
English Whitebeam	p. 147
Grey-leaved Whitebeam	

Cheddar Gorge
Cheddar Whitebeam	
Twin Cliffs Whitebeam	p. 162
Gough's Rock Whitebeam	
Round-leaved Whitebeam	p. 166
English Whitebeam	p. 147

138

SORBUS | INTRODUCTION

Characters of non-pinnate whitebeams (excluding the Isle of Arran)

	Leaf outline	Leaf lobes	Veins	Underside hair	Leaf-margin teeth*	Leaf-base	Leaf-tip
Common	oval	none or rounded (cut to 20%)	18–28	white, dense	uni- or bi-	rounded to wedge	pointed
Rock	oblong	none	12–19	white, dense	uni- or weakly bi-	wedge	rounded
Orange	oval	none or small	18–24	greyish green	bi-, small	wedge	pointed
Lancastrian	broadly oval	none	15–22	greyish white, dense	basal 25% no teeth; rest uni-	wedge	rounded
English	oval	rounded (cut to 30%)	15–23	grey-white	weakly bi-	round or wedge	rounded
Grey-leaved	oval	none	15–20	greyish white	weakly bi-	wedge	rounded
Irish	round to oval	none	18–24	white, dense	uni-, fine, in apical 50%	wedge	rounded
Scannell's	oval	shallow, rounded	15–18	greenish	uni-, small	wedge	rounded to pointed
Least	narrowly oval	rounded (cut to 33%)	12–18	grey-green, thin	uni-	narrow wedge	pointed
Thin-leaved	oval	shallow	19–25	grey-white or yellowish	bi-	wedge to round	pointed
Welsh	broadly oval	none or very shallow	18–24	greyish white	basal 25% no teeth; rest bi-	wedge	rounded
Llanthony	oblong, tapering	none or shallow	14–20	greyish white	basal 25% no teeth; rest bi-	wedge	± pointed
Ley's	rounded	rounded (cut to 75%)	14–22	greyish white	uni-	wedge	round pointed
Llangollen	narrowly oval	rounded (cut to 33%)	16–22	greyish white	weakly bi-	narrow wedge	pointed
Stirton's	oval	none or weak	18–24	white, matted	uni- or bi-	wedge	pointed
Menai Strait	broadly oval	none	18–24	greyish white, dense	uni-	wedge	rounded
Symonds Yat	narrowly oval	none or shallow	15–22	greyish white	basal 35% no teeth; rest weakly bi-	wedge	pointed
Herefordshire	oval	none or shallow	20–25	white, dense	bi-	wedge	rounded
Green's	broadly oval	none or small	14–20	greyish green	strongly bi-	wedge	truncated
Evans'	broadly oval	none	19–23	greyish green	uni- to weakly bi-	wedge	rounded
Doward	broadly oval	none	19–24	greyish white	uni-, regular	wedge	rounded
Ship Rock	broadly oval	shallow (top 50%)	17_23	greyish white	strongly bi-	wedge	rounded
Round-leaved	round	none	17–23	greyish white	uni- to weakly bi-	rounded	rounded
Leigh Woods	oval	blunt, shallow	15–19	greenish white	weakly bi-	rounded to wedge	rounded
Bristol	oval	triangular (top 50%)	16–21	greenish white	weakly bi-	wedge	pointed
White's	oval	none	17–23	greyish green	uni- to weakly bi-	wedge	pointed
Observatory	broadly oval	blunt, shallow	16–20	white to bluish white	weakly bi-	wedge	pointed
Wilmott's	rhomboidal	shallow	14–22	white	fine bi-	wedge	pointed
Rich's	oval	none	18–22	greyish green	uni-	wedge	rounded
Somerset	narrowly oval	triangular (cut to 20%)	12–22	whitish grey	finely bi-	wedge to tapered	pointed
Watersmeet	oval	triangular (cut to 23%)	15–20	greenish white	finely pointed	wedge to rounded	pointed
Margaret's	oval	none	16–20	greyish white, dense	basal 33% no teeth; rest coarsely bi-	wedge	rounded with point
Bloody	oval	none	8–21	greyish white, dense	basal 33% no teeth; rest coarsely bi-	wedge	rounded with point
Devon	broadly oval	pointed (cut to 75%)	14–20	grey	bi-	rounded	sharply pointed
Twin Cliffs	rounded	shallow at most	16–24	greyish green	basal 40% no teeth; rest strongly bi-	wedge	rounded
Gough's Rock	narrowly oval	none	16–21	greyish green	basal 33% no teeth; rest weakly bi-	wedge	pointed
Cheddar	oval	shallow at most	20–24	greenish grey	basal 40% no teeth; rest uni to weakly bi-	wedge	pointed

* uni- = uniserrate; bi- = biserrate (see *p. 59*)

139

SORBUS | WIDESPREAD — Leaves pp. 61, 63, 65 | Flowers p. 70 | Fruit p. 76 | Twigs p. 84

LC LC Rowan — *Sorbus aucuparia*

A medium-sized tree (H to 18 m) with a rounded crown. In its native range it occurs in mountains, woodlands and valleys at elevations of up to 1,000 m, on a wide range of soils, from chalk to acid (even on peat in Scotland). Found throughout Britain & Ireland, it is also widely planted in gardens, parks and streets for its striking red berries and colourful autumn foliage.

Features: BARK smooth and grey with yellowish lenticels; darkening and cracking with age. WOOD strong, hard, and flexible and historically used for carvings and for making tools and shortbows. TWIGS hairy at first, but hairs rub off with age. BUDS large; pointed egg shape; **dark brown and covered in greyish white hairs**. LEAVES **pinnate** (L to 25 cm); with 5–9 pairs of pointed-oval leaflets and 1 terminal leaflet; upperside dark green, underside greyish hairy; leaves arranged alternately along a branch (*cf.* Ash); pair of small stipules at base of leaf-stalk. FLOWERS in large, tight clusters (D to 15 cm) of up to 100; 5-petalled; yellowish white (D 5–10 mm); with an unpleasant aroma, described by some as of 'slightly rotten meat'. FRUIT small; red; berry-like.

J F M A M J J A S O N D

Did you know? Historically, Rowan was planted to keep witches away from houses and churches.
The berries can be turned into a jelly which is excellent with cold game or wildfowl, and a wholesome 'perry' or 'cider' can also be made from them.

Comparison species | other *Sorbus* with pinnate leaves *e.g.* **False Rowan** (*p. 152*), **True Service-tree** (*p. 148*) and the introduced **Hupeh Rowan** *Sorbus hupehensis* which has leaflets more rounded, **margins toothed only near the tip** (INSET), and **white to pinkish red berries**.
Ash (*p. 298*) has somewhat similar leaves but arranged oppositely.

HUPEH ROWAN

BARK smooth and grey; cracking and darkening with age

Older Rowans in open habitats develop a wide crown (*left*); in late summer its bright red fruit and pinnate leaves are distinctive (*right*).

Whitebeams compared *p. 138*   SORBUS | WIDESPREAD

LEAVES **pinnate; 5–9 pairs of leaflets plus 1 terminal leaflet**; small stipules at the base of the leaf-stalk; leaflet margins toothed

FLOWERS in large clusters

FRUIT red; rounded; L 8–12 mm; W 9–14 mm; LENTICELS absent

BUDS **pointed egg shape; dark brown; covered in grey hairs**

TWIGS shiny brown with scattered wart-like lenticels

Associated species | Rowan blossom attracts many insects in spring. **Welsh Wave moth** caterpillars feed on the leaves. The berries are eaten by thrushes such as **Mistle Thrush**, **Blackbird**, **Fieldfare** and **Redwing**. The latter two are thought to time their migration from Scandinavia and Iceland so that they arrive in Britain & Ireland when Rowans are bearing fruit. These thrushes disperse the seed around the countryside, their digestive systems removing the flesh and then excreting the seed in their droppings. The berries are also sought out by **Waxwings**, an irruptive bird species that occasionally arrives in Britain & Ireland in the autumn and winter, often best located simply by finding the nearest Rowan. Mammals, including **Badger**, **Fox**, **squirrels** and even **Pine Marten**, also eat the berries.

MISTLE THRUSH

PINE MARTEN

141

SORBUS | WIDESPREAD   Leaves *pp. 61, 63, 65* | Flowers *p. 70* | Fruit *p. 76* | Twigs *p. 84*

LC LC Wild Service-tree   *Sorbus torminalis*

A medium-sized tree (H usually to 25 m), a rarer relative of Rowan and Whitebeam that seems to germinate only in areas of ancient woodland. Found mainly on chalk and limestone, but also on nutrient-rich clays, it has spectacular autumn colour. Widespread, although uncommon, native of southern Britain, and introduced or planted elsewhere. **Features:** BARK smooth and grey when young; **cracking into small blocks in a chequerboard pattern as the tree ages.** WOOD hard and fine-grained; has been used as a veneer. TWIGS hairless with numerous lenticels. BUDS **almost round; bud scales green with a narrow brown margin.** LEAVES **distinctive with deep triangular lobes** and usually 8–13 veins. FLOWERS 5-petalled; white (D 10–15 mm) in large, somewhat loose clusters (D to 12 cm). FRUIT ± spherical; dark brown and covered in lenticels in a range of sizes.

J F M A M J J A S O N D

Associated species | The **Wild Service Aphid**, which creates characteristic reddish to yellowish rolled pseudo-galls in spring

Did you know? Wild Service-tree is an indicator of ancient woods and hedgerows (unless it has been planted). It can be a large tree and form part of the canopy although, if shaded it will not flower or fruit. In the Weald of Kent and Sussex, Wild Service fruit is known as 'chequers' or 'chequer berries' (from the chequerboard bark). Many of the area's pubs are named 'The Chequers', and it may be that a drink was made from the berries or that they were added to beer.
Currently Wild Service-tree rarely grows from seed. The seeds are thin-walled and it is thought that they are fully digested by birds and hence not dispersed. Therefore, it has been suggested that historically Wild Boar, cattle and possibly even humans were the main dispersal agents of viable seed. However, new trees do develop from root suckers, and although some mature trees have been found to be 200 years old, if they developed from a root sucker, this may be only a fraction of the true age of the tree.
The wood is fine and fine-grained and has been used as a veneer.

BARK smooth and grey, cracking into small blocks

Compared with Rowan, Wild Service-tree has looser flower clusters and differently shaped leaves (*left*); and usually develops into a larger tree (*right*).

Whitebeams compared *p. 138* SORBUS | WIDESPREAD

LEAVES
L 55–100 mm;
W 50–105 mm
OUTLINE **triangular with deep triangular lobes** cut 20–60% of the distance to the midrib; basal lobes at right angles to the leaf-stalk;
CHARACTER shiny; leaf-margin can have sparse, fine, forward-pointing teeth
VEINS typically 8–13

FRUIT **brown, rounded;** L 9–17 mm; W 9–15 mm

TWIGS hairless; with numerous lenticels

BUDS round; scales green with narrow brown margin

SORBUS | INTRODUCED, WIDESPREAD

DD ■ Orange Whitebeam *Sorbus croceocarpa*

Introduced medium-sized tree (H to 21 m; trunk girth to 1·7 m) with a broad, rounded, compact crown. Brought to Britain in 1874, it is widely planted and has naturalized in some places.

LEAVES
L 75–150 mm;
W 50–120 mm
OUTLINE **oval** (1·2–1·6× as long as wide); usually unlobed but can have small lobes halfway up the leaf
LEAF-BASE usually wedge-shaped
LEAF-TIP pointed
CHARACTER dark green; leaf-margins with small biserrate teeth; those at the end of main veins are larger, prominent, broad and rather blunt
VEINS **18–24**
UNDERSIDE greyish green hairy

FRUIT **bright orange** (can be flushed red); typically L 11–22 mm; W 11–16 mm; LENTICELS numerous; larger towards the base

J F M A M J J A S O N D

Comparison species | Devon Whitebeam (*p. 158*) – leaves with 14–20 veins; fruit brown to orange-brown

143

SORBUS | WIDESPREAD   Leaves pp. 61, 63, 65 | Flowers p. 70 | Fruit p. 76 | Twigs p. 84

LC LC Common Whitebeam — *Sorbus aria*

A medium-sized tree (H 15–25 m) with a preference for chalk and lime-rich soils, but also tolerant of other soil types. Its ability to withstand pollution means it has become a widely planted urban tree. **Features:** BARK smooth and grey; cracking with age. WOOD hard and tough; used to make cogs for machinery. TWIGS shiny brown and usually hairless with scattered wart-like lenticels. BUDS cone-shaped; shiny brown to greenish brown; bud-scales have hairy edges and appear to be stuck together. LEAVES **dark green; unlobed** or, at most, shallowly lobed; **underside densely white hairy**. FLOWERS in tight clusters (D to 10 cm); 5-petalled; white (D ± 15 mm); with an unpleasant aroma, described by some as of "slightly rotten fish". FRUIT small; red; berry-like; with a few lenticels at most.

J F M A M J J A S O N D

Associated species | As with all *Sorbus* trees, the berries are eaten by birds, and the flowers attract insects. The white caterpillars of *Argyresthia sorbiella*, a tiny moth, feed on the shoots and flower buds.

ARGYRESTHIA SORBIELLA

Did you know? Whitebeams, like Rowans, can also occasionally be found as an epiphyte – a tree growing on another tree – if their seeds have been dropped by birds. This is encountered most commonly in the damp climate of the west of the region, but can occur anywhere.
The species was often planted as a boundary marker as the underside of the leaves flashing white in the wind would draw attention to it.
Historically, overripe berries were turned into jelly to accompany meats, particularly venison.

The flowers of Common Whitebeam sit in clusters of leaves (*left*); making this and other whitebeams favourites of urban planting schemes (*right*).

144

Whitebeams compared *p. 138* *SORBUS* | WIDESPREAD

LEAVES
L 65–125 mm; W 40–95 mm
OUTLINE **long oval; unlobed** or with shallow lobes (cut max. 10–20% of the distance to the midrib)
LEAF-BASE rounded
LEAF-TIP pointed
CHARACTER dark green; leaf-margins with uniserrate or biserrate teeth
VEINS usually 18–28
UNDERSIDE **densely white hairy**

FRUIT red; large; typically looks wider than long; L 10–15 mm; W 10–15 mm;
LENTICELS **a few; small to medium-sized**

TWIGS shiny brown; scattered wart-like lenticels

BARK smooth and grey; cracking with age

BUDS shiny brown; cone-shaped

LC LC VU **Rock Whitebeam** *Sorbus rupicola*

A shrub or small tree (H to 10 m) widespread in western Britain and Ireland – and also occurring in Norway. In Britain & Ireland the second-most widespread (after Rowan) and fourth-most common *Sorbus*, although the populations are often small and scattered, and on cliff faces. It is one of the last species of whitebeam to come into leaf, flower and fruit – the fruit often being at its best in mid-October.

LEAVES
L 70–150 mm; W 35–75 mm
OUTLINE **oblong** (1·5–2·4× as long as wide); unlobed
LEAF-BASE **wedge-shaped**
LEAF-TIP usually blunt and rounded
CHARACTER dull dark green; leaf-margins with uniserrate to weakly biserrate teeth
VEINS **usually 12–19**
UNDERSIDE **densely white hairy**

J F M A M J J A S O N D

FRUIT deep red; large, look wider than long; L 10–14 mm; W 12–15 mm; LENTICELS moderate number

Comparison species | **Lancastrian Whitebeam** (*p. 149*) is very similar but has more (15–22) veins which are forward-pointing; **Margaret's** (*p. 160*) and **Grey-leaved Whitebeams** (*p. 147*) have broader, biserrate leaves; **Gough's Rock Whitebeam** (*p. 162*) has biserrate leaves, often with a pointed tip. **Common Whitebeam**, **Thin-leaved** (*p. 156*), **Stirton's** (*p. 154*), **Bloody** (*p. 160*), **White's** (*p. 166*) and **Willmott's Whitebeams** (*p. 164*) have fruit longer than wide.

SORBUS | WIDESPREAD   Leaves *pp. 61, 63, 65* | Flowers *p. 70* | Fruit *p. 76* | Twigs *p. 84*

LC Swedish Whitebeam   *Sorbus intermedia*

Medium-sized tree (H to 10 m, but up to 18 m recorded) introduced from Scandinavia and widely planted since 1789, although a few old trees in remote Scottish locations hint at the possibility of the species arriving naturally, dispersed by migrating birds. **Features:** LEAVES oval to oblong (L 5–10 mm; W 30–80 mm); **lobed to deeply lobed** with 12–17 veins and margins with strongly forward-pointing teeth; tip slightly pointed; base straight to wedge-shaped; **underside with greenish white hairs**. FLOWERS white with yellow anthers. FRUIT orange-red to red when ripe; L 11–15 mm; W 10–13 mm – usually longer than wide; at most very few small to medium-sized lenticels.

J F M A M J J A S O N D

FRUITS orange-red; **longer than wide**; a few lenticels at most

LEAVES typically ± oval with **wide, shallow lobes**

LEAF-VEINS 12–17

LEAF UNDERSIDE greenish white hairs

Comparison species | English Whitebeam (*p. 147*) has more veins, shallower lobes, greyish white undersides and red fruit. **Mougeot's Whitebeam** *S. mougeotii* (leaf *right, top*) cultivated since the late 1800s is planted as a landscape tree but naturalizes. It has white hairs on the leaf underside, 20–24 veins, shallower lobes and a wedge-shaped leaf-base. **German Service-tree** *S. ×thuringiaca* (leaf *right, bottom*), a natural and planted hybrid between Rowan (*p. 140*) and Common Whitebeam, typically has 1–3 pairs of free basal leaflets and 16–30 veins. **Common Whitebeam** (*p. 144*) can have shallow lobes but has a whitish underside.

MOUGEOT'S WHITEBEAM

GERMAN SERVICE-TREE

Swedish Whitebeam has the clustered flowers typical of all *Sorbus* (*left*); tree in a typical setting (*right*).

Whitebeams compared *p. 138*   SORBUS | VARIOUS AREAS

NT NT EN English Whitebeam — *Sorbus anglica*

Shrub or small tree (H to 10 m) of woods and rocky places, mostly on Carboniferous limestone, in south-west England, Wales, and Killarney, Ireland. The world population is approximately 1,000 trees across 30 or so sites. It was first collected in 1836 at Craig Breidden and known by a variety of names before it was finally designated as a species in 1914.

LEAVES
L 75–120 mm; W 50–85 mm
OUTLINE **oval, with shallow rounded lobes** (cut max. 10–30% of the distance to the midrib)
LEAF-BASE rounded to wedge-shaped (angle of base 40–59°)
LEAF-TIP rounded
CHARACTER **crinkly undulating appearance**
VEINS usually 15–23
UNDERSIDE grey-white hairy; the main veins at the centre of the leaf have distinct cross veins giving a feathery appearance

Comparison species
Llangollen Whitebeam (*p. 154*) – leaf-base narrower; restricted range from where English Whitebeam is not known.
Ley's Whitebeam (*p. 154*) – leaves narrower with deeper lobes (cut 50–75% of the distance to the midrib).
Least Whitebeam (*p. 156*) – leaves smaller, narrower.
Swedish Whitebeam (*p. 146*) – fruit larger and more orange; leaves with deeper, more even lobes and fewer veins.

FRUIT crimson; typically look wider than long; L 9–12 mm; W 10–13 mm; LENTICELS a few at most at the base of the fruit

VU VU Grey-leaved Whitebeam — *Sorbus porrigentiformis*

A small tree or shrub (H to 5 m) found in scattered locations in south-west England and Wales. There are about 30 populations comprising a total of approximately 500 trees growing in a wide range of habitats, from damp north-facing slopes to dry south-facing slopes.

LEAVES
L 60–100 mm; W 30–70 mm
OUTLINE oval (1·2–1·7× as long as wide); widest at 54–66% of the distance from the leaf-base unlobed
LEAF-BASE wedge-shaped
LEAF-TIP rounded
CHARACTER **dark green; leaf-margins with weakly biserrate outward-pointing teeth**
VEINS 15–20
UNDERSIDE greyish white; hairy

Comparison species
Welsh (*p. 156*) and **Llanthony Whitebeams** (*p. 156*) – leaves with a less rounded tip and more strongly biserrate teeth; fruit narrower. **Margaret's** (*p. 160*) and **Rock Whitebeams** (*p. 145*) – leaves narrower. **Symonds Yat** (*p. 168*) and **Gough's Rock Whitebeams** (*p. 162*) – leaves narrower with a more pointed tip. **Leigh Woods Whitebeam** (*p. 164*) has lobed leaves. **Round-leaved Whitebeam** (*p. 166*) – leaves round. **Rich's Whitebeam** – leaves larger, and relatively wider

FRUIT dark red; L 9–13 mm; W 10–15 mm; usually looks wider than long; LENTICELS a few

ROSACEAE (ROSES) *SORBUS* | TRUE SERVICE-TREE   Leaves *pp. 61, 63, 65* | Flowers *p. 70* | Fruit *p. 76* | Twigs *p. 84*

LC CR True Service-tree (Whitty Pear)   *Sorbus domestica*

Medium-sized tree (H to 20 m or more) that is possibly native or an ancient introduction. It can be found in the Wye Valley, Herefordshire; the Camel Valley, Cornwall; and on the Welsh coast on limestone sea cliffs in Glamorgan. A few have also been found in other locations, including East Anglia.

BARK reddish; flaking to a 'chequerboard' pattern with age

Comparison species | Rowan (*p. 140*) has hairy buds; flowers with 3–4 styles; fruit that is small, round and orange-red; smooth grey bark and leaves with a persistent, smooth-edged stipule.

LEAVES pinnate, with 5–8 pairs of leaflets and a terminal leaflet.

LEAFLETS often pale green or yellowed, with fine greyish green hairs on upper surface when young (lost as the leaf ages). Leaflet-margins with teeth towards the tip, with the inner edge often smooth and lacking teeth.

FRUIT yellow to green or brown (can be partially flushed red); apple- or pear-shaped; L 15–50 mm; W 18–40 mm;
LENTICELS numerous; large

LF-STALK lacks stipules

TWIGS shiny, olive brown, with occasional lenticels; can have greyish bloom

BUDS pointed, green and shiny with dark fringes to the scale edges

Did you know? For many years, the only known wild specimen (known as the 'Whitty Pear') was in a remote part of the Wyre Forest in Worcestershire. It was described in 1678 by Mr Pitt, who says that 'he found it in the preceding year as a rarity growing wild in a forest of Worcester'.

Old inhabitants of the district called it the 'Whitty Pear tree' and used to hang pieces of the bark round their necks as a charm to cure a sore throat. The tree was burnt down in 1862 by a fire kindled by an angry poacher, who wanted revenge on a local tree-loving magistrate.

Fortunately, cuttings had been taken and propagated at nearby Arley Castle; individuals were then planted out over the Wyre Forest. In 1983 a new population of about 80 trees was found on a south-facing limestone sea cliff in Glamorgan. Initially, it was thought they were introduced specimens, but ring counts from dead branches determined the trees were extremely slow growing and could be up to 400 years old. Subsequently other trees have been found elsewhere in Glamorgan and in Gloucestershire.

Given the evidence and the cliff location of the Glamorgan trees, it seems likely that the True Service is in fact a rare native species and that both the Whitty Pear and the Glamorgan trees could be descended from the same 'edge-of-range colonising stock' that has now declined to a few remnant populations.

Whitebeams compared *p. 138* ROSACEAE (ROSES) *SORBUS* | LANCASHIRE

Lancastrian Whitebeam

Sorbus lancastriensis

Small tree or large shrub (H to 6 m) endemic to the limestone screes, rocks and cliffs within 30 km of Morecambe Bay. It is relatively abundant with about 2,300 individuals recorded and most frequent in small populations between Cunswick Scar and Warton Crag. It has a preference for open scrubland; freely flowering and fruiting and colonising well if grazing pressures are removed. In John Ray's *Historia Plantarum* (1688) there is a description which appears to describe this species, although it was only formally named by E. F. Warburg in 1952.

LEAVES
L 60–120 mm; W 40–75 mm
OUTLINE oval (1·40 to 1·75× as long as wide); unlobed
LEAF-BASE wedge-shaped (angle of base 42–47°)
LEAF-TIP rounded
CHARACTER dark green; basal 10–25% of leaf-margins untoothed; remainder uniserrate, unequal, outward-pointing
VEINS usually 15–22
UNDERSIDE densely grey-white hairy

FRUIT dark red; looks wider than long; L 10–12 mm; W 12–15 mm; LENTICELS moderate number near the base, a few scattered over the upper part

Comparison species in Morecambe Bay

Rock Whitebeam (*p. 145*) can be difficult to separate.

LEAVES narrower (1·5–2·4× as long as wide)

fewer (12–19) veins; usually forward-pointing

Common Whitebeam (*p. 144*)

LEAVES more (18-28) veins

FRUIT longer than wide

Lancastrian Whitebeam habitat, White Scar, Cumbria.

ROSACEAE (ROSES) *SORBUS* | IRELAND   Leaves *pp. 61, 63, 65* | Flowers *p. 70* | Fruit *p. 76* | Twigs *p. 84*

VU VU Irish Whitebeam   *Sorbus hibernica*

Small tree (H to 7 m) found in scattered locations throughout Ireland and considered a rare native in Northern Ireland. First recorded in May 1933 in Galway by R. L. Praeger and H. W. Pugsley. Praeger collected further material which was sent to A. J. Wilmott at the Natural History Museum, London, who recognized that it differed from other known *Sorbus* taxa and coined the epithet '*hibernica*' but left it to E. F. Warburg to formally describe the species in 1957.

FRUIT deep red; rounded; looks wider than long; typically L 10–15 mm; W 11–15 mm; LENTICELS few; moderate to large

Comparison species in Ireland |
Common Whitebeam (*p. 144*) – leaves with a rounded base; underside less hairy; teeth swept upward.
Rock Whitebeam (*p. 145*) – leaves narrower, longer, and more oblong and lacking teeth in the basal third.

NOTE: It is closely allied to Common Whitebeam and not always easily distinguished from it.

CR CR Scannell's Whitebeam   *Sorbus scannelliana*

Small tree (H to approximately 7 m) endemic to Ross Island, Killarney, Kerry, Ireland, first found in June 1988 when the population was assessed as five trees. It was confirmed as a new species in September 2008. It has probably evolved as a hybrid between Rowan (*p. 140*) and Rock Whitebeam (*p. 145*) – the same pairing that led to Arran Whitebeam. It grows in the Killarney area of Ireland, where both parents occur. Its name honours the Irish botanist Maura Scannell.

FRUIT small red berries that are rarely found and have not been properly described to date.

Comparison species in Killarney |
Only **English Whitebeam** (*p. 147*), which has leaves with distinct cross veins between the main veins and a whiter underside.
Comparison species (*e.g.* in botanic gardens) |
Arran (*p. 152*), **Least** (*p. 156*) and **Ley's** (*p. 154*) **Whitebeams**.

Scannell's Whitebeam

Whitebeams compared *p. 138*   ROSACEAE (ROSES) *SORBUS* | IRELAND

IRISH WHITEBEAM
LEAVES L 70–110 mm; W 50–80 mm; OUTLINE **rounded to oval** (1·25–1·65 × as long as wide); unlobed
LEAF-BASE wedge-shaped
LEAF-TIP rounded
CHARACTER green; leaf-margins with **straight, symmetrical teeth**, fine-pointed in the apical part of the leaf
VEINS 18–24
UNDERSIDE dense covering of **greyish white** silky hairs

SCANNELL'S WHITEBEAM
LEAVES L 70–90 mm; W 40–50 mm; OUTLINE oval (1·6–1·8 × as long as wide) with **shallow, usually rounded, lobes**
LEAF-BASE wedge-shaped
LEAF-TIP variable, rounded to pointed
CHARACTER dark green; leaf-margins with small, forward-pointing, usually uniserrate, teeth
VEINS 15–18
UNDERSIDE greenish; hairy

Comparison species in Ireland

Common Whitebeam (*p. 144*)

densely white

English Whitebeam (*p. 147*)

Rock Whitebeam (*p. 145*)
LF-VEINS 12–19

FRUIT W > L

Devon Whitebeam (*p. 158*)

FRUIT brown

Swedish Whitebeam (*p. 146*)

FRUIT L > W

NOTE: Devon Whitebeam is designated as **Endangered** on the Irish Red List

ROSACEAE (ROSES) *SORBUS* | ISLE OF ARRAN   Leaves *pp. 61, 63, 65* | Flowers *p. 70* | Fruit *p. 76* | Twigs *p. 84*

EN EN Arran Whitebeam — *Sorbus arranensis*

Small, windswept and stunted slender tree (H typically 1–2 m, although can reach 6 m on good soil). It grows naturally only at the north end of the Isle of Arran where it clings perilously to the steep rocky slopes of two remote glens in association with Arran Service-trees. Approximately 407 have been recorded but are under constant threat as their fragile root systems are easily dislodged from the rocky soil by the strong gales and heavy snowstorms common in their location. Arran Whitebeam was first recorded in the 1830s and is thought to have arisen as a natural hybrid between Rock Whitebeam (*p. 145*) and Rowan (*p. 140*).

Comparison species (*e.g.* in botanic gardens) | Very similar to **Least Whitebeam** (*p. 156*) and almost identical to **Ley's Whitebeam** (*p. 154*) on leaf characteristics alone and probably needs a DNA analysis to be certain of its identity.

CR CR Arran Service-tree — *Sorbus pseudofennica*

Small, slender tree (H usually 1–2 m high but can reach 7 m) is found in two remote glens at the north of the island of Arran where a population of 430 or so are typically found clinging perilously to steep rocky slopes. It was first noted in 1797 but was not named as full species until 1957. This species has arisen from a cross between Arran Whitebeam and Rowan (*p. 140*).

CR CR False Rowan — *Sorbus pseudomeinichii*

Small tree or shrub (H to at least 4 m) found only on the Isle of Arran, Scotland. It is known in the wild from just four specimens, only two of which remain – one beside the main burn in Glen Catacol, and another below a waterfall. A few others are in collections. False Rowan is an example of the continued evolution in *Sorbus* trees and has probably arisen from a cross between Arran Service-tree and Rowan.

Glen Catacol, in the north-west of the Isle of Arran, is the key spot for finding Scotland's rare whitebeams.

ARRAN WHITEBEAM

LEAVES L 60–90 mm; W 30–45 mm
OUTLINE relatively narrow oval (1·5–2·0 × as long as wide) with **rounded lobes cut up to 75% of the distance to the midrib; typically no free leaflets**
CHARACTER yellowish green; leaf-margins with sharp uniserrate teeth pointing mainly forward
VEINS 13–18
UNDERSIDE with greyish white hairs

ARRAN SERVICE-TREE

LEAVES L 50–90 mm; W 40–70 mm
OUTLINE triangular-oval; **partially pinnate with deep lobes** over most of the leaf and **typically with the basal 1–2 pairs of leaflets free**
CHARACTER dark yellowish green; leaflet-margins have sharp teeth, mainly at the tip of the lobes
VEINS 12–20
UNDERSIDE with thin grey hairs

FRUIT scarlet; rounded; L 9–11 mm; W 8–10 mm; LENTICELS a few; inconspicuous

FRUIT scarlet; intermediate between those of Arran Service-tree and Rowan; usually look longer than wide; typically L 8–12 mm; W 7–11 mm; LENTICELS a few; inconspicuous

FALSE ROWAN

LEAVES OUTLINE pinnate; L 75–130 mm; W 45–105 mm; **4–5 pairs of free lateral leaflets and a much larger terminal leaflet**

Comparison species on the Isle of Arran

Rowan (p. 140)

LEAVES terminal and lateral leaflets similarly sized

FRUIT scarlet; intermediate between those of Arran Service-tree and Rowan; look longer than wide; typically L 8–11 mm; W 8–10 mm; LENTICELS few; inconspicuous

Rock Whitebeam (p. 145) also occurs but is very different, having unlobed, oblong leaves.

ROSACEAE (ROSES) *SORBUS* | MID & NORTH WALES Leaves *pp. 61, 63, 65* | Flowers *p. 70* | Fruit *p. 76* | Twigs *p. 84*

CR CR Menai Strait Whitebeam — *Sorbus arvonicola*

Small tree (H to 10 m or more) found on limestone rock in open and tall woodland of Nant Porth Nature Reserve, Wales, where approximately 30 trees grow right down to the rocky shoreline. The site is prone to coastal erosion. However, if other tree species fall in the woods, then Menai Whitebeam appears able to readily colonize areas close to the sea. It was known for many years but only recognized as a full species in 2014.

FRUIT bright red; rounded; L 8–13 mm; W 10–13 mm; LENTICELS a few, near the base

Comparison species in the area
Common Whitebeam (*p. 144*) – narrower fruit and leaves with a densely white hairy underside.
Orange Whitebeam (*p. 143*) – lobed leaves and very orange fruit.

NOTE: Its origin is perplexing – it is a tetraploid related to Grey-leaved (*p. 147*) and Irish Whitebeams (*p. 150*) – from which it is not easily distinguished – neither of which are known from north Wales.

CR CR Stirton's Whitebeam — *Sorbus stirtoniana*

Small tree or shrub (H to 5 m) growing on the cliffs of the west and north crags of Craig Breidden, from where approximately 39 trees are known, often growing close to the rocky cliff faces. It is a tetraploid member of the *Sorbus aria* group. The species was named in honour of Prof. Charles Stirton for his work establishing the National Botanic Garden of Wales.

FRUIT dark red; usually **longer than wide**; L 11–15 mm; W 11–13 mm; LENTICELS few; medium-sized; scattered mainly at the base

Comparison species in the area
Thin-leaved Whitebeam (*p. 156*) – leaves wider above the middle.
Grey-leaved (*p. 147*) and Rock Whitebeams (*p. 145*) have fruits wider than long and leaves which are widest above the middle.
Common Whitebeam (*p. 144*) – leaves and fruit smaller.

EN EN Llangollen Whitebeam — *Sorbus cuneifolia*

Small tree or shrub (H to 5 m, typically less). Endemic to west-facing limestone crags of Eglwyseg, north of Llangollen, Wales where about 315 trees cling to steep cliff faces. Described formally in 2009 by Tim Rich but was first recognized in the 1950s by A. J. Wilmott and E. F. Warburg who, in unpublished records, used the working name '*castelli*' (after its occurrence on the ruined walls of Castell Dinas).

FRUIT crimson; ± spherical; L 9.0–11.5 mm; W 9.0–11.5 mm; LENTICELS few; usually small

Comparison species in the area
English Whitebeam (*p. 147*) – leaf-base broadly wedge-shaped with the angle of the base 40–59°.

CR CR Ley's Whitebeam — *Sorbus leyana*

Shrub or small tree (H to 10 m, taller in cultivation) confined to Wales seemingly with a preference for rocky cliffs and (often inaccessible) slopes. Found in 1896 by Rev. Augustin Ley there are perhaps only 9 trees remaining, although other estimates are of approximately 23 trees in two sites, with a further 6 planted at Penmoelallt. It is a hybrid between Rowan and either Rock or Grey-leaved Whitebeam.

FRUIT red; roundish; L 8–10 mm; W 9–10 mm; LENTICELS few; small; scattered

Comparison species in the area
Sorbus ×motleyi (a backcross between Ley's Whitebeam and Rowan) (N/I) – lowest 2–3 pairs of leaflets free; terminal leaflet large, broad, lobed.
Least (*p. 156*), English (*p. 147*) and Swedish Whitebeams (*p. 146*) – deepest lobes typically cut < 33% (rarely to 50%).

Whitebeams compared p. 138　　　　　　ROSACEAE (ROSES) *SORBUS* | MID & NORTH WALES

MENAI STRAIT WHITEBEAM
LEAVES L 85–125 mm; W 50–90 mm
OUTLINE broadly oval; typically unlobed
LEAF-BASE wedge-shaped
LEAF-TIP rounded
CHARACTER green; leaf-margins with fine-pointed toothed teeth in the apical part
VEINS 18–24
UNDERSIDE densely greyish white hairy

Widespread species in the area for comparison

Common Whitebeam

densely white

STIRTON'S WHITEBEAM
LEAVES L 60–110 mm; W 40–75 mm
OUTLINE oval (1·3–1·8× as long as wide; **widest nearer the leaf-base**, at approx. 38–55% of the distance from the base; typically unlobed but can be rarely weakly lobed
LEAF-BASE tapering to wedge-shaped
LEAF-TIP pointed
CHARACTER green; leaf-margins with uniserrate or biserrate teeth
VEINS 18–24; those halfway up the leaf angled at 29–47° to the midrib
UNDERSIDE matted, white hairs

Grey-leaved Whitebeam

FRUIT W > L

LLANGOLLEN WHITEBEAM
LEAVES L 80–115 mm; W 50–75 mm
OUTLINE narrowly oval (1·50–1·85× as long as wide); **with shallow lobes** (cut max. 33% to the midrib) in the apical ⅔ of the leaf
LEAF-BASE narrowly wedge-shaped (angle of base 32–44°)
LEAF-TIP pointed
CHARACTER dark green; leaf-margins with weakly biserrate teeth
VEINS 16–22, with distinct cross-veins
UNDERSIDE greyish white hairy with **'feathery' intermediate veins between the main veins**

Rock Whitebeam

FRUIT W > L

LEY'S WHITEBEAM
LEAVES L 65–100 mm; W 45–70 mm
OUTLINE rounded (1·8–2·5× as long as wide); **with rounded lobes** (cut 30–75% of the distance to the midrib)
LEAF-BASE wedge-shaped
LEAF-TIP roundly pointed
CHARACTER distinctive lobe shape
VEINS 15–21
UNDERSIDE greyish white

Swedish Whitebeam

ROSACEAE (ROSES) *SORBUS* | BANNAU BRYCHEINIOG

EN EN Welsh Whitebeam
Sorbus cambrensis

Small tree or shrub (H to 8 m) endemic to the eastern Bannau Brycheiniog, west of Abergavenny. The total population is probably approximately 180 trees across eight locations. It has been recorded scattered widely within Cwm Clydach, at Blackrock, at Craig-y-Cilau, and at Coed Pantydarren.

FRUIT deep red; ± spherical; L 10·5–13·0 mm; W 11·5–14·0 mm; LENTICELS moderate number; medium-sized

Comparison species in the area |
Grey-leaved Whitebeam (*p. 147*) – tends to be smaller; fruit wider than long; leaves somewhat more rounded, with a more rounded tip and less strongly biserrate teeth.
Llanthony Whitebeam (*p. 156*) – leaves narrower with a narrower wedge-shaped base and deeper teeth.
Thin-leaved Whitebeam (*p. 156*) – larger leaves; fruits longer than wide.

EN EN Thin-leaved Whitebeam
Sorbus leptophylla

Small tree or shrub (H to 5 m), often with pendulous branches and most typically found as a shrub with its trunk pressed against a rock face. Endemic to two sites (Craig-y-Cilau and Craig-y-Rhiwarth) comprising approximately 75 trees and first described in 1952 by E. F. Warburg.

FRUIT dark red; rounded; L 12–16 mm; W 12–16 mm; LENTICELS few; small

Comparison species in the area |
Grey-leaved Whitebeam (*p. 147*) – leaf-base more rounded; fruit smaller and usually wider than long.
Common Whitebeam (*p. 144*) – leaves with less prominent teeth and a densely white hairy underside.

VU VU Least Whitebeam
Sorbus minima

Small, twiggy tree or shrub (H to 9 m) endemic to the Llangattock escarpment. Believed to have arisen as a cross between Rowan (*p. 140*) and possibly Rock (*p. 145*) or Grey-leaved Whitebeams (*p. 147*), it was first found by Rev. Augustin Ley in 1893 at Craig y Cilau, from where approximately 778 trees have been recorded, although this population has been reduced as a result of quarrying.

FRUIT scarlet; small, usually ± rounded; L 5–10 mm; W 8–11 mm; LENTICELS few; small

Comparison species in the area |
English Whitebeam (*p. 147*) – leaves much larger; fruit larger.
Ley's Whitebeam (*p. 154*) – leaves broader with deeper lobing in the lower half of the leaf.

NOTE: In September the leaves turn a distinctive yellow colour, making it easy to spot at a distance

EN EN Llanthony Whitebeam
Sorbus stenophylla

Shrub or small tree (H to 8 m) restricted to the Llanthony Valley in Wales, where there are about 100 trees growing at three sites: Cwmyoy (± 50 trees), Tarren yr Esgob (± 50 trees) and Darren Lwyd (± 7 trees) with other possible specimens at Craig-y-Cilau and Pwll-du Quarry. First found in 1874 by the Rev. Augustin Ley.

FRUIT red; rounded; L 11–13 mm; W 12–14 mm; LENTICELS moderate number; variable in size

Comparison species in the area |
Welsh Whitebeam (*p. 156*) – leaves wider, with a wider wedge-shaped base, and shallower toothing.

Whitebeams compared *p. 138* — ROSACEAE (ROSES) *SORBUS* | BANNAU BRYCHEINIOG

WELSH WHITEBEAM
LEAVES L 60–110 mm; W 45–80 mm
OUTLINE **broadly oval, at its widest just below half leaf-length**; usually unlobed or with small, shallow lobes at most
LEAF-BASE wedge-shaped
LEAF-TIP flatly rounded
CHARACTER dark green; leaf-margins with forward-pointing biserrate teeth although untoothed from the base to ±25% of leaf length
VEINS usually 18–24
UNDERSIDE greyish white hairy

Other species in the Bannau Brycheiniog area

Common Whitebeam

densely white

THIN-LEAVED WHITEBEAM
LEAVES L 90–150 mm; W 60–80 mm (large for a whitebeam)
OUTLINE oval (1·5–1·9 × as long as wide); with shallow lobes
LEAF-BASE wedge-shaped to somewhat rounded
LEAF-TIP pointed
CHARACTER dull green; leaf-margins with forward-pointing biserrate teeth
VEINS usually 19–25
UNDERSIDE thinly grey-white or yellowish hairy

Grey-leaved Whitebeam

FRUIT W > L

LEAST WHITEBEAM
LEAVES L 40–80 mm; W 20–40 mm
OUTLINE **narrowly oval** (1·8–2·5 × as long as wide); **with shallow rounded lobes** (cut max. 20–33% of the distance to the midrib)
LEAF-BASE narrowly wedge-shaped
LEAF-TIP pointed
CHARACTER dark green; leaf-margins with forward-pointing uniserrate teeth
VEINS 12–18
UNDERSIDE even and somewhat thinly grey-green hairy

English Whitebeam

LLANTHONY WHITEBEAM
LEAVES usually L 70–115 mm; W 45–70 mm
OUTLINE **oblong, tapering to base**; widest nearer the leaf-tip, at approx. 50–66% of the distance from the base; usually unlobed but a few can have shallow lobes
LEAF-BASE **wedge-shaped**
LEAF-TIP rounded to shallowly pointed
CHARACTER dark green; basal 10–25% of leaf-margins untoothed; remainder with deeply incised, acute, **biserrate teeth that usually curve towards the leaf-tip**
VEINS 14–20
UNDERSIDE greyish white hairy

Ley's Whitebeam

ROSACEAE (ROSES) *SORBUS* | NORTH DEVON & SOMERSET 1/2

VU VU EN Devon Whitebeam — *Sorbus devoniensis*

A medium-sized tree (H to 20 m) that grows in hedges and woodland, particularly in Devon (hence the name), but also in eastern Cornwall, south Somerset and south-eastern Ireland. It has also been planted in Northern Ireland and has become naturalized. In Devon this species shows a preference for soils over shales, grits and slates over those that are lime-rich. The large berries have been sold for eating and even turned into wine.

Comparison species in the area | **Somerset Whitebeam** – leaf-base narrower; fruit smaller, more orange. **Watersmeet Whitebeam** – leaves more glossy, deeper lobed (cut 10–23% of the distance to the midrib), and with more obvious teeth on the margins.

EN EN Watersmeet Whitebeam — *Sorbus admonitor*

Medium-sized tree (H to 16 m) found along the North Devon coast around Watersmeet (where it is most frequent, with at least 110 trees) and Lynmouth in open woodland or among Bracken on rocky ground. Watersmeet Whitebeam was first noted to be different from the more widespread Devon Whitebeam in the 1930s.

NOTE: One tree at Watersmeet car park had a 'No Parking' notice nailed to it. Consequently, the species has also gone by the name No-Parking Whitebeam.

Comparison species in the area | **Devon Whitebeam** (*p. 158*) – leaves less glossy and less deeply lobed (6–18% to the midrib at the centre). **Somerset Whitebeam** (*p. 158*) has a denser crown and narrower leaves that usually have a tapered base.

EN EN Somerset Whitebeam — *Sorbus subcuneata*

A small tree (H 10–15 m) endemic to rocky oak woodland along the coast of North Devon and Somerset. The distribution of Somerset Whitebeam was first described in 1939. Approximately 300 trees are known from nine locations, the largest population (almost 270) is at East Lyn near Watersmeet.

The heavily wooded, steep cliffs of north Devon are the home of many species of rare whitebeam.

Comparison species in the area | **Devon Whitebeam** – leaves larger, more rounded base; fruit, on average, larger, duller orange/brown when ripe. **Watersmeet Whitebeam** – leaves broader, rounded at the base. **Grey-leaved** (*p. 147*), **Rock** (*p. 145*), **Bloody** and **Margaret's Whitebeams** (*p. 160*) – fruit red; leaves unlobed with a white hairy underside.

Whitebeams compared *p. 138* — ROSACEAE (ROSES) *SORBUS* | NORTH DEVON & SOMERSET 1/2

DEVON WHITEBEAM

LEAVES L 60–110 mm; W 40–80 mm
OUTLINE broadly to narrowly oval (1·2–1·7 × as long as wide) with **shallow lobes** (cut 6–18% of the distance to the midrib)
LEAF-BASE usually ± rounded
LEAF-TIP sharply pointed
CHARACTER leaf-margins with forward-pointing biserrate teeth; those at the end of main veins prominent
VEINS 14–20
UNDERSIDE grey; hairy

FRUIT orange-brown to brown; ± spherical; L 12–17 mm; W 12–17 mm; **LENTICELS** numerous; large, becoming smaller towards the top of the fruit

Selected species in the area for comparison

Grey-leaved Whitebeam

FRUIT W > L

Rock Whitebeam

WATERSMEET WHITEBEAM

LEAVES L 70–100 mm; W 35–50 mm
OUTLINE oval to pointed-oval (1·2–1·5 × as long as wide); with **shallow lobes** (cut 10–23% of the distance to the midrib)
LEAF-BASE broadly wedge-shaped to rounded
LEAF-TIP pointed
CHARACTER dark green; leaf-margin with short forward-pointing fine teeth
VEINS 15–20
UNDERSIDE greenish white hairs

FRUIT brown; ± spherical; L 12–17 mm; W 13–17 mm; **LENTICELS** numerous, of varying size

Bloody Whitebeam

SOMERSET WHITEBEAM

LEAVES L 60–115 mm; W 35–60 mm
OUTLINE narrowly oval (1·6–2·4 × as long as wide); with **shallow, broadly triangular lobes** (cut 10–20% of the distance to the midrib) which are usually absent from the basal 33–50% of the leaf
LEAF-BASE shallowly **wedge-shaped** to tapered
LEAF-TIP pointed
CHARACTER bright green; leaf-margins with fine biserrate teeth; those at the end of main veins have a sharply pointed projection
VEINS 12–22
UNDERSIDE dense greenish white (when young) to whitish grey hairs

FRUIT brownish orange, ripening brown; ± spherical; L 10–15 mm; W 12–16 mm; **LENTICELS** numerous; large; smaller towards the top of the fruit

Margaret's Whitebeam

FRUIT W > L

159

Margaret's Whitebeam — *Sorbus margaretae*

A shrub or small tree (H to 6 m) endemic to the south-west coast of England from Culbone, South Somerset to Combe Martin, North Devon. There are at least 100 known trees in woodland on cliffs and rocky places on acidic soils. It was first recognized as a distinct species by Margaret E. Bradshaw in 1984, and named after her.

Comparison species in the area | **Bloody Whitebeam** (*see box below*). **Rock Whitebeam** (*p. 145*) – leaves oval to pointed-oval; teeth uniserrate. **Grey-leaved Whitebeam** (*p. 147*) is similar but has broader leaves.

NOTE: grows together with the very similar Bloody Whitebeam and Rock Whitebeam.

Margaret's and **Bloody Whitebeams** are very difficult to separate. There are minor leaf differences but they are not reliable. The best time to separate the two is when in fruit (Sep–Oct), when the broader, darker red berries of **Margaret's** looks quite different to the lighter red, rounder berries of **Bloody**.

Bloody Whitebeam — *Sorbus vexans*

A small tree (H to 10 m) endemic to the North Devon and Somerset coast. There are some 70 known trees in small, scattered populations, although there could be many more. Some sites suffer an invasive spread of Rhododendron, which creates shady conditions the whitebeams do not grow well in. First recorded in 1939 as a Devon form of Rock Whitebeam, it was identified as a species in 1957 by E. F. Warburg.

Comparison species in the area | **Margaret's Whitebeam** (*see box above*). **Rock Whitebeam** (*p. 145*) – leaves oval to pointed-oval; teeth uniserrate.

NOTE: Warburg puzzled over these trees for many years – he reflected this in the names: *vexans*, and 'Bloody', which mean 'annoying' in Latin and English respectively (NOTE: the berries are also blood red).

Rich's Whitebeam — *Sorbus richii*

A medium-sized tree (H to 15 m or more, although usually much less in exposed situations) found along the shores of the Severn Estuary from Portishead to Clevedon. It is a tetraploid (see *p. 137*), related to the *Sorbus eminens* group. First described in 2014, there are more than 40 known trees at five sites.

Rich's Whitebeam on low cliffs near Clevedon, North Somerset.

Comparison species in the area | **Common Whitebeam** (*p. 144*) – leaves with a rounded leaf-base; more veins (18–28); and uniserrate teeth. **Round-leaved Whitebeam** (*p. 166*) – leaves much rounder. **Grey-leaved-leaved Whitebeam** (*p. 145*) has smaller and narrower leaves.

ROSACEAE (ROSES) *SORBUS* | NORTH DEVON & SOMERSET 2/2 | SEVERN ESTUARY

MARGARET'S WHITEBEAM
LEAVES
L 80–100 mm; W 40–70 mm
OUTLINE oval (1·5–2·0× as long as wide); unlobed
LEAF-BASE tapering down from near the middle of the leaf length into a wedge shape
LEAF-TIP rounded with a point
CHARACTER yellowish green; basal 33% of leaf-margin untoothed, remainder with coarse, unequal biserrate teeth that are curved and slightly forward-pointing
VEINS 16–20
UNDERSIDE dense greyish white hairs

FRUIT red; very broad; L 9–12 mm; W 12–15 mm; looks much wider than long
LENTICELS variable; from few to many

Selected species in the area for comparison

Grey-leaved Whitebeam

FRUIT W > L

Rock Whitebeam

BLOODY WHITEBEAM
LEAVES
L 80–100 mm; W 40–70 mm
OUTLINE oval (1·5–2·0× as long as wide); unlobed
LEAF-BASE tapering down from near the middle of the leaf length into a wedge shape
LEAF-TIP rounded with a point
CHARACTER yellowish green; basal 33% of leaf-margin untoothed, remainder with coarse, unequal biserrate teeth that are curved and slightly forward-pointing
VEINS 8–21
UNDERSIDE dense greyish white hairs

FRUIT scarlet; ± rounded; L 12–15 mm; W 11–16 mm;
LENTICELS a few medium-sized near the base with smaller ones in the top half

Common Whitebeam

densely white

Round-leaved Whitebeam
LF-TEETH fine

RICH'S WHITEBEAM
LEAVES
L 85–120 mm; W 60–90 mm
OUTLINE oval (1·30–1·55× as long as wide); **widest at ±50% of the distance from the leaf-base; unlobed**
LEAF-BASE **wedge-shaped**
LEAF-TIP **broadly rounded**
CHARACTER mid-green; basal 25–33% of leaf-margins untoothed; remainder with slightly forward-pointing uniserrate
VEINS 18–22
UNDERSIDE greenish grey; hairy

FRUIT dark red; obviously broader than long;
L 10–12 mm; W 13–15 mm; LENTICELS few; mainly at the base of the fruit

ROSACEAE (ROSES) *SORBUS* | CHEDDAR GORGE   **Leaves** *pp. 61, 63, 65* | **Flowers** *p. 70* | **Fruit** *p. 76* | **Twigs** *p. 84*

CR CR Cheddar Whitebeam — *Sorbus cheddarensis*

A shrub or small tree (H to 7 m or more) endemic to Cheddar Gorge, Somerset where it grows on Carboniferous limestone slopes and rocks. At least 23 trees have been recorded since it was first found in 2005 by Libby Houston, who gave it the nickname 'Pinstripe' because it has prominent parallel veins, especially visible on the greenish underside of the leaf.

FRUIT red; usually look ± spherical; L 10–14 mm; W 10–13 mm; LENTICELS few; small to medium-sized

Comparison species Cheddar Gorge | **Common Whitebeam** (*p. 144*) – leaves with densely white hairy underside, variable teeth and lobes; fruit usually longer than wide. **Grey-leaved Whitebeam** (*p. 147*) – leaves obovate; fruits wider than long. **English** (*p. 147*) and **Swedish Whitebeams** (*p. 146*) – both of which have deeper lobed leaves. **Twin Cliffs Whitebeam** (*p. 162*) – leaves broadly elliptic with large, coarse toothing. **Gough's Rock Whitebeam** (*p. 162*) – leaves much narrower.

CR CR Twin Cliffs Whitebeam — *Sorbus eminentoides*

A shrub or medium-sized tree (H to 20 m) endemic to the limestone cliffs of Cheddar Gorge from where 20 trees are known. The species was first identified in 2006 and grows in association with Common, English, Round-leaved, Grey-leaved and Gough's Rock Whitebeams.

FRUIT red; usually look ± spherical; L 12.0–14.5 mm; W 12–15 mm; LENTICELS few; small to medium-sized

Comparison species Cheddar Gorge | **Round-leaved Whitebeam** (*p. 166*) – leaves with fine, regular teeth. **Common Whitebeam** (*p. 144*) – leaves narrower with whitish undersides; fruits usually longer than wide. **English** (*p. 147*) and **Swedish Whitebeams** (*p. 146*) – leaves more deeply lobed. **Cheddar Whitebeam** has elliptic leaves with fine teeth. **Grey-leaved Whitebeam** (*p. 147*) – leaves obovate; fruits wider than long. **Gough's Rock Whitebeam** – leaves much narrower.

CR CR Gough's Rock Whitebeam — *Sorbus rupicoloides*

Shrub or small tree (H to 7 m or more) usually with downward-arching branches. Endemic to the limestone cliffs of the south side of Cheddar Gorge where it grows in association with Common (*p. 144*), English (*p. 147*), Round-leaved (*p. 166*) and Twin Cliffs Whitebeams. It was first found in 2006 and is known from approximately 13 trees.

FRUIT ripens unevenly to red; usually look wider than long; L 11–14 mm; W 13–17 mm; LENTICELS few; mainly towards the base

Comparison species in Cheddar Gorge | Narrow leaves with greenish white undersides and large, broad fruit distinguish Gough's Rock Whitebeam from other British *Sorbus* except the narrow-leaved **Rock** and **Bloody Whitebeams**. **Rock Whitebeam** (*p. 145*) – leaves more oblong with ± uniserrate teeth which curve towards the leaf apex; fruits, on average, smaller. **Bloody Whitebeam** (*p. 158*) – leaf widest beyond the middle; tip shallow-angled.

Whitebeams compared *p. 138* ROSACEAE (ROSES) *SORBUS* | CHEDDAR GORGE

CHEDDAR WHITEBEAM
LEAVES L 90–120 mm; W 60–80 mm
OUTLINE oval; widest at 50% of the distance from the leaf-base; unlobed or with shallow lobes
LEAF-BASE curved taper
LEAF-TIP rounded
CHARACTER matt green; basal 15% of leaf-margin untoothed, remainder uniserrate to weakly biserrate (teeth small, acute, forward-pointing)
VEINS 20–24; prominent, whitish
UNDERSIDE greenish grey hairy

Selected Cheddar Gorge *Sorbus* for comparison

Common Whitebeam

densely white

TWIN CLIFFS WHITEBEAM
LEAVES L 80–110 mm; W 55–85 mm
OUTLINE **rounded** (1·6 to 1·8× as long as wide); **unlobed or very weakly lobed**
LEAF-BASE broadly wedge-shaped
LEAF-TIP rounded
CHARACTER green; basal 20–33% of leaf-margin untoothed, remainder strongly, **coarsely biserrate**
VEINS 16–24; angled at 30–40° to the midrib
UNDERSIDE greyish green hairy

Grey-leaved Whitebeam

FRUIT W > L

GOUGH'S ROCK WHITEBEAM
LEAVES L 70–120 mm; W 45–75 mm
OUTLINE **narrowly oval**; widest at 50% of the distance from the leaf-base; unlobed
LEAF-BASE wedge-shaped
LEAF-TIP pointed
CHARACTER dark green; coarsely wrinkled; basal 33% of leaf-margin untoothed, remainder weakly biserrate and pointed outwards
VEINS 16–21
UNDERSIDE greenish white hairy

Rock Whitebeam
LF-VEINS point forward

FRUIT W > L

Round-leaved Whitebeam
LF-TEETH fine

Cheddar Whitebeams grace the steep craggy slopes of Cheddar Gorge.

ROSACEAE (ROSES) *SORBUS* | AVON GORGE 1/2 | **Leaves** pp. 61, 63, 65 | **Flowers** p. 70 | **Fruit** p. 76 | **Twigs** p. 84

CR CR Observatory Whitebeam — *Sorbus spectans*

A shrub or small tree (height to 9 m) endemic to the rocks, cliffs and slopes below the Observatory at the south end of the Avon Gorge, near Brunel's suspension bridge. Approximately 60 trees are known, and it was first recognized as a distinct species in October 1999 although not fully described until 2014. The name *spectans* relates to 'looking out or observing', due to its proximity to the Observatory.

FRUIT dark red; look spherical to longer than wide; L 9–14 mm; W 9·0–13·5 mm; LENTICELS few; small; mainly at the base

Comparison species Avon Gorge | **Common Whitebeam** (*p. 144*) – leaves larger, usually unlobed, base rounded, 18–28 veins; fruits larger. **Willmott's Whitebeam** – leaves larger, narrower, more deeply lobed; fruits L > W. **White's Whitebeam** (*p. 166*) – leaves larger, obovate; fruits longer than wide. **Leigh Woods Whitebeam** – leaves obovate, more or less truncate. The hybrid *Sorbus ×avonensis* (N/I) – leaves obtuse.

EN EN Leigh Woods Whitebeam — *Sorbus leighensis*

A shrub or small tree (H to 10 m or more), endemic to the Somerset side of the Avon Gorge, where the population of approximately 130 trees can be locally frequent on open Carboniferous limestone rocks and scree. It is a triploid, derived from hybridization between Common and Grey-leaved Whitebeams known since at least the 1980s.

FRUIT red; ± spherical; L 10–12 mm; W 10–14 mm; LENTICELS few; small to medium-sized

Comparison species Avon Gorge | **Common Whitebeam** (*p. 144*) – leaves broader, rounder with 18–28 veins and a densely white hairy underside. **Grey-leaved Whitebeam** (*p. 147*) – leaves with sharper, more strongly biserrate teeth. **Willmott's Whitebeam** – leaves oval with a white hairy underside. **White's Whitebeam** – leaves unlobed with more pointed leaf-tip; fruits L > W.

EN EN Willmott's Whitebeam — *Sorbus wilmottiana*

A small tree (H to 10 m or so), endemic to both sides of the Avon Gorge on vertical rock faces, rocky slopes, screes and shallow soils, and short open scrub and grassland. It is a light-demanding species and grows best away from dense woodland. Approx. 90 trees are known, at least 5 of which were cut down in error during management work. However the species does respond to coppicing, and most have survived.

FRUIT bright red; rounded, look longer than wide; L 12–13 mm; W 10·5–13·0 mm; LENTICELS medium-sized; scattered

Comparison species Avon Gorge | **Common Whitebeam** (*p. 144*) – leaves broader, rounder with more (18–28) veins and a densely white hairy underside. **Grey-leaved Whitebeam** (*p. 147*) – leaves unlobed; fruits W > L.

NOTE: It was first found by E. F. Warburg on 14th September 1933 and confirmed later by genetic tests. The name *wilmottiana* was first used in 1962.

EN EN Bristol Whitebeam — *Sorbus bristoliensis*

Medium-sized tree (H typically to 15 m; up to 22 m recorded) of the rocky limestone woods and scrub on both sides of the Avon Gorge in Bristol. The estimated population of approximately 330 trees make it probably the second-most abundant Whitebeam in the area after Common Whitebeam.

FRUIT yellow, orange or red; rounded but can look wider than long; L 9–12 mm; W 9·5–12·0 mm; LENTICELS numerous, mostly towards the base

Comparison species Avon Gorge | **Common Whitebeam** (*p. 144*) has red fruits and leaves that are unlobed or with rounded lobes and a densely white hairy underside.

NOTE: It is thought to have originated as a hybrid between Wild Service-tree (*p. 142*) and Round-leaved Whitebeam (*p. 166*).

Whitebeams compared *p. 138*   ROSACEAE (ROSES) *SORBUS* | AVON GORGE 1/2

OBSERVATORY WHITEBEAM
LEAVES L 75–100 mm; W 65–75 mm
OUTLINE **broadly oval**; widest at ± the mid-point; with **shallow blunt lobes** or unlobed
LEAF-BASE wedge-shaped
LEAF-TIP usually **sharply pointed**
CHARACTER mid- to dark green; thick and 'waxy'; leaf-margins weakly biserrate
VEINS **16–20; smaller secondary veins branch off from main veins**
UNDERSIDE **densely white to 'bluish white' hairy**

Other Avon Gorge *Sorbus*
Common Whitebeam

densely white

LEIGH WOODS WHITEBEAM
LEAVES L 70–105 mm; W 50–70 mm
OUTLINE **broadly oval** (1·2 to 1·5 × as long as wide); widest at ± the mid-point; with shallow blunt lobes from mid-length
LEAF-BASE narrowly rounded to wedge-shaped
LEAF-TIP truncated to rounded
CHARACTER light green; basal 10–15% of leaf-margins untoothed; remainder with **small, outward-pointing strongly biserrate teeth**
VEINS **15–19**
UNDERSIDE white to greenish white hairy

Grey-leaved Whitebeam

FRUIT W > L

WILLMOTT'S WHITEBEAM
LEAVES L 60–110 mm; W 35–60 mm
OUTLINE **± rhomboidal** (1·6–2·1 × as long as wide); widest at 46–58% of the distance from the leaf-base; with **shallow lobes**
LEAF-BASE wedge-shaped
LEAF-TIP pointed
CHARACTER green; leaf-margins with fine biserrate teeth
VEINS usually **14–22**; angled at 20–37° to the midrib
UNDERSIDE white hairy

Non-native *Sorbus* in the area which can be confused
ORANGE WHITEBEAM (*p. 143*)
0–few lobes

FRUIT larger, orange

BROAD-LEAVED WHITEBEAM
deeper lobes

BRISTOL WHITEBEAM
LEAVES L 50–100 mm; W 35–70 mm
OUTLINE **oval with acute, shallow triangular lobes**, especially in the apical half
LEAF-BASE broadly wedge-shaped
LEAF-TIP pointed
CHARACTER glossy dark green, almost translucent in bright light, especially when young
VEINS 16–21
UNDERSIDE **greenish white hairy**

Sorbus decipens – leaves more deeply lobed; fruit larger, orange-brown, wider than long.

ROSACEAE (ROSES) *SORBUS* | AVON GORGE 2/2 + WYE VALLEY 1/2

CR CR White's Whitebeam — *Sorbus whiteana*

A shrub or small tree (H to 10 m or more) endemic to the Wye Valley and Avon Gorge on Carboniferous limestone rocks. It grows on open cliffs, screes and grassland edges, scrub, and developing woodland, with a population of about 100 trees at four sites. White's Whitebeam is triploid.

FRUIT dark red; look spherical to longer than wide; L 10–13 mm; W 10–13 mm; LENTICELS small to medium-sized

Comparison species | Common Whitebeam (*p. 144*) – leaves usually oval leaves with a densely white hairy underside. **Willmott's Whitebeam** – leaves elliptic, weakly-lobed, with a whitish underside. **Grey-leaved Whitebeam** (*p. 147*) – fruits wider than long; leaves whitish underneath.

VU VU Round-leaved Whitebeam — *Sorbus eminens*

A small tree or shrub (height to 10 m) endemic to the limestone woodlands of the Avon Gorge, Cheddar Gorge and the Wye Valley, with a population comprising 500 trees across about 15 locations.

FRUIT red; looks wider than long; L 10–14 mm; W 12–16 mm; LENTICELS a few; mostly towards the base

Comparison species | Common Whitebeam (*p. 144*) – leaves generally not round, with blunter, biserrate teeth. **Grey-leaved Whitebeam** (*p. 147*) and **White's Whitebeam** – leaves oval. **Rock Whitebeam** (*p. 145*) – leaves oblong. **Doward Whitebeam** – leaf-base narrower.

EN EN Doward Whitebeam — *Sorbus eminentiformis*

A medium-sized tree or shrub (H to 15 m) endemic to the Wye Valley. The total population is probably fewer than 100 trees, most of which occur on the Great Doward.

FRUIT orange or red, dull; look wider than long; L 10–13 mm; W 12–15 mm; LENTICELS few, large

Comparison species in the area | Round-leaved Whitebeam – leaf-base rounded.

EN EN Herefordshire Whitebeam — *Sorbus herefordensis*

A medium-sized tree (H to 20 m) endemic to the Great Doward, centred on the Herefordshire Wildlife Trust's Miners Rest and Woodside nature reserves in tall, closed woodlands (not found on open cliffs like other rare whitebeams). 118 trees were mapped during 2010–2012. It was first found by D. Green in 2010 when it was noticed that the woodland trees had a consistent leaf shape.

FRUIT red; look wider than long; L 12–13 mm; W 12–14 mm; LENTICELS moderate number

Comparison species in the Wye Valley | Common Whitebeam (*p. 144*) – leaves paler green, less glossy and with a less densely hairy underside with less prominent veins. **Round-leaved Whitebeam** and **Doward Whitebeam** – both also occur in tall woodland but have round to broadly oval leaves.

NOTE: A triploid clone which may have originated via hybridization between the diploid Common Whitebeam and one of the tetraploid Doward or Round-leaved Whitebeams.

Whitebeams compared *p. 138*   ROSACEAE (ROSES) *SORBUS* | AVON GORGE 2/2 + WYE VALLEY 1/2

WHITE'S WHITEBEAM
LEAVES
L 85–150 mm; W 50–95 mm
OUTLINE **oval** (1·4 to 1·9× as long as wide); **usually unlobed**
LEAF-BASE **wedge-shaped**
LEAF-TIP **pointed**
CHARACTER green; basal 10–15% of leaf-margins untoothed; remainder uniserrate to biserrate
VEINS 17–23
UNDERSIDE **greyish green hairy**

Selected species in the area for comparison
Common Whitebeam

densely white

ROUND-LEAVED WHITEBEAM
LEAVES
L 70–140 mm; W 50–120 mm
OUTLINE **rounded to broadly oval** (1·0–1·3× as long as wide); unlobed
LEAF-BASE **rounded to very broadly wedge-shaped**
LEAF-TIP **rounded**
CHARACTER dark green; leaf-margins with **regular, neat uniserrate teeth that point forward**
VEINS 19–24
UNDERSIDE white to greyish white hairy

Grey-leaved Whitebeam

FRUIT W > L

DOWARD WHITEBEAM
LEAVES
L 65–150 mm; W 45–125 mm
OUTLINE rounded to broadly oval (1·0–1·3× as long as wide); unlobed
LEAF-BASE **broadly wedge-shaped**
LEAF-TIP **rounded**
CHARACTER light to mid-green; leaf-margins with regular, neat uniserrate teeth that point forward
VEINS 19–24
UNDERSIDE white to greyish white hairy

Rock Whitebeam
LF-VEINS 12–19

FRUIT W > L

HEREFORDSHIRE WHITEBEAM
LEAVES
L 90–130 mm; W 65–90 mm
OUTLINE **oval; usually widest beyond the midpoint**; can have shallow lobes
LEAF-BASE wedge-shaped to broadly rounded
LEAF-TIP usually pointed
CHARACTER **dark green, glossy, thick, leathery**; leaf-margins biserrate
VEINS ± 20–25
UNDERSIDE **dense white hairs and prominent veins**

English Whitebeam

167

Green's Whitebeam — *Sorbus greenii*

A medium-sized tree (H to 18 m) endemic to the Great Doward, Herefordshire. There are at least 59 individuals, mainly found on the open sides of quarries, with a few larger trees scattered in Ash-dominated woodland. The trees grow with Common, Round-leaved (*p. 166*), Evan's, Doward (*p. 166*) and Symonds Yat Whitebeams. It is a triploid related to the Grey-leaved Whitebeam group, and was named after David Green, who found the species for the first time in 2009.

FRUIT dark red; ± spherical; L 10–11 mm; W 10–12 mm; LENTICELS variable; few to a moderate number

Comparison species in the area | Common Whitebeam (*p. 144*) – leaves with blunter, more weakly biserrate teeth and more (18–28) veins. Grey-leaved Whitebeam (*p. 147*) – leaves with a rounded tip and weakly biserrate teeth. White's Whitebeam (*p. 166*) – leaves oval with more (17–23) veins. Rock Whitebeam (*p. 145*) – leaves oblong, blunt-tipped. Doward Whitebeam – leaves with more (19–24) veins and a narrower leaf-base.

Symonds Yat Whitebeam — *Sorbus saxicola*

Small tree or shrub (H to 5 m) endemic to the Wye Valley around Symonds Yat and the Great Doward on Carboniferous limestone cliffs and cliff edges in Beech/Ash woodland. First found in 1999, 40 individuals are currently known from seven small populations, but it is likely there are more on the surrounding cliffs, growing with many other *Sorbus* taxa.

FRUIT dark red; ± spherical; L 9–11 mm; W 9·5–11·5 mm; LENTICELS few; small to medium; mainly near base

Comparison species in the area | Rock Whitebeam (*p. 145*) – leaves larger, with fewer veins and an obtuse tip; fruit averages larger and looks wider. Grey-leaved Whitebeam (*p. 147*) – leaves broader (1·25–1·65 mm; length ± width); fruits that are obviously wider than long.

Evans' Whitebeam — *Sorbus evansii*

A small tree or shrub (H to 6 m or more) endemic to the Great Doward and Coldwell Rocks. There are ± 70 known trees in Ash, oak and Large-leaved Lime woodlands on Carboniferous limestone cliffs and rocks growing in association with Common, Doward (*p. 166*), Green's, Rock and Symonds Yat Whitebeams. Named after a Monmouthshire botanist, Trevor G. Evans, who showed the first located specimen at a BSBI *Sorbus* meeting in September 1982.

FRUIT red to dark red; ± spherical or **slightly wider than long**; L 9–13 mm; W 11–15 mm; LENTICELS few at most, mainly towards base of fruit

Comparison species in the area | Common Whitebeam (*p. 144*) – fruit narrower; leaves more rounded with a densely white hairy underside. Grey-leaved Whitebeam (*p. 147*) – leaves smaller with biserrate teeth. Rock Whitebeam (*p. 145*) – leaves oblong, blunt-tipped; fruit larger. Symonds Yat Whitebeam – leaves narrower, more strongly lobed.

Ship Rock Whitebeam — *Sorbus parviloba*

A small tree (H to 8 m) endemic to Coldwell Rocks – from Quarry Rock to Ship Rock on Carboniferous limestone cliffs and cliff edges in open Beech and Ash woodland. First found in 1999, 8 trees are currently known, but there are likely more in the area. The most accessible trees suffered serious damage in 2019.

FRUIT red; looks slightly wider than long; L 9–13 mm; W 11–14 mm; LENTICELS a few; small and scattered

Comparison species in the Wye Valley | Round-leaved (*p. 166*) and Doward Whitebeams (*p. 166*) – leaves unlobed. Twin Cliffs Whitebeam (*p. 162*) – fruit larger. Common Whitebeam (*p. 144*) – leaves with a densely white hairy underside; fruit longer than wide.

Whitebeams compared *p. 138*   ROSACEAE (ROSES) *SORBUS* | WYE VALLEY 2/2

GREEN'S WHITEBEAM

LEAVES L 70–100 mm; W 50–70 mm

OUTLINE **broadly oval** (1·2–1·6× as long as wide); widest just above the mid-point); **unlobed or with small, forward-pointing acute lobes**

LEAF-BASE wedge-shaped

LEAF-TIP **truncated**; typically with a short, sharp point (although this can be not that pronounced)

CHARACTER green; leaf-margins with forward-pointing, strongly biserrate teeth

VEINS **14–20**

UNDERSIDE greyish green hairy

Selected species in the area for comparison

Common Whitebeam

densely white

SYMONDS YAT WHITEBEAM

LEAVES L 70–110 mm; W 50–70 mm

OUTLINE **narrowly oval** (1·50–1·85× as long as wide); widest at the mid-point; at most shallow lobes in the apical 66% of the leaf

LEAF-BASE wedge-shaped (angle of base 30–43°)

LEAF-TIP pointed

CHARACTER dark green; basal 10–35% of leaf-margins can be untoothed; remainder with small, acute, weakly biserrate teeth that point forward

VEINS **15–22**

UNDERSIDE **greyish white hairy**

Grey-leaved Whitebeam

FRUIT W > L

EVANS' WHITEBEAM

LEAVES L 140–250 mm; W 95–125 mm

OUTLINE **broadly oval to oval** (1·3–1·7× as long as wide); **widest nearer the leaf-tip**, at approx. 56–70% of the distance from the base; usually unlobed

LEAF-BASE wedge-shaped

LEAF-TIP **rounded**

CHARACTER green; leaf-margins with neat **uniserrate**, or weakly biserrate, teeth

VEINS 19–23

UNDERSIDE white or greyish green hairy

Rock Whitebeam

LF-VEINS 12–19

FRUIT W > L

SHIP ROCK WHITEBEAM

LEAVES L 70–120 mm; W 55–100 mm

OUTLINE **± rounded to broadly oval** (1·1–1·5× as long as wide); widest at ± the mid-point; with **shallow lobes from beyond the midpoint**

LEAF-BASE wedge-shaped

LEAF-TIP rounded with a point

CHARACTER dark green; basal 25–40% of leaf-margins untoothed; remainder with outward-pointing strongly biserrate teeth

VEINS 17–23; angled at 29–38° to the midrib

UNDERSIDE **greyish green hairy**

Round-leaved Whitebeam

LF-BASE rounded

ROSACEAE (ROSES) | *PRUNUS* (PLUMS) Leaves *p.65* | Flowers *p.70* | Fruit *p.75* | Twigs *pp.80, 85*

Blackthorn and plums *Prunus* species

Prunus can be recognized from other fruiting woody species in the rose family by their simple leaves and thin-skinned, fleshy fruit (drupe) containing a single large, hard seed ('stone').

CHERRY PLUM — LEAVES broadly oval; ± shiny; L to 70mm. FRUIT D 20–25mm; yellow-reddish; slightly sweet. TWIGS green. Flowers usually 2–3 weeks before Blackthorn. FLOWERS ± with leaves; D 15–22mm. SEPALS **reflexed**.

WILD PLUM — LEAVES ± broadly oval; dull; L to 80mm (smaller in some forms). FRUIT variable by type; D 10–50mm; green, yellow or purplish; sweet to bitter. TWIGS grey-brown; can have a few spines. FLOWERS with leaves; D 15–25mm. SEPALS usually not reflexed.

BLACKTHORN — LEAVES ± broadly oval; dull; L to 30mm. FRUIT D 8–15mm grey-black with bluish bloom; very bitter. TWIGS blackish grey; sharply spined. Flowers usually 2–3 weeks after Cherry Plum. FLOWERS well before leaves; D 12–16mm. Smallest flowers of the three. SEPALS **not reflexed**.

Blackthorn (*left*), just coming into bloom with Cherry Plum in full flower (*right*) 21st March, Hampshire.

ROSACEAE (ROSES) | *PRUNUS* (PLUMS)

Blackthorn

Prunus spinosa

A spiny deciduous shrub or small tree (H to 7 m), Blackthorn is a widespread native species that is found in hedges and at woodland edges, and in coastal scrub, on a wide range of soils. It produces numerous suckers by which it can rapidly colonize areas. Typically found as a thick shrub but it can grow into a small tree. It is also widely planted as its spiny twigs make it a good hedging plant. **Features:** TWIGS blackish grey; spiny. BUDS small (L 1–2 mm) egg-shaped; reddish brown. LEAVES **small, oval** (L to 30 mm) with a finely toothed margin; dull green and hairless although the underside veins can have some hairs. FLOWERS small (D 10–15 mm); white; **appearing before the leaves**. FRUIT **small purple to black oval plum** (known as a sloe) with a distinctive **bluish white bloom**, which rubs off (also see *facing page*).

J F M A M J J A S O N D

in fruit

LEAVES ± broadly oval; dull; smaller than other *Prunus*

FRUIT D 8–15 mm; grey-black with bluish bloom; TASTE very bitter

TWIGS blackish grey; **sharply spined**

In flower – *facing page*

Did you know? Blackthorn thorns can easily break off the plant and can deeply penetrate human flesh. Blackthorn spines can be covered in bacteria, which can cause acute medical problems, including cellulitis and sepsis, which can develop rapidly. Caution should always be taken when around Blackthorn, and if a rash begins to develop around a wound caused by a thorn then medical attention should be sought without delay.

Associated species | The rare **Brown** and **Black Hairstreak** butterflies can be found around Blackthorn, as their caterpillars feed on the leaves. Blackthorn is susceptible to the fungus *Taphrina pruni*, which causes 'Pocket Plum'. An affected sloe develops into a distorted gall, with the surface covered in a white mass of the fungus. This occurs in May–June and can lead to a failure of the sloe crop.

Brown Hairstreak (*left*) is almost wholly dependent on Blackthorn; sloes ripening in late summer (*right*).

ROSACEAE (ROSES) | *PRUNUS* (PLUMS) — Leaves *p. 65* | Flowers *p. 70* | Fruit *p. 75* | Twigs *pp. 80, 85*

DD Cherry Plum — *Prunus cerasifera*

A deciduous shrub or small tree (H to 10 m) that is similar in form to Blackthorn although not usually very spiny – even lacking spines altogether. It is a native of south-eastern Europe and western Asia and was introduced into Britain & Ireland by 1597 and has been widely planted since the start of the 20th century, frequently becoming naturalized in woods, hedges and scrubby areas. **Features:** TWIGS inner branches can be spiny, dark brown to grey, contrasting with the green to purplish 1st-year twigs. BUDS relatively pale brown; broadly conical (L 1–2 mm) with a relatively broad base. LEAVES broadly oval (L 30–70 mm); ± glossy. FLOWERS white to pinkish (D 15–22 mm) **appearing with the opening leaves early in the spring.** FRUIT small (D 20–25 mm), ± globular 'large cherry'-like yellow to purple plum.

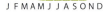
J F M A M J J A S O N D

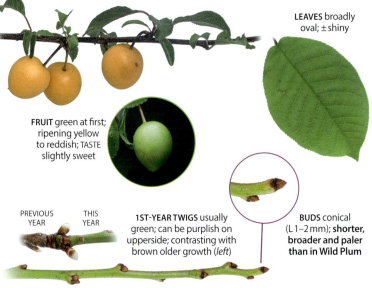

LEAVES broadly oval; ± shiny

FRUIT green at first; ripening yellow to reddish; TASTE slightly sweet

PREVIOUS YEAR — THIS YEAR

1ST-YEAR TWIGS usually green; can be purplish on upperside; contrasting with brown older growth (*left*)

BUDS conical (L 1–2 mm); **shorter, broader and paler than in Wild Plum**

FLOWERS larger than Blackthorn's; sepals reflexed

In flower – *p. 170*

The small fruits come in a wide range of colours (*left*); in the open, trees develop a spreading form (*right*).

ROSACEAE (ROSES) | *PRUNUS* (PLUMS)

Wild Plum *Prunus domestica*

A deciduous large shrub or small tree (H to 8 m) which has developed from Cherry Plum, or perhaps from a cross between Cherry Plum and Blackthorn. It has been known in Britain & Ireland in the wild since 1777, but had been cultivated since *c.* 1000. Also known as Domestic Plum, the species contains many subspecies and varieties which can be hard to differentiate. **Features:** TWIGS grey-brown; can also be greenish; can be sparsely spiny. BUDS dark brown; narrowly conical (L 2–6 mm) with a broad base. LEAVES ± **broadly oval** (L 40–80 mm); dull. FLOWERS white to pinkish (D 15–25 mm); in clusters of 2–3; appearing with the leaves. FRUIT *see below*.

J F M A M J J A S O N D

Damson and Bullace; two of the Wild Plum types more commonly encountered in woods and hedgerows

DAMSON

BULLACE

FRUIT D 20–40 mm green, yellow or purplish; TASTE sweet

1ST-YEAR TWIGS usually grey-brown; can be greenish, especially on underside

LEAVES ± broadly oval; dull; L to 80 mm (smaller (L to 40 mm), more Blackthorn-like in some forms)

BUDS conical (L 2–6 mm); **longer, narrower looking and darker than those of Cherry Plum**

FRUIT D 10–15 mm blue-black with bluish bloom; TASTE very bitter

FLOWERS sepals usually not reflexed

In flower – *p. 170*

Plum varieties | The many forms (over 2,000 worldwide) of plums grown over the centuries have fruits that occur in a wide range of colours, tastes and sizes – up to 8 cm wide in varieties like Victoria Plum. Some examples are shown below. The flesh is usually sweet and contains a single large seed (the stone), the shape of which can be useful for assigning the fruit to subspecies group.

P. spinosa (p. 169)	*P. domestica* ssp. *insititia*	ssp. ×*italica*	ssp. *domestica*
 BLACKTHORN STONE rounded	 BULLACE DAMSON 'Shropshire Prune' STONE longer than wide; somewhat flattened	 GREENGAGE STONE longer than wide; slightly flattened	 'Victoria' STONE much longer than wide; obviously flattened

173

ROSACEAE (ROSES) | *PRUNUS* (CHERRIES) — Leaves *p. 65* | Flowers *p. 70* | Fruit *p. 76* | Twigs *p. 84*

LC LC Bird Cherry ☠ — *Prunus padus*

A medium-sized deciduous tree (H to 19 m) with a natural range that has a northern and western bias, growing to elevations of ± 600 m in Britain & Ireland (up to 2,200 m in Europe). It is a familiar sight growing by lime-rich streams in *e.g.* the Scottish Glens, the Lake District, and the Pennines. It has been planted in streets and gardens, particularly in the south of England, for its attractive white flowers. **Features:** BARK dark grey-brown with a distinctive strong aroma of tannin; **horizontal lenticels can produce a yellow resin if the bark is damaged.** TWIGS shiny, dark brown; can be covered with fine hairs when fresh. BUDS ± egg-shaped to conical (L 5–11 mm); dark brown with a paler, sharply pointed tip. LEAVES oval, large (L to 10 cm; W to 6 cm) with a finely toothed margin, sharply pointed tip and a rounded leaf-base; typically hairless but hairs may be present along the main vein on the leaf's underside. FLOWERS fragrant, almond aroma; white (D 10–20 mm), **in erect or drooping cylindrical spikes** of up to 40. FRUIT small (D 6–8 mm); **globular**; glossy dark purple to black when ripe.

J F M A M J J A S O N D

Similar species | **Rum Cherry** (*p. 179*) which has smaller, more tightly packed flowers in shorter spikes and leaves with tufts of brown hairs along the midrib of the underside.

Did you know? One wood where Bird Cherry grows wild in the south is Waylands Wood, Norfolk, mentioned in the Domesday Book and famous as the location of the *Babes in the Wood* children's tale.
The reddish brown heartwood was used in cabinet making and wood turning, whilst extracts from the bark were used for medicinal purposes. The bitter black cherries were used to flavour brandies and wines.

BARK dark grey-brown with distinctive unpleasant aroma

Bird Cherry can be a large obvious tree (*left*) but can also be inconspicuous in woodland if not in flower (*right*).

ROSACEAE (ROSES) | *PRUNUS* (CHERRIES)

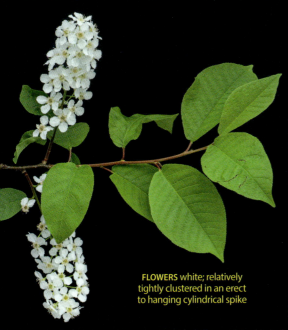

FLOWERS white; relatively tightly clustered in an erect to hanging cylindrical spike

BIRD CHERRY ERMINE MOTH

CHERRY ERMINE WEB

POCKET PLUM GALL

LEAVES light green; oval with a finely toothed edge; usually hairless but can have white hairs along the main vein on underside

FRUIT cluster of **dark purple to black, very bitter cherries** each on a short stalk attached to a main fruit-stalk

BUDS egg-shaped; dark brown with paler tip

TWIGS shiny, dark brown; can be finely hairy when young

Associated species | The flowers provide food for many insect species, and the cherries are eaten by many birds and small mammals. Twenty-eight species of insects have been recorded feeding on Bird Cherry in Britain & Ireland, four of which are specific to Bird Cherry: two aphids, *Rhopalosiphum padi* and *Myzus padellus*, the **Bird Cherry Ermine Moth**, and a sawfly, *Pristiphora retusa*. Both *R. padi* and the ermine moth can cause widespread damage to trees, and the moth can defoliate the tree as well as cover the tree in a grey silk web. In addition, large deposits of **sooty mould** can develop on honeydew excreted by the aphids.

Bird Cherry can also suffer from the effects of *Taphrina padi*, a 'tongue fungus' which causes a form of **Pocket Plum Gall** – resulting in swollen and distorted hollow cherries without a seed.

175

ROSACEAE (ROSES) | *PRUNUS* (CHERRIES)   Leaves *p. 65* | Flowers *p. 70* | Fruit *p. 76* | Twigs *p. 86*

LC LC Wild Cherry   *Prunus avium*

A medium to large tree (H to 30 m). It is fast-growing and occurs widely across Britain & Ireland where it can be common in woodlands and hedgerows. It has often been planted in parks, gardens and streets for its spring flowers. It prefers deep moist soils, particularly those that are lime-rich, but it can tolerate fairly acid soils. **Features:** BARK smooth and often shiny reddish brown to grey when young; **peeling off in horizontal strips, with horizontal lenticels which create a distinctive stripy effect on the bark** as it ages. The bark of older trees becomes deeply fissured and thick at the base of the trunk. TWIGS grey to reddish brown with raised lenticels; hairless. BUDS egg-shaped; shiny red-brown; end-bud (L 2–7 mm) hairless. LEAVES variably shaped, oval to oblong (L to 15 cm; W to 7 cm), usually with a **long pointed tip**; upperside dull green; underside paler green; leaf-margin with forward-pointing irregularly shaped teeth; leaves arranged alternately on long shoots; can be clustered on short shoots or spurs. FLOWERS white (D 15–30 mm), in clusters of 2–6 on long (L 15–45 mm) stalks; appear at the same time as the leaves, usually on short shoots. FRUIT globular cherry (D 20–30 mm); shining red to very dark red.

J F M A M J J A S O N D

Did you know? Humans have eaten Wild Cherries for many thousands of years, with stones being found on Bronze Age archaeological sites. In Ely, Cambs, people visited cherry orchards on 'Cherry Sunday', paying 6d (2½p) a person to eat as many cherries straight from the tree as they liked. Curiously, the fruit of Wild Cherry is the only non-poisonous part of the tree. Even cherry stones contain amygdalin, a chemical compound that releases cyanide when chewed. The leaves also contain prunasin, another plant compound which can produce hydrogen cyanide. In spite of this, the fruit's flesh makes excellent jams and preserves.

The fine-textured wood of Wild Cherry is used for veneers, musical instruments and cabinet making and was extensively used for making smokers' pipes. It makes good firewood when green, producing perfumed smoke. Historically the leaves were also used to flavour liquors and custards, while the bark produced a fine yellow dye.

The scientific name *avium* is thought to be derived from the Latin *avis* (bird) referencing the fact that the birds eat and disperse the fruit.

BARK horizontal lenticels create a distinctive stripy pattern with age

Wild Cherry flowers are one of the early signs of spring with their distinctive clusters of loose flowers (*left*) which conspicuously brighten the leafless woodlands in spring (*right*).

ROSACEAE (ROSES) | *PRUNUS* (CHERRIES)

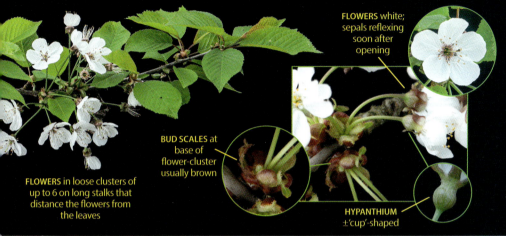

FLOWERS white; sepals reflexing soon after opening

BUD SCALES at base of flower-cluster usually brown

FLOWERS in loose clusters of up to 6 on long stalks that distance the flowers from the leaves

HYPANTHIUM ±'cup'-shaped

LEAF-STALK usually with a pair of prominent **red glands near the leaf-base**

FRUIT shining bright red cherries; each with their own stalk

LEAVES typically **long oval to oblong with a long tip** and irregular forward-pointing teeth on the margin

TWIGS grey to red-brown

BUDS egg-shaped; shiny red-brown with some usually occurring in clusters; tip of end-bud hairless

Similar species | Many other cherry species and cultivars are planted; the two most likely to be found in a naturalized state are **Dwarf Cherry** which differs in flower-cluster and leaf details; and **Japanese Cherry** which differs primarily in details of the leaf – see *p. 179* for details of both.

BRIMSTONE

PLATYCHEIRUS HOVERFLY

Associated species | Wild Cherry is very important for wildlife, as the cup-shaped white flowers are used by pollinating insects such as **bees** and **hoverflies** as sources of early pollen and nectar. The sweet, edible cherries also provide food for **birds** and **small mammals** in the summer. The leaves provide food for the caterpillars of the **Cherry Fruit**, **Cherry-bark** and **Brimstone** moths among many others. The prominent red glands on the leaf-stalk attract **ants** by producing small quantities of sugar-rich nectar. Ants will defend the leaf from any caterpillar that is encountered.

ROSACEAE (ROSES) | PLANTED/NATURALIZED CHERRIES Leaves p.65 | Flowers p.70 | Fruit p.76 | Twigs p.86

Cherry Laurel ☠ *Prunus laurocerasus*

LC LC

Small to medium-sized evergreen tree or large shrub (H to 15 m) that resembles Bay Laurel, hence the common name. Native to the Black Sea region and the Caucasus, it has been planted in Britain & Ireland for hedging and in gardens since 1611 and has naturalized in woods and scrubby areas via birds dispersing the seed. Its thick foliage creates shade and can lead to a loss of native flora and fauna in its vicinity.

Features: BARK smooth and reddish brown. LEAVES large (L 10–25 cm); **leathery; shiny dark green**; leaf-margin finely toothed margin; can have **almond aroma when crushed.** FLOWERS creamy-white (D 8–12 mm); sweet-smelling; in upright cylindrical spikes of 30–40. FRUIT inedible, bitter-tasting small cherry (D 10–12 mm); green when young; ripening red then black.

J F M A M J J A S O N D

LF-STALK green

LEAVES large; shiny; leathery; dark green with finely toothed margin

coarse teeth

red

PORTUGAL LAUREL

FRUIT small, black, bitter cherry

Similar species | The closely related **Portugal Laurel** *Prunus lusitanica* (*box, left*) differs in its coarsely toothed leaves and red leaf-stalks. **Bay** (*p. 331*) differs in having pale yellow-green ♂ and ♀ flowers on separate trees; smaller leaves (L to 12 cm); and shiny black berry-like fruit.

Did you know? Cherry Laurel contains cyanogenic glycoside compounds and enzymes that convert these compounds to hydrogen cyanide gas when the plant cells are damaged by chewing, crushing, cutting or burning. This chemical reaction produces the smell of almonds. **The leaves and seeds contain similar levels of cyanide to Apricot and Peach seeds and, if eaten, will cause severe discomfort to humans. The smoke from burning Cherry Laurel should not be inhaled.**

The distinctive flower spikes (*left*); when in flower Cherry Laurel can be covered in masses of blossom (*right*).

FLOWERS upright clusters of 30–40 white, sweet-smelling flowers

178

ROSACEAE (ROSES) | PLANTED/NATURALIZED CHERRIES

Rum Cherry ☠
Prunus serotina

LC

An introduced species from the eastern USA that is similar to Bird Cherry, but has **FLOWERS** smaller (D 6–8 mm), in tighter clusters and **LEAVES** with **tufts of brown to whitish hairs along the underside midrib** and extending along the leaf-stalk in many. Found occasionally in hedges, woodland edges, roadsides and gardens, mostly in southern and central England.

Similar species | **Bird Cherry** (*p. 174*) which has larger (D 10–20 mm) flowers and hairless leaf-stalks.

LEAVES usually with brown to whitish hairs on underside, and leaf-stalk

FLOWERS smaller and in a tighter cluster than those of Bird Cherry

Dwarf (Sour) Cherry
Prunus cerasus

LC

Shrub or small tree (H to 10 m) introduced from Asia and naturalized in hedges, copses and field corners. Similar to Wild Cherry, but differs in its smaller clusters of 2–4 flowers on shorter (L 10–40 mm) stalks; green flower bud scales; 'bowl'-shaped hypanthium; and smaller leaves with more regular round-tipped teeth. The red fruit is sour and is cultivated as Morello Cherry.

Japanese Cherry
Prunus serrulata

Small tree (H to 12 m) introduced from Asia *c.* 1822, and widely planted in streets, parks and gardens for their variety of flower forms – the best known being 'Kanzan', with striking pink flowers in the spring. Separating Japanese from Wild Cherry is straightforward when in flower, but harder when not, however there are subtle differences in the leaves and buds (see *below*).

FLOWERS clusters closer to the leaves than in Wild Cherry

BUD SCALES usually green

HYPANTHIUM ±'bowl'-shaped

LEAVES much smaller than Wild Cherry's

LF-TEETH ± equal-sized, rounded

HYPANTHIUM usually elongated

LEAVES like Wild Cherry's but usually with a longer pointed tip;

WILD CHERRY

LF-TEETH much longer, pointed tips

TWIGS typically lack the clustered buds of Wild Cherry

BUDS longer than Wild Cherry's; end-bud tip with buff hairs

Similar species | **Wild Cherry** (*p. 176*) has a leaf-margin with irregular round-tipped teeth.

ROSACEAE (ROSES) | APPLES Leaves *p. 65* | Flowers *p. 70* | Fruit *p. 75* | Twigs *pp. 80, 85*

DD Wild Apple (Crab Apple)   *Malus sylvestris*

A relatively uncommon medium-sized tree (H to 14 m) of hedges, copses, wood pasture and ancient woodlands, thriving in fertile and heavy soils. Although not renowned for being particularly long-lived, some trees have been found to be ± 120 years old. Wild Apples often grow singly, with individual trees some distance apart. They have very dense, irregular and complex crowns (like large hawthorns) and are a distinctive dark slate-grey colour in winter. Wild Apples were one of the ancestors of domesticated apple varieties, which arose thousands of years ago, from hybridization between *Malus sieversii* (from China and Kazakhstan) and European Wild Apples. This ability to hybridize has now resulted in probably one-third of British and Irish Wild Apples being hybrids, which makes it vital to protect populations of known pure Wild Apples (such as in the Highlands of Perthshire and the Borrowdale area of the Lake District). It can be very difficult to distinguish Wild Apple from Domestic and hybrid Apples (see *opposite*), and sometimes DNA analysis is needed to be certain.

Features: BARK greyish brown; becoming scaled with age. TWIGS grey-brown; can have spines. BUDS pointed egg-shape (L 2–8 mm); red- to purple-brown; hairy or hairless. LEAVES small; shiny; oval (L to 6 cm) with **small rounded teeth**; **leaf-stalk and underside hairless** or with a few sparse, stiff hairs at most. FLOWERS pink when young, becoming white (D 25–50 mm); **anthers yellow**. FRUIT small (D typically < 35 mm) **green-yellowish apples**.

J F M A M J J A S O N D

LEAF-SCARS
stepped

TWIGS of both species smooth; those of **Domestic Apple** usually hairy (at least near tip); those of pure **Wild Apple** usually hairless. Twigs of pure Wild Apple can develop spines

BUDS generally more pointed than those of Domestic Apple

BARK greyish brown; with rectangular scales on older trees

A typical 'hawthorn-like' Wild Apple (*left*), and its distinctive winter dark slate-grey appearance (*right*).

ROSACEAE (ROSES) | APPLES

Hybrid Apple *Malus sylvestris* × *domestica*

FLOWERS pink at first (in bud) turning white

ANTHERS yellow

Similar species | **Wild** (*p. 184*) and **Plymouth Pears** (*p. 183*) both have purple anthers.

Associated species
Wild Apple has a relatively high wildlife value and attracts some 90 species, particularly insects, to its spring flowers. The trees also provide excellent growing conditions for **Mistletoe** and lichens, and moss communities also develop on the bark. Its small, bitter fruits provide birds and mammals with food in the autumn. However, surprisingly, there is still little detailed information available about Wild Apple and further research is needed to understand the true biodiversity values of this native species.

WILD APPLE

SEPALS hairless

LEAF-STALKS + LEAF UND **hairless or, at most, a few sparse hairs** on leaf-stalk and leaf-margin at base

LEAVES smaller (L to 6 cm); shiny; oval; with small, rounded teeth (larger than those of pears)

FRUIT small (D usually < 3·5 cm); greenish yellow

DOMESTIC APPLE
Malus domestica

SEPALS hairy

LEAF-STALKS + LEAF UND **largely covered with felted hairs**; Hybrid Apple is hairy for ⅓–½ leaf length.

LEAVES larger (L to 15 cm) and more strongly toothed than those of Wild Apple

FRUIT larger (D usually > 4 cm); highly variable, green yellow and red

Did you know? It has been suggested that the name 'Wild Apple' (rather than 'Crab Apple') should be used so as to distinguish the pure native species from the many hybrids and cultivars of Domestic Apple that use the Crab Apple name.

Wild Apples have been used for thousands of years and have been found as food remains from the late Neolithic (*c.* 2500 BCE). They have been recorded along with hazel nuts from a Mesolithic site on Colonsay (*c.* 7000 BCE) and found in a bucket in a Norwegian Viking ship from 820 CE. A 'Crab Fair' in Egremont, Cumbria has taken place annually since 1267.

The small 'Crab' Apples can be used for jams and wine and a fermented juice called Verjuice.

Wild Apples are usually at their highest densities in ancient wood pasture systems, with up to 10 mature trees per hectare. Historically, this may be because the seeds pass unscathed through the stomachs of cows, horses and pigs (although seemingly not sheep), helping with their distribution and germination. A Danish study even found that more than 90% of Wild Apple seedlings were growing in cow pats, suggesting that the modern-day relative rarity of Wild Apple may be down to a lack of large grazing mammals. Around lakes such as Loch Lomond, large populations of trees can also be found at the high-water mark, suggesting that they spread via floating apples, meaning water may also be important for distribution.

Unfortunately, hybridization with Domestic Apple does appear to threaten the long-term survival of Wild Apple as a pure species to the extent that Wild Apple is flagged on the IUCN Red List as needing further study. It seems vital that Domestic Apple should no longer be planted in areas important for pure Wild Apple, and seeds of pure Wild Apple should be collected for seedbanks and nurseries to ensure the continued survival of the species.

ROSACEAE (ROSES) | APPLES

Domestic apples

Historically, there were more than 2,500 varieties of cultivated apples in Britain & Ireland, all named and selected for the taste, size and texture of their fruit. They are collectively called *Malus domestica* – the Domestic Apple – but they go by a huge range of fantastic names, ranging from Newton Wonder and Beauty of Kent to Pitmarston Pineapple, to name but three.

The origins of Domestic Apples stretch back to the Tien Shan Mountains in Kazakhstan, from where they arose from hybridization between *Malus sieversii* (from China and Kazakhstan) and Wild Apple many thousands of years ago. Domestic Apples were probably brought to Britain & Ireland by the Romans as grafted fruit trees.

One complexity of apple reproduction is that the ovary of any single apple flower has ±5 compartments, which each contains 2 ovules – meaning that up to 10 seeds may develop. However, the fertilization of every ovule is not essential for fruit to develop, although generally the more seeds that develop in the apple, the larger it is, and approximately 6 or 7 seeds are necessary for good fruit set. Each one of these 10 seeds was pollinated by a different pollen grain that could have come from any neighbouring apple tree. Therefore, each seed is potentially unique and will produce a unique fruit. To ensure that the fruit that develops is the same as the parent tree, a process called grafting is required.

There are many different methods of grafting; however, the basic principle is to join the graft material or scion – a shoot – from one tree with the rootstock from another (usually of the same species) to make a new tree. Both are cut, and the two cut surfaces are bound together until they fuse and become one. The resulting tree will then grow with some of the characteristics of its rootstock (in terms of height and spread) but the fruit will be the same as the original tree, *e.g.* a Bramley.

Through the process of grafting, humans have selected apples that were liked for their fruit and then continued to propagate those varieties through time.

Ananas Reinetter — Ashmeads Kernel — Crispin

Lamb Abbey Pearmain — Falstaff — Crown Gold

Leaves *p.65* | Flowers *p.70* | Fruit *p.75* | Twigs *pp.80, 85*   ROSACEAE (ROSES) | PEARS

LC EN Plymouth Pear ILLEGAL TO PICK   *Pyrus cordata*

A large shrub or small tree (H to 8 m) first recognized as different in 1870 before being formally described in 1871 by T. R. Archer Briggs, a Plymouth naturalist. He wrote '*What is very remarkable is the late period – the beginning of May – at which it flowers, corresponding not with our pears generally, which are in blossom quite a fortnight or three weeks before, but with the apple and crab.*' Always rare, the species has been **known from Plymouth** (one of few cities to have given its name to a tree), although it was also found in 1989 **near Truro**. It is mostly found in hedgerows. Plymouth Pear has an estimated population of between 40 and 60 trees and is legally protected (WCA 1981, Schedule 8). **Features:** BARK **greenish grey with a cracked and fissured appearance** breaking up into small square plates. TWIGS purplish; hairless; can have spines. BUDS pointed-oval (L 4–6 mm); pink-brown to purplish brown. LEAVES heart-shaped or oval (L to 4·5 cm), with a straight to wedge-shaped base. FLOWERS white to pale pink (D 20–25 mm) with a slightly 'fishy' aroma; anthers purple. FRUIT small (D 10–20 mm), hard, **globular brownish** pears on long stalks.

J F M A M J J A S O N D

LEAVES oval (or heart-shaped) with straight to wedge-shaped base

BUDS pointed-oval; pink- to purplish brown

FRUIT small (D <2 cm); globular

BARK greenish grey with cracks and fissures

TWIGS purplish; hairless; can be spiny

Similar species | Plymouth Pear is rare and has a limited range and so is unlikely to be confused. However, **Wild Pear** and **cultivated Pear** both have yellowish to dark brown twigs with reddish to orange-brown buds; flower much earlier (April–early May); and have fruit that is at least a little more pear-shaped. **Wild Apple** (*p. 180*) has yellow anthers.

Close-up of Plymouth Pear flowers (*left*), which can cover a tree in late spring (*right*).

ANTHERS purple

183

ROSACEAE (ROSES) | PEARS   Leaves *p. 65* | Flowers *p. 70* | Fruit *p. 75* | Twigs *pp. 80, 85*

LC LC Wild Pear   *Pyrus pyraster*

A medium-sized tree or large shrub (H to 15 m). The origin of the Wild Pear is uncertain, but it was probably brought to the country thousands of years ago, as it has found in Neolithic peat from Essex. It can be found scattered alongside road verges, in hedgerows and on woodland edges, most commonly in the south and east. It is now an uncommon tree and may actually be rare, as many supposed Wild Pears are actually 'wildings' (*i.e.* trees descended from domesticated pears) as well as the probability that a lot of Wild Pear records are actually naturalized, Cultivated Pear *Pyrus communis*, or hybrids of the two (see *opposite*), and all three taxa are not usually differentiated by botanical recorders. **Features:** BARK smooth grey or brown; breaking into **distinctive small rectangular plates with age**. WOOD pale pink; easily stained and used for veneers, carving and for musical instruments such as flutes and – when stained – black piano keys. TWIGS yellowish to dark brown; usually **spiny and hairless**. BUDS egg-shaped to conical (L 3–5 mm); shiny; typically reddish brown although can be dark brown to almost black; usually hairless. LEAVES glossy dark green; oval or rounded with a sharp point at the tip (L to 7 cm); hairy when young; becoming hairless with age; leaf-margins finely toothed. FLOWERS white (D 20–30 mm) in clusters of 6–12; **anthers purple**; slightly 'fishy' aroma. FRUIT small (D 15–40 mm), greenish or yellowish, **typically quite rounded, pears**.

J F M A M J J A S O N D

Similar species | Wild Pear and Cultivated Pear can be difficult to identify with certainty. They are best separated as follows: **Wild Pear** typically has spines on the branches combined with small fruit (usually 1–4 cm long); **Cultivated Pear** generally has larger leaves, twigs and buds; usually spineless branches; and pears greater than 5 cm long. **Plymouth Pear** (*p. 183*) typically has purple-brown twigs and buds; small round fruit and flowers later. In flower **Wild Apple** (*p. 182*), which has yellow anthers.

BARK grey or brown with **small rectangular plates**

A verified mature Wild Pear at Danebury, Hampshire (*left*), which has rounded fruit but spineless branches and Perry Pears (*right*) in Worcestershire.

ROSACEAE (ROSES) | PEARS

Cultivated Pear *Pyrus communis*

FLOWERS white; in clusters of 6–12

ANTHERS purple

Did you know? One often overlooked tree in the British treescape is the Perry Pear (*facing page, right*). Thought to be wild hybrids between cultivated Pear and Wild Pear, these were selected to produce perry, the pear equivalent of cider. Perry Pears are widespread in the landscape of Gloucestershire, Herefordshire and Worcestershire. Both Wild and Perry Pears provide excellent growing conditions for Mistletoe.

Historically there were more than 100 named Perry varieties with wonderful names such as 'Huffcap', 'Merrylegs', 'Stinking Bishop', 'Bosbury Scarlet' and 'Bartestree Squash'. One of the most remarkable trees in Britain & Ireland is the Perry Pear at Much Marcle in Herefordshire, an ancient tree from 1776 covering half an acre of ground. It achieved this through the branches growing laterally and then rooting where they touched the ground.

CULTIVATED PEAR – all parts usually larger than those of Wild Pear

LEAVES L to 12 cm

TWIGS hairless; brown; **typically spineless**; usually thicker (D 3–5 mm)

BUDS L usually 4–8 mm

FRUIT L >5 cm; pear-shaped

WILD PEAR

LEAVES L to 7 cm

LEAF-TEETH smaller teeth than those of Crab Apple

TWIGS hairless; brown; **usually spiny**; slender (D ± 3 mm)

BUDS L usually 3–5 mm

FRUIT L < 5 cm; ± globular

A selection of the wide variety of cultivated Perry Pear fruit shapes.

ROSACEAE (ROSES) | HAWTHORNS   Leaves *p. 63* | Flowers *p. 70* | Fruit *p. 77* | Twigs *p. 80*

LC LC Hawthorn   *Crataegus monogyna*

Usually a much branched thorny shrub, but can be a medium-sized tree (H to 15 m) with a rounded, densely twiggy crown. Found throughout Britain & Ireland (except for extreme north-west Scotland) and tolerates a very wide range of soils. It is probably best known as a hedgerow shrub although it is also common in scrub, thickets and woodland edges. **Features:** BARK grey-brown, smooth becoming pink-brown and scaly with age. TWIGS spiny; thorn L to 20 mm. BUDS egg-shaped to spherical (D 2–6 mm). LEAVES **5–7-lobed** (L 1–6 cm); lobes cut at least halfway to the midrib; upperside shiny, dark or bright green (can be slightly leathery); underside grey-green; usually hairless except for hair-tufts in the lower vein axils. FLOWERS in loose clusters of 10–18; white (D 10–15 mm); typically **1 style**, rarely 2. FRUIT bright red tough-skinned berry (D 5–10 mm); inner flesh yellowish with a **single stone**.

J F M A M J J A S O N D

> **Similar species** | Other hawthorn relatives which may be encountered as planted trees include **Cockspurthorn** *C. crus-galli* (N/I) which has hairy leaves and 2–3 fruit stones; and **Large-sepalled Hawthorn** *C. rhipidophylla* (N/I) which has more deeply toothed leaves, larger deeper red fruits and very obvious sepals which are longer than wide. The hybrid between **Hawthorn** and **Midland Hawthorn** (*C. ×media*) is common and has a wide range of variable features intermediate between the parents.

TWIGS grey-brown and spiny with thorns up to 20 mm long

BUDS ± spherical; dark reddish brown

BARK pink-brown; scaly when mature

LC LC Midland Hawthorn   *Crataegus laevigata*

Usually a multi-branched shrub, although can develop into a small tree (H to 12 m) with a broad, dense, rounded crown. The rarer of the two native hawthorns. It is a shade-tolerant species found on heavy soils and closely associated with ancient woods to the extent that it is one of the 'indicator plant' species for that habitat. It can also be found in old hedges. Found wild in the south-east of England north to the Midlands but generally absent elsewhere, except where planted. Ornamental forms, including the pink-flowered 'Paul's Scarlet' (which arose from a branch 'sport' in the 1880s as the result of a chance genetic mutation), are frequently seen in towns and cities. **Features:** BARK grey-brown to orange-brown; becoming cracked and fissured with age. TWIGS grey-brown; usually shiny; fewer shorter (L 6–15 mm) thorns than Common Hawthorn. BUDS ± spherical (L 1–2 mm). LEAVES **1–3-lobed**; lobes shallow, cut less than halfway to the midrib; upperside shiny dark green; underside pale green; at most scattered hairs along the main veins when young. FLOWERS in clusters of 3–11; white (D 15–18 mm); typically with **2 styles**, but up to 5 recorded. FRUIT bright red tough-skinned berry (D 8–14 mm) inner flesh yellowish flesh with **2–5 stones**.

J F M A M J J A S O N D

TWIGS grey-brown and often shiny, with fewer, shorter spines than Hawthorn

BARK grey-brown to orange-brown; cracked when mature

ROSACEAE (ROSES) | HAWTHORNS

Did you know? Hawthorn can reach a considerable age. In hedgerows that have been managed, a 100-year-old tree may only have the stem girth of a fizzy drinks can. This is because hedge management can be akin to what happens to a tree when it is bonsaied; thus Hawthorn trees may sometimes be the oldest plants in a landscape, but because they are still relatively small in girth or stature, their age and importance can be overlooked. Indeed, the oldest Hawthorn in Britain & Ireland is thought to be the Hethel Old Thorn in Norfolk and is considered to be more than 700 years old. Fortunately, this tree also lives in what is our smallest nature reserve at 0.025 hectares.

Another well-known Hawthorn is the Glastonbury Thorn, which has a second flowering during winter. This is a form of Hawthorn called 'Biflora', famously known (according to the legend) to have grown from the staff of Joseph of Arimathea as he brought Christianity to England. It is genuinely intriguing to see this famous tree flowering around Christmas Day.

The wood of both hawthorns has been used for tool handles and walking sticks, while the berries are an excellent source of vitamin C. The stems of Midland Hawthorn are also often contorted, improving the walking sticks that can be made from the tree.

HAWTHORN

in fruit

LEAVES
5–7 deep lobes; hair-tufts in the axils of lower veins

FLOWERS
1 style

FRUIT
1 seed

broadly similar in terms of form and flowering; differing in leaf shape and details of the flowers and fruit

MIDLAND HAWTHORN

LEAVES
1–3 shallow lobes; lacks hair-tufts on underside

FLOWERS
2 styles

FRUIT
2 seeds

Known as 'May Blossom', the vast numbers of Hawthorn flowers can entirely cover bushes (*left*) resulting in vast numbers of berries in autumn (*right*).

ROSACEAE (ROSES) | HAWTHORNS

PHYLLONORYCTER OXYACANTHAE MINES

SMALL EGGAR CATERPILLARS

BLACK-SPOTTED LONGHORN

Associated species – hawthorns

Hawthorn has a high wildlife value as its flowers provide nectar for spring insects and its berries provide excellent food for small mammals and birds, especially **thrushes**. By the late summer whole shrubs can become a dark red colour under the weight of the berries.

As the major component of hedgerows throughout Britain & Ireland, Hawthorn is a vital element of one of our most important habitats. Generally, hedges are composed of two main species, Hawthorn and Blackthorn, but ancient hedges can support many more species, including Hazel, Dogwood and Field Maple. These Hawthorn hedgerow networks allow species to travel through the countryside, using the hedges for food and shelter. **Bats** in particular use hedges, and the **Greater Horseshoe** and **Lesser Horseshoe Bats**, **Brown Long-eared Bat** and **Barbastelle**, for example, use Hawthorn hedges for foraging. Hedgerows also play a vital role for the **Dormouse**, which replaces body fat lost during hibernation by feeding on Hawthorn flowers in the spring.

Hawthorn in hedges also provides song posts and perches for territorial and breeding birds such as **Dunnock**, **Yellowhammer** and **Whitethroat**.

When Hawthorn flowers are abundant in the spring, **honeybees** and many other species use the flowers as vital source of nectar and pollen. In ancient wood-pasture systems like the New Forest, a single Hawthorn can support a myriad of **beetles** and **bees**, including rarities such as the **Oak Mining Bee** and many longhorn beetles such as the **Four-banded**, **Six-spotted** and **Black-spotted Longhorns**.

Hawthorns are also used as the caterpillar food plant of many butterfly and moth species, such as the **Small Eggar**, and in total well over 200 insect species have been reported to use Hawthorn.

Coleophora coracipennella, a moth, burrows into the flower buds to feed on pollen. In one study, as many as 20% of flowers of Midland Hawthorn in an Oxfordshire woodland were sterilised by this species. Another moth, *Phyllonorycter oxyacanthae*, creates distinctive mines on the leaves. **Mistletoe** and the slime mould *Reticularia lycoperdon* have also been found growing on Midland Hawthorn.

Many bird species, including Whitethroat (*left*) use the tangle of spiny Hawthorn branches to nest in; while thrushes like Redwing and Fieldfare (*right*) consume large quantities of berries during the autumn.

Leaves *p.65* | Flowers *p.73* | Fruit *p.77* | Twigs *p.80* ROSACEAE (ROSES) | COTONEASTERS

CR Wild Cotoneaster *Cotoneaster cambricus*

A low, spreading deciduous shrub (H to 1·5 m; W to 2 m when grown in botanical gardens). First discovered in 1783, Wild Cotoneaster, or the Great Orme Berry, is a long-lived, endemic native species, **globally restricted to the Great Orme peninsula of north Wales**. There has been much debate about the status of the species, with suggestions that the plants may be a form of *Cotoneaster integerrimus*. The Royal Botanic Gardens at Kew investigated the plant's DNA and revealed there is low genetic variability in the original Great Orme cotoneasters, and the plants form a distinct genetic cluster, closely related but separated from other plants within the *C. integerrimus* group.

Features: TWIGS grey-brown with dense yellow hairs, especially near the tip. BUDS naked (L ± 5 mm), pressed against twig. LEAVES oval (L 1–4 cm); grey-green with a few scattered hairs on the upperside and a dense layer of white hairs on the underside. FLOWERS **solitary or in clusters of 2–4**; white to pale pink (D ± 3 mm); petals erect. FRUIT small (D ± 3 mm); red; berry-like.

J F M A M J J A S O N D

WINTER TWIG grey-brown; densely yellow hairy; buds naked; **dark brown narrow stipules obvious**

Did you know? The Wild Cotoneaster population on Great Orme was reduced by 19th-century collectors who took plants for gardens. The remaining plants have suffered from overgrazing by rabbits, goats and sheep, which has reduced plant size and suppressed flowering, thus preventing new shrubs from establishing. Only 6 plants are known in the wild, which appear unable to spread naturally from seed, and so it has full legal protection under Schedule 8 of the Wildlife and Countryside Act 1981. To attempt to improve the chances for the species, 33 additional plants, taken from seeds and cuttings, were planted on Great Orme but, to date, only 11 appear to have survived. The locations of plants are now kept confidential, to restrict the impacts of collectors, and the existing plants are monitored.

One of the 6 known native Wild Cotoneasters on Great Orme.

ROSACEAE (ROSES) | COTONEASTERS   Leaves *p. 65* | Flowers *p. 73* | Fruit *p. 77* | Twigs *p. 80*

Introduced Cotoneasters
Cotoneaster species

Although found across temperate Asia, Europe and North Africa, the greatest diversity of species can be found in southwestern China and the Himalayas. Up to 300 species have been described, of which more than 100 have been recorded in Britain & Ireland, the vast majority of these as naturalized occurrences. Many of these species are included in the Wildlife and Countryside Act Schedule 9 list of invasive non-native species, which means that it is illegal to cause them to spread in the wild.

Features of Cotoneaster
Form: low and prostrate to more erect shrubs and trees.
Leaves: arranged alternately; evergreen or deciduous; oval to pointed-oval.
Flowers: white to pinkish; 5 petals (open or closed depending on the species); single or in groups.
Fruit: small, berry-like (pomes); bright red through orange.

Cotoneaster Identification
Due to the sheer number of species, varieties, cultivars and hybrids, and the similarities between many of them, Cotoneaster identification is not straightforward. Specialist keys are available but, even so, correctly naming an encountered plant may not be possible without expert analysis. Important features are the plant's form; the size, shape, texture and details of the upper- and undersides of the leaves of sterile (non-flowering) shoots; whether the flowers are open or closed; the number of flowers in an flower-cluster; the size of the fruit and how many seeds it contains.

The following are some of the most widely encountered species – arranged by leaf size.

Large-leaved species (L > 50 mm)

■ Bullate Cotoneaster   *C. rehderi*

Form H 2–5 m; deciduous, arching shrub. **Leaves** oval to broadly oval; L 70–210 mm; W 45–90 mm; appear very blistered; VEINS very deep, 8–11 pairs; ABOVE shiny mid-green; BELOW light green, yellowish hairs when young. **Flowers** 10–30; PETALS erect, incurved, red or maroon with pink border; STAMENS 20; ANTHERS white. **Fruit** L 8–11 mm; shiny; cardinal red.

LF

UNDERSIDE yellowish hairs   UPPERSIDE **8–11 pairs of very deep veins**

■ Hollyberry Cotoneaster   *C. bullatus*   SCHEDULE 9

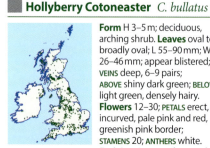

Form H 3–5 m; deciduous, arching shrub. **Leaves** oval to broadly oval; L 55–90 mm; W 26–46 mm; appear blistered; VEINS deep, 6–9 pairs; ABOVE shiny dark green; BELOW light green, densely hairy. **Flowers** 12–30; PETALS erect, incurved, pale pink and red, greenish pink border; STAMENS 20; ANTHERS white. **Fruit** L 7–8 mm; ± shiny; red to rich red.

FRUIT

LF

UNDERSIDE dense hairs   UPPERSIDE **6–9 pairs of deep veins**

ROSACEAE (ROSES) | COTONEASTERS

Medium-leaved species (L 19–50 mm)

Franchet's Cotoneaster *C. franchetii*

Form H 2–3 m; evergreen, erect or arching shrub. **Leaves** widely spaced on twigs; oval to broadly oval; L 25–37 mm; W 13–19 mm; ABOVE slightly wrinkled; ± shiny grey-green; BELOW silvery to yellowish, densely hairy. **Flowers** 5–25; PETALS erect, incurved, dark red with off-white border; STAMENS 20; ANTHERS pink to purple. **Fruit** L 8–10 mm; shiny orange-red softly hairy.

UPPERSIDE slightly wrinkled
FRUIT
LVS
UNDERSIDE silvery to yellow; densely hairy

Himalayan Cotoneaster *C. simonsii* SCHEDULE 9

Form H 3–4 m; deciduous or semi-evergreen, erect shrub. **Leaves** broadly oval; L 19–33 mm; W 13–24 mm; ABOVE shiny mid- to dark green; BELOW light green, hairy when young. **Flowers** 2–6; PETALS erect, incurved, dark red with white border; STAMENS 20; ANTHERS white. **Fruit** L 10–12 mm; shiny; bright orange to orange-red.

UPPERSIDE
FRUIT
LVS
UNDERSIDE light green; hairy when young

Small-leaved species (L < 18 mm)

Entire-leaved Cotoneaster *C. integrifolius* SCHEDULE 9

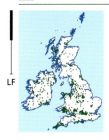

Form H 0·5–1·5 m; evergreen, procumbent to arching. **Leaves** oval to narrowly oval; L 8–17 mm; W 3–8 mm; ABOVE shiny dark to blue-green; BELOW greyish white, densely hairy. **Flowers** 1–2; very short-stalked; PETALS spreading, white; STAMENS 20; ANTHERS red-purple. **Fruit** L 7–9 mm; dull; dark rich red to crimson; can be sparsely hairy.

UPPERSIDE
FRUIT
UNDERSIDE greyish with dense hairs

Wall Cotoneaster *C. horizontalis* SCHEDULE 9

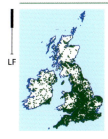

Form H 0·5–1·0 m; **deciduous**, arching shrub; often against walls; **herringbone branching**. **Leaves** broadly oval; L 5–12 mm; W 5–9 mm; ABOVE shiny dark green; BELOW paler, with sparse long hairs. **Flowers** 1–3; PETALS erect, incurved, dark red with blackish red base and pale crimson border; STAMENS 10(–13); ANTHERS white. **Fruit** L 5–6 mm; shiny; orange-red.

UPPERSIDE shiny
FRUIT
LVS
UNDERSIDE paler with sparse long hairs
distinctive herringbone branching

RHAMNACEAE | BUCKTHORNS Leaves *p. 65* | Flowers *p. 72* | Fruit *p. 76* | Twigs *p. 81*

LC LC Alder Buckthorn ☠ *Frangula alnus*

A spineless, deciduous shrub or small tree (H to 6 m) found on peaty soils in damp woods and bogs. **Features:** BARK OUTER dark blackish brown; INNER yellow. TWIGS dark brown with many lenticels; can have corky projections. BUDS end-bud egg-shaped (L 3–7 mm); brown; bud scales absent – new leaves protected by covering of dense hairs; side-buds are smaller (L 2–5 mm) and pressed close to the branch. LEAVES oval (L to 6 cm); dark green, with an **untoothed margin** and **6–10 pairs of prominent veins** which can be slightly hairy; turn yellowish red in the autumn. FLOWERS small (D 3–4 mm); star-shaped; 5-petalled; greenish white; bisexual. FRUIT small (D 6–10 mm), 2–3-seeded berries; green when young, ripening red then purple/black.

FLOWERS in clusters along branch

FLOWERS greenish white; star-shaped; **5-petalled**

J F M A M J J A S O N D

LEAVES oval with **untoothed margins** and **6–10 pairs** of veins

TWIGS dark brown; leaf-buds arranged **alternately**

BUDS scales absent; new leaves covered in dense hairs

INNER BARK yellow

Similar species | Both buckthorns are somewhat similar to the dogwoods (*p. 304*), particularly in leaf-shape.

Did you know? Charcoal from this tree was an important part of gunpowder. This species had a wide range of uses: the wood was used to prepare charcoal, due to its slow-burning nature; the berries, which ripen from green to red to black, were used for dyes and also as a purgative, along with its bark. The tree is a food plant for the Brimstone butterfly.

FRUIT loose cluster of berries; green ripening red then black

Leaves *p. 66* | Flowers *p. 72* | Fruit *p. 76* | Twigs *pp. 80, 83, 84*   RHAMNACEAE | BUCKTHORNS

Buckthorn   *Rhamnus cathartica*

A moderately fast-growing, short-lived, deciduous shrub or small tree (H to 10 m), found on chalk and limestone across southern and central England, and central Ireland. It is shade-tolerant and grows in hedgerows or as scrub on downland. **Features:** BARK OUTER grey, cracking with age; INNER orange. TWIGS dark brown with a greyish surface. BUDS conical (L 3–7 mm); blackish brown with pale fringes; end-bud replaced by a thorn in many. LEAVES variously arranged; simple; green; oval (L to 6 cm) with **3–5 pairs of conspicuous main veins which curve strongly towards the leaf-tip**; turn yellow in the autumn. FLOWERS small (D 4–5 mm), inconspicuous yellow/green with 4 (can have 5) petals; ♂ and ♀ on separate plants. FRUIT small (D 6–10 mm), purplish black, 3–4-seeded berries.

J F M A M J J A S O N D

FLOWERS in clusters along branch

♂ FLOWERS   ♀ FLOWERS

LEAVES oval with finely toothed margins and **3–5 pairs** of forward-pointing veins

TWIGS darkish brown with a greyish covering; can be spiny

BUDS ± conical; side-buds usually arranged **oppositely** but can be alternate on some twigs

END-BUD **replaced by a thorn in many trees**

INNER BARK orange

Did you know? Historically the wood was used in making charcoal. Another name for the species is the Purging Buckthorn, in reference to the use of the inner bark and berries to create a violent laxative. Buckthorn is one of the favoured foodplants of the Brimstone butterfly, and the berries are eaten by birds, which disperse the seeds in their droppings.

FRUIT clustered berries; green ripening black

ELAEAGNACEAE | SEA-BUCKTHORNS   Leaves *p.67* | Flowers *p.69* | Fruit *p.77* | Twigs *p.80*

Sea-buckthorn *Hippophae rhamnoides*

A densely branched, deciduous shrub (H to 9 m), native to Britain and introduced into Ireland. Sea-buckthorn is very hardy and both salt and drought resistant. It has developed a symbiotic relationship with *Frankia*, a bacteria which fixes nitrogen, helping the shrub survive in difficult conditions. The species has an extensive root system and readily develops suckers, allowing it to spread easily. Because of this it has been used to stabilise extensive sand-dune areas, particularly in its native range on the east coast of England. Elsewhere in the country it is often planted and naturalized. **Features:** BARK rough; brown or black. TWIGS dark brown; spiny. BUDS small (L 2–5 mm), egg-shaped and covered in a layer of small, shield-shaped, bronze scales (♀ with 2–4; ♂ with 6–8). LEAVES **long and very narrow** (L to 80 mm; W to 15 mm); upperside dark grey-green; underside silver-grey, densely hairy. FLOWERS tiny (D 3–4 mm); greenish; on separate ♂ and ♀ plants; ♂s in a short brownish spike, which fall off after pollen release; ♀s appear with the first leaf shoots, in small clusters along the length of a branch that is at least 2 years old. FRUIT distinctive **egg-shaped orange berries** (D 6–8 mm) on shrubs at least 7 years old.

J F M A M J J A S O N D

Did you know? In ancient Greece, Sea-buckthorn was fed to horses to help make their coats shiny (*Hippophae* – *hippo* = horse; *phaos* = to shine).

Juice from the nutrient-rich berries, which are high in proteins, vitamins B12, C and E, and the pigment beta-carotene (all antioxidants) has been used for *e.g.* ice cream, jam and as an additive to sparkling wine. Sea-buckthorn oil has been used medicinally, including for eczema, sunburn and chemical burn treatments. In Russia the first Sea-buckthorn factory in Bisk manufactured dietary products and cream for protecting against cosmic radiation for use by Russian cosmonauts.

The shiny orange colour from the berries can also be extracted using hot water and has been used as a wool dye. Shrubs in commercial orchards can produce fruit for 30 years.

Sea-buckthorn is thought to be native in the sand dunes of eastern England (*left*), perhaps having arrived as seed from Europe dispersed by migrating thrushes that feed on the nutritious berries (*right*).

ELAEAGNACEAE | SEA-BUCKTHORNS

IN FRUIT ♀

FRUIT distinctive egg-shaped orange berries

LEAVES thin; willow-like; dark green upperside; dense silver hairs on underside

♂ ♀

FLOWERS tiny, inconspicuous, greenish appearing just as the leaves appear

BUDS egg-shaped; ♀ with 2–4 scales; ♂ with 6–8

TWIGS dark brown; spiny

Associated species | The bracket fungus *Phellinus hippophaeicola* grows on the main stem and older branches. The fungus fruit body is a semicircular or hoof-shaped bracket, with a soft reddish brown velvety upper surface on young growth, becoming harder and greyer with age.

One rare native moth, the **Seathorn Groundling** is found in scattered sites, especially in Norfolk. It flies from mid-August to the end of September. The larvae initially mine the leaves and, when larger, spin individual webs which bind the leaves together at the ends of the shoots.

The berries are attractive to birds and are an important foodstuff for migrating **thrushes** arriving along the east coast in the autumn.

PHELLINUS HIPPOPHAEICOLA

SEATHORN GROUNDLING

JUGLANDACEAE | WALNUTS Leaves *p.61* | Flowers *p.68* | Fruit *p.74* | Twigs *p.87*

LC Walnut *Juglans regia*

A large tree (H to 30 m), probably introduced by the Romans. Nowadays widespread in southern and central England, growing best on well-drained soils, especially clays. Further north and west it is found largely in parks and gardens. **Features:** BARK smooth and grey when young, becoming fissured. TWIGS grey to purplish brown with distinctive leaf-scars; **pith chambers 8–12 per cm**. BUDS terminal large (L 5–10 mm); laterals smaller (L 3–5 mm). LEAVES large (**L to 45 cm**); **pinnate**; dark green; aroma of furniture polish if crushed. FLOWERS ♂s tiny, in catkins (L 5–15 cm); ♀ tiny. FRUIT globular (D to 50 mm); green; smooth.

J F M A M J J A S O N D

♂ FLOWERS green; in catkins

♀ FLOWERS small; green; STIGMAS yellowish, tinged purple

FLOWERS just before leaves

LEAVES 5–9 oval leaflets which get larger towards the leaf-tip

LEAF-SCARS large, distinctive 'Y'-shape

FRUIT round, smooth and green; containing a stony seed, the inner part of which is the familiar wrinkled brown 'nut'

FRUIT   SEED

TWIGS quite stout; BUDS egg-shaped to globular; end-bud larger than side-buds

BARK smooth, silvery grey becoming fissured with age

Similar species | Black Walnut
Juglans nigra differs in its leaves having 9–23 longer, more narrowly pointed and more irregularly toothed leaflets and slightly hairy twigs which have smaller **pith chambers (14–18 per cm)**. Introduced into Britain & Ireland by 1629.

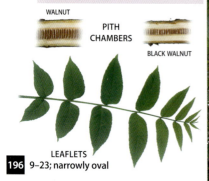

WALNUT

PITH CHAMBERS

BLACK WALNUT

LEAFLETS 9–23; narrowly oval

Introduction to the elms

Obtaining a confident, conclusive identification of an individual elm tree in the field is difficult in some cases, and impossible in others. Dr Oliver Rackham (a Cambridge academic and writer) once referenced the complexity of elms by describing them as *'the most critical genus in the entire British flora'*. This complexity is evidenced by the vigorous identification debate that has occurred over the last 100 years, which has seen elm taxonomy classified in a number of ways: from one very variable species; through two, three, five or seven (often with named hybrids and varieties); and upwards of 60 species in recent British floras and keys (*e.g.* Sell 2018 and Eversham 2021 – see *p. 346*). Indeed, at a meeting on the conservation of elm trees in 2022, when the author asked the question "How many elm species are there in Britain & Ireland?", none of the attending experts were prepared to state a view.

In the early 2000s an analysis using molecular markers suggested that there are essentially only two native species in Britain & Ireland – the relatively consistent-looking Wych Elm (*p. 200*) and the highly variable Field Elm (*p. 201*), which comprises a range of subspecies and varieties that have caused much confusion for centuries. In addition, the two species may hybridize, producing a range of trees with variable characters.

As of early 2025 the classification of British elms remains complex and unresolved with the addition of two intriguing possibilities: the widely planted White Elm has now been found in a Suffolk woodland, with circumstances hinting at it being an additional native species; and a small number of trees with consistent morphological differences that were discovered recently around Tasburgh in Norfolk appear, after genetic analysis, to be significantly distinct from other elms.

A planted Camperdown Elm (*above*) and the size differential between its very large leaves and those of Atinian Elm (*below*).

CAMPERDOWN ELM

ATINIAN ELM

The approach taken in this book is therefore to describe in detail the 'two native species', 'Wych Elm' and 'Field Elm', and the commonly encountered, possibly native, European White-elm. In the field this distinction is pragmatic and sufficient for most people, unless they have a deeper interest in the taxonomy and identification of elms as a group. To cover this, a wider list of potential elm taxa is set out in lesser detail. This list is based on a September 2024 Natural England workshop in which interim genetic analysis data were presented exploring whether a larger number of distinct forms within the diversity of British elms could be treated as new 'species'. As a result, it is likely that over the next few years elm taxonomy and field identification will take on a different form.

Whatever the genetic data may eventually show, the 'in-the-field' reality is that there is enormous physical variability in British elm trees, with locally distinct tree forms/species being found around the country as is clearly shown by the picture above, which shows the massive differences in leaf size and shape that occur between one of the smallest leaved elms, the 'Atinian' Elm, and one of the largest, the 'Camperdown' Elm. These trees (elms) are certainly worth exploring!

ULMACEAE | ELMS   Leaves *pp. 65, 67* | Flowers *p. 69* | Fruit *p. 75* | Twigs *p. 86*

Associated species | Elms support a wide range of other species and host approximately 80 species of invertebrate. The early pollen is sought by many insects, and the rare **White-letter Hairstreak** butterfly larvae feed on the flowers and leaves, normally towards the top of the tree. The white lines on the underside of a White-letter Hairstreak's hindwing are the same width as the veins on an elm leaf, making the butterflies well camouflaged when they lie sideways to sun themselves. Moth caterpillars such as those of the **Light Emerald**, **Dusky-lemon Sallow**, **Clouded Magpie** and **White Spotted Pinion** also feed on the leaves, along with the caterpillars of the beautiful **Large Tortoiseshell** butterfly. Elms also host bark beetles, particularly **elm bark beetles** (especially *Scolytus scolytus* and *S. multistriatus*), which have acted as the transmitters of the Dutch elm disease which killed off most of Britain's approximately 30 million mature elms. Elm bark beetles spread the infection through wounds in elm twigs and then breed in the bark of the dying elms. The larvae of **Elm Zigzag Sawfly**, introduced from Asia, creates distinctive feeding patterns, which can cause some defoliation. Elm seeds are also eaten by birds, including **Chaffinch**, **Greenfinch**, **Bullfinch** and **Hawfinch**.

BARK BEETLE BREEDING GALLERIES

ZIGZAG SAWFLY DAMAGE

WHITE-LETTER HAIRSTREAK

DUSKY-LEMON SALLOW

LARGE TORTOISESHELL

Did you know? Elms produce attractive, tough timber which is durable if kept dry or continuously wet. Small-dimension wood has been used to make furniture, coffins and flooring, and larger timber has been used for underwater docks, harbours, docks and wooden boats, including parts of the *Mary Rose* (1512), often sourced from mature hedgerow trees.

Although mature elm trees are now rare, elms have spread as root suckers and some have formed into hedges. They remain healthy until the trees reach a size that the beetles can re-infest.

ULMACEAE | ELMS

DD CR European White-elm *Ulmus laevis*

A tall tree (H to 30 m), similar in form to Wych Elm. It is found in floodplain woodland and trees develop buttresses as an adaptation to wet ground conditions. It has been known in Britain since the early 19th century and has long been planted in streets and parks. Historically, it was regarded as an introduction from Europe although a recent discovery in a Suffolk wood opens up the possibility that it may be native. One key identification feature is the dangling flowers and fruits have long stalks, which persist, looking like tufts of blackish 'wires'. **Features:** BARK smooth and grey when young, becoming increasingly fissured and cracked with age, creating a network of grey-brown scales; under-bark reddish brown. TWIGS thin; greyish brown; usually with tiny whitish hairs. BUDS narrowly egg-shaped (L 5–8 mm), pointed; bud scales 6–8, purple-brown to reddish brown with darker margins. LEAVES oval (L to 12 cm; W to 6 cm); **thin and papery**; upperside smooth; underside hairy; leaf-base strongly asymmetrical (AS: 9–17 mm); leaf-margin with fine-pointed hooked teeth; leaf-stalk (L 7–17 mm). FLOWERS appear before the leaves; in clusters of up to 30; **flower-stalks long** (L to 20 mm); anthers purplish red. FRUIT winged seeds (wing L to 15 mm); wings pointed with a hairy margin; points converge; seed located halfway along wing length.

J F M A M J J A S O N D

BARK like Wych Elm, although the scales on older trees can be thinner

SHORT SHOOT

LEAVES longer leaf-stalk and more asymmetrical than most elms

FLOWERS on long stalks

TWIGS greyish brown; usually with tiny hairs

BUDS reddish brown with darker margins

FRUIT hairy + long-stalked

A large European White-elm at Dinefwr, Wales (*left*); the long-stalked fruit (*right*) with hairy margins is unlike any of the native elms.

ULMACEAE | ELMS　　　　　　　　　　　　　Leaves *pp. 65, 67* | Flowers *p. 69* | Fruit *p. 75* | Twigs *p. 86*

DD LC Wych Elm　　　　　　　*Ulmus glabra* agg.　GROUP 1 *p. 204*

A tall, non-suckering tree (H to 45m) with a rounded spreading canopy and long spreading branches. It has a preference for heavy moist clays and loams although will grow on chalky soils. It is moderately shade-tolerant, wind-resistant and copes with coastal exposure. Most common in hillside woods in Scotland; widespread but much less common elsewhere. **Features:** BARK smooth and grey when young, becoming darker and heavily ridged with ± parallel furrows with age; can have burrs; underbark grey-brown. TWIGS stout; grey and flexible; hairy or hairless; lacking corky wings. BUDS **broadly egg-shaped** (L 3–6mm), pointed; bud scales 4–6, dark brown with **distinct rufous to whitish hairs**. LEAVES oval to narrowly oval (L to 16cm; W to 11cm), typically abruptly pointed, or with 3 pointed lobes, particularly on the most vigorous shoots; **upperside rough sandpaper-like texture**; underside typically hairy; leaf-base asymmetrical (AS: 2–8mm); leaf-margin roughly double-toothed; leaf-stalk (L 3–7mm). FLOWERS appear before the leaves; ± sessile; in clusters of up to 25; anthers purplish red. FRUIT winged seeds (wing L 15–20mm); small notch between wings; **seed located halfway along wing length**.

J F M A M J J A S O N D

LEAVES generally much larger than those of other elms

FLOWERS largest of the elms; unstalked

LEAF-TIP long drawn-out; can have 3 pointed 'horn'-like lobes (*right*)

LEAF-STALK short; covered by lobe in many

FRUIT hairless; seed located centrally

FRUIT CLUSTER

BUDS distinct whitish to rufous hairs

TWIGS greyish brown; hairy or hairless; **corky ridges absent**

BARK smooth and grey, grey furrowing with age; paler than in other elms

An ancient Wych Elm growing in the Bannau Brycheiniog National Park near Llangorse, Wales.

Did you know? The word 'wych' has its origins in Old English and means pliant or supple.

The wood is tough, and resists splitting, even when wet, which led to its use in boatbuilding, for underground water pipe and for wheel hubs.

Like other elms, this species has suffered from Dutch elm disease, and its population is much reduced.

200

ULMACEAE | ELMS

DD LC Field Elm *Ulmus minor* agg. GROUPS 2–6 *pp.204–211*

Field Elm (or what is currently known as Field Elm) is potentially a complex group of species, subspecific taxa and hybrids. Genetic analysis is ongoing but there appears to be some basis for differentiation by 'type' based on morphology and genetics which reflect the geographical groupings found in the '*minor*' aggregate in Britain and neighbouring countries.

The Field Elm complex comprises typically vigorously growing, suckering trees (H to 30m, rarely more) found in a wide range of habitats, but predominantly in small copses and hedgerows. **Features:** BARK grey-brown with long, deep vertical fissures with age. TWIGS hairy or hairless; older twigs **with or without corky ridges** (or 'wings'). BUDS egg-shaped (L 2–6mm); **hairless**; typically dark brown to purplish black with paler margins. LEAVES highly variable between types (see following pages), although generally less than 7cm long; W to 6cm); upperside of optimal leaves (see *p. 203*) smooth or rough – leaves growing in shade and on suckers coarse and hairy in most taxa and may differ in texture from the optimal leaves; **leaf-base weakly to strongly asymmetrical**; leaf-margin variably toothed; leaf-stalk variable in length. FLOWERS appear before the leaves; ± sessile; in clusters of up to 20; anthers purplish red. FRUIT winged seeds (wing L 10–12mm); small notch between wings; **seed located towards the tip of the wings**.

J F M A M J J A S O N D

OLDER TWIGS can have corky ridges

FLOWERS slightly smaller than those of Wych Elm; unstalked

FRUIT hairless; **seed located near tip**

BARK has long, deep, vertical fissures

TWIGS greyish brown; hairy or hairless

BUDS hairless; dark brown with paler margins

LEAVES highly variable in shape, basal asymmetry and leaf-stalk length – see pages *204–211*

A mature 'English' Elm* (*right*) is an exceedingly rare sight as they almost entirely succumbed to Dutch elm disease, although Field Elms can still be found as smaller plants growing from root suckers in e.g. hedgerows.
* NOTE English Elm *U. procera* s.s. (together with Plot's Elm *U. plotii*) is included here within the '*minor*' aggregate.

Author's note | Another Jonathan Stokes, an 19th-century doctor and botanist, described Plot's Elm in his 1812 book of medicinal plants as growing at North Wingfield in Derbyshire, naming it *Ulmus surculosa argutifolia*. As the 18th-century Jonathan Stokes lived in the same village as the author's family, albeit 20 years earlier, the chances of a family connection seem worthy of further exploration!

ULMACEAE (ELMS) | IDENTIFICATION

Elm complexity – a field approach

Translating the ongoing research into the genetics, distribution and characteristics of elms into a guide to field identification, while linking historical elm 'species' with the more recent treatments, is not straightforward. There do appear to be consistent morphological differences between the genetic groupings, although these are not necessarily identification features that have been used in the past. What follows is an attempt to reconcile the new genetics work with observable characteristics to create a pragmatic approach that can be used in the field. However there are some limitations: taxa that have been recently regarded as separate species may prove to have a different taxonomic designation (*i.e.* subspecies or forms); there is the possibility that a number of the currently named species may prove to be part of a single species with very variable morphology; and the ongoing work may find potentially new species (*e.g.* the 'Tasburgh Elm' – see *p. 212*).

The broad framework that seems to elicit the widest agreement is as follows:
- **'Wych' Elm** *U. glabra* and similar-looking taxa form a genetically distinct group;
- **'Field' Elm** is a complex of species, subspecies and forms, as yet undetermined;
- **'Wych' Elm × 'Field' Elm hybrids** show a highly variable mix of features and may include species that have arisen from hybridization events (*cf.* Whitebeams (*p. 134*));
- **European White-elm** *U. laevis* is a distinct, possibly native, species as evidenced by its lack of hybridization with native British & Irish elms.

The following is an overview that combines current genetic information from the ongoing Natural England project (2025) with work by Sell (2018) and Eversham (2021). It presents the key identification information for optimal leaves (see *facing page*) within 6 broader groups based on leaf morphology. In addition, several taxa placed in these groups may be transferred to the 'basal elm species' (Group 8) if they prove to be significantly genetically distinct.

The taxa that follow are referred to by only their English names as used in Sell (2018) and Eversham (2021) as genetic work is still ongoing. Their scientific names have been omitted in this section, but are included for reference in Appendix I (*p. 346*).

Leaf upperside texture

The first and most crucial feature is whether optimal leaves (from July onwards) are hairy on the upperside. When rubbed gently (using a fingertip) hairy leaves have a '**rough**' texture noticeably different from the '**smooth**' feel of those lacking hairs. This is best done in the field. NOTE: leaves of 'rough' species are always rough; finding a smooth leaf in 'smooth' species may take some searching.

LEAF UPPERSIDE TEXTURES

rough (hairy)   smooth (hairless)

best assessed by gently rubbing a finger tip on the leaf surface

The elm groups

After the initial separation by texture the groups are further differentiated by shape and other features as described in the accounts in the broad genetic groups (to be confirmed) as below.

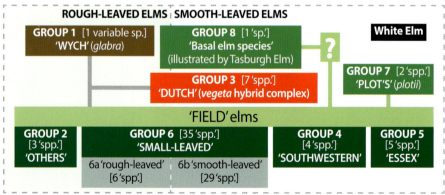

ULMACEAE (ELMS) | IDENTIFICATION | ID

Elm identification

Many of the features normally used in tree identification (*e.g.* mature shape, bark, winter twigs and flowers) are largely not useful for elms. In addition, many of the 'species' have not been studied enough to provide accepted detailed descriptions of winter twigs, flowers and fruits. Also, mature tree shape is largely impossible to use as there are so few left after the ravages of Dutch elm disease. Currently, therefore, elm identification is best undertaken using leaves from particular shoots in late summer.

'English' Elm branch showing the location of leaves suitable for identification

BRANCH FROM A SUNNY POSITION

Optimal leaf for identification – a fully developed leaf on a small side-shoot (L usually <10 cm) in full sun

2nd-year twig

short side-shoots

'ENGLISH' ELM

undamaged short side-shoots usually have 3–6 leaves

Which leaves to look at – optimal leaves:
- Leaves for elm identification are best collected from **July to the end of September**.
- Leaves of **the smallest side-shoots (L usually <10 cm) on second-year twigs growing in full sun**, are usually ideal. NOTE these are normally found in fine-branched canopy growth which can be difficult to reach.

Which leaves not to look at:
- **Leaves collected before July** – early in the summer 'smooth-leaved' elms still may have hairs and feel rough if stroked. From July onwards these hairs are lost on 'smooth-leaved' elms but retained on 'rough-leaved' elms.
- **Leaves produced from suckers, epicormic growth and clipped hedges** – the shape and form of these leaves is more variable (often rough in 'smooth' species), and therefore they are not optimal for identification.
- **Leaves growing in shade** – these are often 'hairy' (rough), even on 'smooth-leaved' species.

Identification features of optimal elm leaves – those on a sunny side-shoot and fully developed

Check and measure:
- Whether, in the field, the leaf upperside is rough (hairy) or smooth when gently rubbed
- **L** – leaf length (from base of blade to tip)
- **W** – leaf width at the widest point
- **AS** – length of leaf-base asymmetry
- **LS** – length of leaf-stalk (from join with twig to where it first joins the leaf-base)
- **LR** – Leaf length to width ratio
- **LVC** – Leaf vein count (on longest side)

Appendix I (*p. 346*) gives the leaf measurements described above as well as the English and scientific names for the taxa covered in this section.

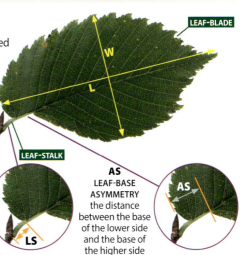

LEAF-BLADE

LEAF-STALK

LS
LEAF-STALK LENGTH
the length between the stalk base and the lowest part of the connected leaf-blade

AS
LEAF-BASE ASYMMETRY
the distance between the base of the lower side and the base of the higher side

203

ULMACEAE | ELMS   Leaves *pp. 65, 67* | Flowers *p. 69* | Fruit *p. 75* | Twigs *p. 86*

ROUGH-LEAVED ELMS

WYCH ELMS | **Features:** BUDS hairy; FRUIT seeds located centrally in wings

GENETIC GROUP 1 'WYCH' ELMS [1 variable sp.]

LEAVES rough, typically large (L >8 cm), 'horned' (see *p. 200*); FORM non-suckering.

1 LC **Northern Wych Elm** | Differs from Southern Wych Elm in its proportionately longer (L 10–18 cm), narrower leaves (L 2–3 × W). The wych elm of northern and western Britain and Ireland.

1a LC **(Southern) Wych Elm** – full description *p. 200*; leaves L 8–17 cm; L 1·6–2·1 × W.

1b DD **Moss's Elm** | Similar to **1a** Southern Wych Elm but smaller-leaved (L 10–15 cm) and with glandular hairs on young twigs. Hedgerows, roadside, copses. Widespread in England from Cornwall to Derbyshire; East Anglia and Wales.

1c DD **Camperdown Elm** | Similar to **1a** Southern Wych Elm but larger-leaved (L 14–18 cm) – the largest-leaved British & Irish elm. Originating from a park in Dundee, this tree is occasionally planted in its weeping form in parks, gardens and churchyards, rarely in semi-natural habitats.

1d DD **Exeter Elm** | Leaves are the smallest and broadest (L 8–11 cm; L 1·3–1·5 × W) and with the largest, longest, 'claw-like' teeth of the wych elms. Apparently brought into cultivation by a nursery in Exeter in 1829 and widely planted in parks and gardens, although now only occasional and widely scattered on road verges. This form is reasonably resistant to Dutch elm disease and has been used in breeding resistant variants.

Northern Wych Elm		AS	3–13 mm
L	10–18 cm	LS	3–6 mm
W	4–9 cm	LVC	17–27
LR	2–3		

Southern Wych Elm		AS	2–8 mm
L	8–17 cm	LS	3–7 mm
W	4–11 cm	LVC	16–26
LR	1·6–2·1		

Moss's Elm

Exeter Elm

ULMACEAE | ELMS

FIELD ELMS | **Features:** BUDS hairless; FRUIT seeds located centrally in wings

GENETIC GROUP 2 'OTHER' ELMS [3 'spp.']

MORPHOLOGICAL GROUP A | **Features:** FORM branches and/or stems/trunk with cork (though not a feature unique to this group); LEAVES rough, generally round, mite bumps often present (see *inset right*); OTHER very vulnerable to Dutch elm disease.

bumps caused by mites

LEAVES SMALLER, ROUNDER
(L 6–9 cm; L ≤ 1·6 × W); TWIGS CORKY;
BARK WITH VERTICAL FISSURES

2 LC **English Elm** | Often referred to as *Ulmus procera*, this suckering tree with a straight trunk is the tallest of the elms (H to 35 m), although most mature trees have been lost to Dutch elm disease. The leaves are green, broadly oval to circular (L 5–9 cm; L 1·2–1·6 × W) and 'floppy' with a rather buckled appearance; harshly rough on both sides; leaf-base asymmetry **to 7 mm**; leaf-stalk L 5–9 mm; longer side leaf-vein count 10–16. The leaves are attacked by the leaf-gall mite *Aceria campestricola*. Abundant in much of southern and central England. Road and track sides, hedges, field margins, copses, usually as self-coppicing shrubs. Similar to **51** Western Elm, which has larger, narrower, more pointed leaves which are only very slightly rough at most.

English Elm			
L	6–9 cm	AS	3–7 mm
W	4–7 cm	LS	5–9 mm
LR	1·2–1·6	LVC	10–16

3 LC **Atinian Elm** | Widespread and locally abundant in southern half of England and parts of west Wales. There is some evidence that it was introduced by the Romans to grow vines against and can certainly be found growing around Portchester Castle (a Roman shore fort in Hampshire). Very similar to **3** English Elm, and differs in having **leaves that are smaller, yellower and harder**; leaf-base asymmetry **no more than 4 mm**; leaf-stalk slightly shorter (L 3–7 mm) and **on average fewer veins (10–13)**. Road and tracksides, hedges, field margins, copses, often in river valleys.

Atinian Elm			
L	5–8 cm	AS	1–4 mm
W	3–6 cm	LS	3–7 mm
LR	1·1–1·6	LVC	10–13

LEAVES LARGER (L 7–12 cm);
TWIGS NOT CORKY; BARK CORKY

4 LC **Corky-barked Elm** | Leaves quite narrow with prominent jagged teeth. Very similar to **21** Western Elm which is only slightly rough at most and has broader, less tapered leaves with smaller teeth. Widespread in Norfolk, Suffolk and Leics, scattered in Cambs and Essex, with outliers in Wiltshire, Sussex, Herefordshire.

Corky-barked Elm

205

ULMACEAE | ELMS   Leaves pp. 65, 67 | Flowers p. 69 | Fruit p. 75 | Twigs p. 86

GENETIC GROUP 6a 'SMALL-LEAVED' ELMS (part) – rough-leaved spp. [6 of 35 'spp.']
MORPHOLOGICAL GROUP B | Features: LEAVES rough; variously shaped; hugely variable.

Morphological group B contains 'rough-leaved' taxa that are genetically within the 'small-leaved' group. Those marked with an asterisk* are of interest due to genetic differentiation that may be significant.

LEAVES ± RHOMBIC TO ROUNDED

5 EN **Bassingbourn Elm*** | Similar to, but larger-leaved than, **12** Huntingdon Elm (smooth-leaved; GENETIC GROUP 3). Field margins and hedgerows in Bassingbourn, East Hatley and Bottisham, Cambs.

6 EN **Woodland Elm*** | Leaves neat, diamond-shaped and almost symmetrical at the base. Mainly in ancient woodlands in Cambs and known from about 6 sites.

7 CR **Hayley Elm*** | Similar to **6** Woodland Elm which has more jagged teeth and to **27** Bonhunt Elm (GENETIC GROUP 6b, p. 209) which has rounded teeth. Only in a few boulder-clay woods in Cambs.

8 CR **Dark-leaved Elm** | Leaves long, tapered and crenate, similar to **54** Plot's Elm (GENETIC GROUP 7) which has shorter, less tapered leaves. By roads, tracks, hedges, field margins, around Halstead- Braintree-Colchester, Essex. Also occasionally on fen droves in north Cambs.

9 CR **Madingley Elm** | Rather lacking in distinctive features. Leaves fairly symmetrical with fairly small sideways-jutting bulges at the leaf-base. Currently known only from the ancient Madingley Wood in Cambs.

10 CR **Sacombe Elm*** | Rather like **9** Madingley Elm and almost impossible to differentiate. Roadsides, hedges. Currently known only from the area around Dane End and Sacombe, between Ware and Stevenage, Hertfordshire.

Bassingbourn Elm*

Woodland Elm*

Sacombe Elm*	L 5–7 cm	AS 2–8 mm
	W 3–4 cm	LS 3–9 mm
	LR 1·8–2·0	LVC 11–20

Hayley Elm*

ULMACEAE | ELMS

SMOOTH-LEAVED ELMS

FIELD ELMS | **Features:** BUDS hairless; FRUIT seeds located centrally in wings

GENETIC GROUP 3 'DUTCH' ELMS

Wych Elm × non-Wych Elm hybrids **Features:** LEAVES large, ± oval (L 7–16 cm; L 1·5–2·0 × W); teeth pointed and forward-curved; many with a bare section of vein at the base of the shorter side (most prominent in Huntingdon Elm – see *inset below*).

11 DD **Dutch Elm** | Introduced and widely planted by roads, in hedges, wood margins and parks in towns and cities as well as the countryside. Widely distributed from Cornwall to East Anglia, north to Yorkshire and south Wales. Large-leaved with pointed, forward-curved teeth. Leaves can be rather rough early in the season but generally smooth and fairly glossy by mid-summer.

12 DD **Huntingdon Elm** | Close to **11** Dutch Elm but with leaves generally longer and narrower (some twice as long as wide). Distinctive exposed vein at the base on the short side. Said to have originated in Hinchingbrooke Park, Huntingdon, in 1747, now widely planted by roads, hedges and wood margins, in towns and cities as well as the countryside.

Dutch Elm — L 7–14 cm | W 4–8 cm | LR 1·5–1·8 | AS 4–13 mm | LS 6–13 mm | LVC 13–21

Huntingdon Elm — L 8–16 cm | W 5–9 cm | LR 1·5–2·0 | AS 6–18 mm | LS 9–16 mm | LVC 13–22

The following five taxa probably arose through the partial genetic merging of two lineages and hybridization events. These five are recognized by Eversham (2021) but there are likely many other taxa which all look very similar. NOTE **15**–**17** probably arose as Wych Elm × genetic group 4 hybrids.

Features: LEAVES ± broadly oval (L 5–12 cm; L 1·2–3 × W); teeth blunt and/or quite evenly spaced.

13 CR **Laxton Elm** | A broad-crowned tree; leaves like **1a** Southern Wych Elm but shorter (L 8–12 cm) with distinctive, rather blunt, teeth. Currently known only from around Laxton, north Northants, south Norfolk (where some trees look cherry-like), and an old record from Hertfordshire.

14 EN **Cut-leaved Elm** | Rather lacking in distinctive features. Leaves with rather even squarish teeth which are noticeably spaced. Several mature trees in the ancient Madingley Wood, Cambs, and two trees in Leics.

15 LC **Davey's Elm** | Leaves broadly oval with blunt teeth. By roads, hedges and in valleys in Cornwall, extending into Devon, Dorset and Hampshire. Frequent in south Norfolk with scattered records from other counties. NOTE trees in Cornwall are probably genetically distinct from those elsewhere.

16 VU **Fat-leaved Elm** | Leaves ± rhombic- to round-leaved; rough early in the season but quickly becoming smooth and glossy. Leaves often broadly round, almost circular, with deep-impressed veins and fairly large, neat and even teeth. Roadsides and hedges. Frequent in Cornwall, also found in Devon and Dorset.

17 EN **Scabrid Elm** | Very similar to **22** Western Elm, which has slightly smaller and broader leaves. In hedgerows and copses by roads, tracks and field margins, and woods along beaches. Frequent in southern Cornwall, scattered records elsewhere in Cornwall, one record from Devon and a few in Ceredigion in Wales.

Davey's Elm — L 5–10 cm | W 5–7 cm | LR 1·5–1·7 | AS 6–9 mm | LS 6–10 mm | LVC 15–22

207

ULMACEAE | ELMS Leaves pp. 65, 67 | Flowers p. 69 | Fruit p. 75 | Twigs p. 86

GENETIC GROUP 4 'SOUTHWESTERN' ELMS [4 'spp.']

Features: LEAVES thin, smaller, broad and rounded (L 3–8 cm; L 1·4–2·5 × W) with an abruptly pointed tip; FORM trees often upright growth habit.

SMALLER LEAVES (L 2–7 cm)

18 DD **Cornish Elm** | The smallest-leaved (L 2–4 cm; W 1·5–2·0 cm) smooth elm. Found by roads, around farms and in hedges, infrequent in south-west Cornwall and scattered across the rest of Cornwall and Devon. **21** Small-leaved Elm (GENETIC GROUP 6b) is similar (and the most common elm in Cornwall) but has larger (L 5–7 cm) leaves.

19 EN **Goodyer's Elm** | Broader crown and steeply ascending branches, not unlike **18** Cornish Elm but with larger, broader leaves (L 3–6 cm; W 2–4 cm) with a shortly asymmetrical base and with fine regular teeth. First found in the 17th century and now best seen around the pub at the village of Rockford in the New Forest.

Goodyer's Elm
L 3–6 cm	AS 1–4 mm
W 2–4 cm	LS 4–11 mm
LR 1·5–2·0	LVC 9–15

20 EN **Jersey Elm** | Rather broad, rounded leaves and neat, rather square, blunt teeth. Somewhat asymmetrical, with a rounded bulge on the longer side, and a relatively long leaf-stalk. The crown is narrow with the main branches getting progressively shorter towards the top of the tree. Mostly a hedgerow and roadside tree in the Channel Islands, and occasionally in southern England (sometimes planted elsewhere).

21 LC **Small-leaved Elm** | Very similar to **18** Cornish Elm, but with larger leaves (L 5–7 cm; W 3–4 cm). Common throughout Cornwall and Devon, extending into Somerset and Dorset, with scattered records in Sussex, Middlesex and Bucks.

LARGER LEAVES (L 6–11 cm); TWIGS CORKY

22 VU **Western Elm** | Leaves very slightly rough, broadly oval, with a noticeably tapering tip and fairly large, forward-pointing teeth in the forward half. Similar to **3** English Elm which has smaller, broader, less pointed, much rougher leaves. Common in hedgerows in Cornwall and west Devon. A few records from Ceredigion, Wales.

Western Elm

GENETIC GROUP 5 'ESSEX' ELMS [5 'spp.']

Features: LEAVES similar in shape to GENETIC GROUP 4; FORM often broader in growth habit.

23 EN **Peninsula Elm** | Leaves (L 5–7 cm; W 3–4 cm) reminiscent of Wild Apple (p. 180), neat and fairly regularly-toothed. Mainly known from hedgerows and roadsides on the Dengie peninsula in south Essex, also found at a site in Bedfordshire.

24 CR **Smooth-leaved Elm** | Neat leaves less than twice as long as wide, with fairly large regular teeth. By roads and in hedges, so far known from a few localities in north Essex.

25 EN **Long-toothed Elm** | Despite the English name, the teeth of this elm are not long, but actually strongly curved towards the leaf-tip. Currently known from a few sites in south Essex south and east of Chelmsford, and a few woodlands in Ceredigion.

26 CR **Assington Elm** | A unique jagged-leaved elm, with proportionately the longest teeth of any British elm. No really similar species, and confined to roadsides, copses and hedgerows around the village of Assington Green in Suffolk.

27 EN **Essex Elm** | Very similar to **54** Plot's Elm (GENETIC GROUP 7), differing in its larger leaves (L 4–7 cm; W 3–4 cm) that have a longer leaf-stalk (L 4–9 mm). Roadsides, hedges and field margins, frequent in north Essex, especially near Waltham Cross. Outlying records from Suffolk.

Assington Elm

ULMACEAE | ELMS

Peninsula Elm

Long-toothed Elm

GENETIC GROUP 6b 'SMALL-LEAVED' ELMS (part) – smooth-leaved spp. [28 of 34 'spp.']
MORPHOLOGICAL GROUP C | Features: LEAVES rough; variously shaped; hugely variable due to the presence of poorly recognized species and complex hybridization arising from partial genetic merging of two lineages within the group as well as some external influences.

Morphological group C contains 'smooth-leaved' taxa that are genetically within the 'Small-leaved' group. Those marked with an asterisk* are of interest due to genetic differentiation that may be significant.

28 CR **Bonhunt Elm*** | Leaves widest near the base and with rounded (rather than pointed) rather deep teeth and a crimped appearance. Currently known only from roadside and hedges in the Wicken Bonhunt area of north Essex and a roadside tree and suckers near Bartlow, Cambs. Similar to **7** Hayley Elm and **8** Dark-leaved Elm (both GENETIC GROUP 6a, *p. 206*) which both have rough leaves that are widest at the middle and with more pointed teeth.

29 EN **Fat-toothed Elm*** | Leaves with shallow, broad-based teeth and a broadly rounded lobe on the long side. Roadsides and hedges on Cambs-Essex border and at one site in Bedfordshire.

Continued overleaf

Bonhunt Elm*

Fat-toothed Elm*

209

ULMACEAE | ELMS   Leaves *pp. 65, 67* | Flowers *p. 69* | Fruit *p. 75* | Twigs *p. 86*

Continued from previous page

GENETIC GROUP 6b 'SMALL-LEAVED' ELMS (part) – smooth-leaved spp. [28 of 34 'spp.']

MORPH GROUP C | Features: LEAVES rough; variously shaped; hugely variable due to the presence of poorly recognized species and complex hybridization arising from partial genetic merging of two lineages within the group as well as some external influences.

Morphological group C contains 'smooth-leaved' taxa that are genetically within the 'Small-leaved' group. Those marked with an asterisk* are of interest due to genetic differentiation that may be significant.

LEAVES TRAPEZOIDAL

30 VU **Wedge-leaved Elm*** | A distinctive elm with reddish corky stems and leaves distinctly asymmetrical. Roadsides, hedges and stream banks in Suffolk, Cambs, the east Midlands and outliers in Wales and Herefordshire.

LEAVES WITH DISTINCTLY JAGGED EDGES

31 VU **Long-tailed Elm** | Long lanceolate leaves that have a long drawn-out top part (hence the name) and long prominent broadly triangular teeth. Scattered in north Essex, south-east Cambs, west Suffolk and east Norfolk and one site in Wales.

32 EN **Narrow-crowned Elm** | Similar to **31** Long-tailed Elm but generally less asymmetrical and with less acute teeth. Hedges and copses in north Essex and west Suffolk.

33 EN **Prominent-toothed Elm** | Most likely to be confused with **31** Long-tailed Elm but differs in having jagged teeth only in the upper part of the leaf. Hedgerows and riverside, so far known from scattered sites in Suffolk and Cambs.

34 CR **Chaters' Elm** | Similar to **31** Long-tailed Elm but with shorter asymmetry. Forming a small wood west of Aberaeron, Ceredigion, and possibly also in Dorset.

35 EN **Burred Elm** | The burrs on the bark are an unusual feature of this elm, but otherwise similar to many of the elms in this group. By tracks, field margins, ditches and roads, and in hedges, copses and parks. Widespread in Huntingdonshire, Cambs and Bedfordshire and extending into Northants.

36 EN **Jagged-leaved Elm** | Leaves relatively symmetrical with prominent triangular teeth. Found on roadsides, hedges, copses and field margins, mainly in Huntingdonshire, also extending into Cambs fenland.

37 CR **Pebmarsh Elm** | Leaves neatly spear-shaped with only slight asymmetry and rather flat-topped, broad teeth. Hedges, woodland, scattered in Essex and Hertfordshire. Currently known from a few roadside trees around Pebmarsh in north Essex and Navestock Heath in south Essex.

38 CR **Tall Elm** | The leaves are unusual in being parallel-sided for 30% of their length. This tall elm is only found in hedgerows, streamsides and copses around Boxted and Hawkedon in west Suffolk.

LEAVES SPEAR-SHAPED, STRONGLY TAPERING

39 EN **Pointed-leaved Elm** | Leaves oval with a long pointed tip and variable, blunt-ended teeth; medium to large asymmetry. Hedgerows, scattered in north Essex, single records from Cambs and Norfolk.

40 EN **Narrow-leaved Elm** | Leaves rhombic to narrowly spear-shaped, with a fairly long pointed tip. Roadsides, field margins. Frequent in Huntingdonshire, Cambs, and the east Midlands.

41 EN **Cambridge Elm** | Leaves nondescript but fairly long and narrowly spear-shaped. Thought to be the commonest elm in Cambs, especially along the valley of the River Cam, and also common in Huntingdonshire, and extending into Bedfordshire, Leics, and Northants. NOTE this is *U. minor* s.s.

Wedge-leaved Elm*

Prominent-toothed Elm

ULMACEAE | ELMS

42 [EN] **Luffenham Elm** | Leaves narrowly oval with an almost symmetrical base and quite small, neat teeth; underside with triangles of white or fawn felty hairs in the vein axils – the most narrowly oval of this group of elms. Currently known from 3 or 4 regrowing stumps at South Luffenham, Leics, and 4 or 5 trees in a wood near the coast of Ceredigion.

43 [CR] **Hatley Elm** | A very striking elm with leaves rather resembling a broad-leaved asymmetrical White Willow (p. 241). Currently known only from a few ancient woodlands on boulder clay, and a fen drove, in Cambs.

Hatley Elm

LEAVES FAIRLY NONDESCRIPT; VARIOUS SHAPES

44 [VU] **Anglo-Saxon Elm** | Leaves fairly small; neatly rhomboidal leaves with prominent regular triangular teeth in the upper half; not very asymmetrical; leaf-stalk quite long. Common near the coast in Norfolk and Suffolk, less frequent inland, and a handful of Cambs records.

45 [VU] **Curved-leaved Elm** | Leaves fairly small; neat longish oval; very smooth with fine regular teeth in upper half. Barely asymmetrical; leaf-stalk quite long. Roadsides in north Herts, Cambs and Essex.

46 [LC] **Pale-leaved Elm** | Leaves broad; spear-like with a pointed tip; pale yellow-green in spring, when trees look 'Silver Birch'-like. Distribution uncertain: but mainly in Norfolk (where common), Suffolk, Cambs with outliers in the east Midlands.

47 [NT] **Large-toothed Elm** | Leaves broad with prominent triangular teeth. Roadsides, hedges, copses and field margins, scattered records in North Essex, East Suffolk, East Norfolk and Huntingdonshire.

48 [EN] **Rhombic-leaved Elm** | Leaves medium-sized; rhombic to rounded with regular shallow teeth near the base and rather jagged teeth towards the tip. Slightly asymmetrical at most with a narrow bulge on the long side. Hedges, woodland, scattered in Essex and Herts, with a few records from Hunts and Northants.

49 [VU] **East Anglian Elm** | Leaves fairly small; neat, rhomboidal with prominent regular triangular teeth in the upper half. Not very asymmetrical. Widespread in Norfolk, Suffolk, Essex and Herts, occasional in Cambs.

50 [EN] **Leathery-leaved Elm** | Rather lacking in distinctive features. Leaves fairly symmetrical with little bulging at the base. A few scattered records on roadsides in Cambs and Huntingdonshire.

51 [VU] **Coritanian Elm** | Leaves broadly oval with blunt teeth; strongly asymmetrical with a large bulge on the longer side. By roads, tracks, footpaths and in hedges and copses, in East Anglia and the east Midlands and possibly further north.

52 [VU] **Round-leaved Elm** | Some trees have small leaves, others have larger. Leaves generally fairly broad, strongly asymmetrical with a few fairly large, ±irregular teeth. Roadsides, hedges and field margins: widespread in Norfolk and Suffolk, extending into east and west Midlands.

53 [EN] **Dwarf-leaved Elm** | Leaves reminiscent of Wild Apple (p. 180); very smooth with a moderately asymmetrical base and small, neat, fairly fine teeth. Scattered throughout Essex, where it is commonest in the south but extending into Cambs and Hertfordshire.

Anglo-Saxon Elm L 4–7 cm W 2–4 cm LR 1·5–2·0 AS 1–4 mm LS 4–8 mm LVC 11–15

Pale-leaved Elm L 5–8 cm W 3–5 cm LR 1·5–2·0 AS 5–15 mm LS 10–20 mm LVC 14–17

ULMACEAE | ELMS   Leaves *pp. 65, 67* | Flowers *p. 69* | Fruit *p. 75* | Twigs *p. 86*

GENETIC GROUP 7 'PLOT'S' ELMS [3 'spp.']

Features: LEAVES ± oval (L 7–16 cm; L 1·5–2·0 × W), teeth rounded and forward-pointing; SHORT SHOOTS rarely formed as they continue to grow through the season; FORM slender with drooping tip.

54 EN **Plot's Elm** | Another historically long-known elm; also referred to as *Ulmus plotii*. A distinctive narrow tree (H to 20 m) with pendulous branches that have the leading shoot floppy and one-sided, with the uppermost 5–6 m of the crown growing more on one side of the tree than the other, giving a slender floppy-topped look. Leaves appear to be smooth on both sides but are minutely hairy above and can be hairy below, with white hair tufts in the axils. Found in hedgerows with a distribution centred around the East Midlands.

55 NT **Midland Elm** | Very similar to Plot's Elm but, as a mature tree, with a round-topped conical crown and larger leaves (L 5–7 cm; W 3–5 cm). Roadsides, hedges and field margins, frequent in Leics and Rutland, occasional in Suffolk, outlying records from Gloucestershire and Edinburgh.

56 DD **Sowerby's Elm** | Leaves broadest just above the middle with teeth towards the tip being larger and more prominent. Distribution uncertain: thought to be widespread in eastern England and eastern Midlands.

Plot's Elm
L	4–6 cm	AS	4–10 mm
W	2–4 cm	LS	1–6 mm
LR	1·3–2·3	LVC	8–14

GENETIC GROUP 8 BASAL ELM SPECIES (illustrated by 'Tasburgh' Elm) [1 'sp.']

A group of significantly genetically distinct elms, pending the results of ongoing analysis. Most are included in other groups (indicated by *) but may be transferred into this group in the future.

57 **'Tasburgh' Elm** | A bit of an oddity; a few trees recently discovered in the village of Tasburgh, Norfolk, have a highly distinctive appearance on account of their thick waxy leaves. They also have large thick hairy buds (like a Wych Elm). The recent genetic work undertaken by Natural England seems to suggest this tree is very distinct from other elms.

'Tasburgh' Elm
NOTE *single leaf sample only*

L	11 cm	AS	8 mm
W	5·5 cm	LS	7·5 mm
LR	2	LVC	13

FAGACEAE | BEECH, OAKS AND CHESTNUTS

Introduction to Beech, oaks & chestnuts

Worldwide the Fagaceae family of trees and shrubs (which includes beeches, chestnuts and oaks) contains almost 1,000 species. The vast majority are deciduous, although there are evergreen species in the tropics. All have simple leaves; unisexual flowers on the same tree (♂s typically numerous in catkins and ♀s typically in small, few-flowered groups); and fruit which is a nut or nut-like fruit. Many have fruits which develop within a scaly or spiny case (*e.g.* Beech); others in cases (cupules) that may not fully enclose the seed (*e.g.* the acorns of oaks).

Collectively the Fagaceae family represents many of the main tree species in northern Europe forming much of the broadleaved woodland cover, with their seeds providing a significant supply of food for many birds and mammals. They are also an important part of the timber supply chain and are used for many items, from walls to floors, cupboards to chairs and for alcohol barrels.

British & Irish Fagaceae family members have quite different-looking flowers, leaves and fruit.

BEECH • OAKS • SWEET CHESTNUT

FAGACEAE | BEECH Leaves *p.65* | Flowers *p.69* | Fruit *p.74* | Twigs *p.83*

Beech   *Fagus sylvatica*

A tall tree (H to 40 m), often with a huge domed crown. It can be found almost anywhere in Britain & Ireland as it is widely planted and extensively used for hedging although, as a native, it is probably restricted to the south of England where it is the dominant tree in many woods on chalk and limestone. It has a preference for well drained soils, but will tolerate heavy clay. It appears to have arrived after the last glaciation, with remains having been found in Neolithic deposits in Essex, Hampshire and Norfolk. The distinctive 'Copper Beech' cultivar with deep purple leaves has been planted since the 17th century. **Features:** BARK distinctive; **smooth silver-grey**; commonly covered in lichens and algae. TWIGS dull purple-brown at first, greyer by their second year. BUDS **long, pointed** (L 11–25 mm); reddish brown; held away from the twig on short stalks. LEAVES oval (L 4–10 cm) with 5–9 pairs of veins (particularly prominent on the underside); vibrant lime green; **very hairy when fresh becoming darker and hairless with age; leaf-margin wavy, untoothed.** FLOWERS ♀ flowers in pairs surrounded by a 'cup'; ♂ flowers in rounded flower heads. FRUIT brown spiny husk, containing 2 roughly triangular edible seeds (known as beechnuts or beech mast) with a mild nut-like taste.

J F M A M J J A S O N D

> **Did you know?** Beech trees are long-lived and are regarded as 'ancient' when more than 225 years old. The oldest trees in Britain & Ireland are 350–400 years old, although these are mostly pollards.
> Beech trees are adapted to survive on relatively dry soils, often with little topsoil. Their widespread shallow roots are in complex patterns which can be readily seen if there is soil erosion around a tree. Bats (*e.g.* Barbastelle) may use these exposed root masses as a day roost.
> Shrivelled leaves may persist throughout winter (known as marcescence).
> The origins of Beech in Britain have been questioned, because Julius Caesar appeared to suggest that 'Fagus' did not occur in England. However, it is probable that he was referring to Sweet Chestnut (*p.216*), which may have been introduced by the Romans.
> The strong, tough wood has been used for diverse purposes including tool handles, kitchen utensils, children's toys, furniture framing and flooring.

BARK distinctive; smooth; silver-grey 'elephant skin' appearance

A typical Beech wood showing the tall clean stems of the species (*left*); the Meikleour Beech hedge south of Blairgowrie, Perthshire is a world record hedgerow (*right*).

FAGACEAE | BEECH

LEAVES wavy untoothed margin hairy when young; underside with prominent veins

BUDS long, pointed; reddish brown; **held away from the twigs**

♂ FLOWERS yellowish; in hanging rounded catkins;

♀ FLOWERS ± upright to straight; surrounded by bracts; usually in pairs

FRUIT brown husk, covered in spines; opening when ripe to reveal two nuts

NUT roughly triangular; edible

TWIGS dull purple-brown to grey; distinctive 'zig-zag' form

Similar species | Two southern beech species (*Nothofagus*: Nothofagaceae) can be encountered as planted trees. **Roble** *N. obliqua* (N/I), which has smaller oval leaves (L 2–5 cm) with 7–11 pairs of veins and a hairless underside; and **Rauli** *N. alpina* (N/I), which has larger leaves (L to 15 cm) with 15–18 pairs of veins.

Associated species | The dense shade cast by Beech woods makes for a restricted but specialised flora, including orchids such as the reasonably widespread **Bird's-nest Orchid**, the extremely rare, sporadic, **Ghost Orchid** and various **helleborines**. Some 94 invertebrate species have been found on Beech, including moths such as **Clay Triple-line**, **Olive Crescent**, **Barred Hooked-tip** and **Lobster**. Beech mast provides food for **mice**, **voles**, **squirrels** and **birds** and, in the past, pigs and boar. As Beech become ancient, they develop holes and cavities which can be used by hole-nesting birds like **Redstart** and **Pied Flycatcher**. Around their roots, many mycorrhizal fungi also develop, including truffle species such as the **Summer Truffle** and **Autumn (or Burgundy) Truffle**. Other fungi that can be found growing with Beech are two rare *Hericium* species with downward-pointing spines: **Coral Tooth** is extremely rare and composed of delicate coral-like structures; **Bearded Tooth** can grow to the size of a large white football and is one of only four non-lichen fungi that are illegal to pick.

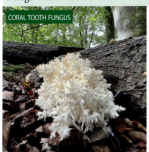

CORAL TOOTH FUNGUS

NARROW-LEAVED HELLEBORINE

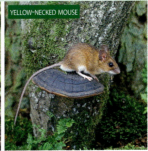

YELLOW-NECKED MOUSE

FAGACEAE | SWEET CHESTNUT — Leaves *p.67* | Flowers *p.69* | Fruit *p.74* | Twigs *p.87*

LC LC Sweet Chestnut — *Castanea sativa*

A tall tree (H to 30 m; trunk girth up to 6 m). It has long been thought that it was introduced by the Romans, but recent research has begun to cast doubt on this, with the first physical evidence being from 650 CE, and the earliest written records from 1113 CE. Sweet Chestnut is now widely established in Britain & Ireland, often actively managed as coppice woodland (especially in the south), but also found in historic gardens, deer parks, parklands, as well as in formal avenues and as high forest. It has a preference for deep, moist, sandy soils and drained clays and does not thrive in very wet or lime-rich soils. **Features:** BARK grey-brown; **fissuring deeply (often spirally)** with age. TWIGS olive to red-brown with raised oval to heart-shaped leaf scars and a rough texture caused by whitish resin granules. BUDS ± egg-shaped (L 5–10 mm); orange/red to purplish (often greenish at the base). LEAVES **large (L 10–25 cm); spear-shaped;** leaf-margins with obvious, regular teeth. FLOWERS in ♂ only or bisexual catkins (L 15–25 cm) with a **distinctive, unpleasant ammonia-like aroma.** ♀s small; softly spiky; green with yellow stigmas at the base of a catkin in clusters of 3; ♂s pale greyish yellow, in long catkins (L to 20 cm). FRUIT **sharply spiny case** containing 1–4 two-skinned nuts – an external, thicker shiny brown skin, and an internal thinner skin which adheres to the edible creamy-white nut.

J F M A M J J A S O N D

BARK often with deep fissures that 'spiral' around the trunk

> **Did you know?** Sweet Chestnut can live to be 500–600 years old and some may even exceed 1,000 years. The tree, and its edible nuts, have been cultivated since at least Roman times and its diverse genetics have led to different cultivars with a variety of nut characteristics, with cultivated trees grown in coppices and orchards. Chestnuts are rich in carbohydrates and are used for many culinary uses including baking, jams, puddings, sweets (candied as *marrons glacés*) and stuffings. In Corsica, a polenta (called *pulenta*) is made with Sweet Chestnut flour, and there are Corsican chestnut beers.
>
> Although its historical status in Britain & Ireland is still uncertain, Oliver Rackham, a 20th-century academic and writer, once described it as an 'honorary native', which seems the perfect description.
>
> Individual trees cannot self-pollinate, making cross-pollination necessary.

The unpleasant smelling flowers attract pollinating flies (*left*); developing nuts in their spiky seed cases (*right*).

FAGACEAE | SWEET CHESTNUT

FAGACEAE | OAKS   Leaves p.63 | Flowers p.69 | Fruit p.74 | Twigs p.87

LC LC Pedunculate (English) Oak   *Quercus robur*

A tall tree (H to 37 m; trunk girth to 12 m) with a broadly spreading crown (W to 20 m). Some of the largest oaks in Britain & Ireland are thought to be more than 1,000 years old. It tolerates a wide range of soils, although grows best on deep fertile clays and loams. Found throughout most of Britain & Ireland in woodlands, hedgerows and parkland from sea level to approximately 450 m elevation. **Features:** BARK silvery grey, when young becoming dark grey-brown and deeply ridged with vertical cracks as the tree ages. TWIGS brown with a reddish, greenish or greyish hue and 3–5 buds in a cluster at the tip. BUDS egg-shaped (L 3–8 mm); reddish brown; usually shorter, darker and with fewer (usually <20) bud scales than those of Sessile Oak. LEAVES thin; on short stalks (L < 10 mm); outline oval (L 8–12 cm; widest towards the tip) with 3–6 pairs of rounded lobes that are divided less than halfway to the midrib; **leaf-base with 'ear'-like auricles**; underside usually hairless, although some can have a few hairs on the main vein. FLOWERS ♂ flowers tiny, greenish yellow, in drooping clusters; ♀ flowers small, inconspicuous pinkish, in the leaf-axils. FRUIT acorn (L 15–30 mm) in a cupule; **stalk long** (L 20–90 mm); up to 4 acorns per stalk; green when young, ripening pale brown prior to falling from the cupule.

J F M A M J J A S O N D

> **Did you know?** It is estimated there are 170 million oak trees in woodlands and 2·3 million oaks outside of woodlands in Britain & Ireland (plus millions of additional seedlings and saplings). England is thought to have more ancient native oaks than all other European countries combined, with more than 65,000 ancient, veteran and notable oak trees having been recorded.
>
> Oak has always been the most widely used hardwood in Britain & Ireland with its durable timber traditionally used for buildings, ships, furniture, panelling and coffins.
>
> Raw acorns have a high concentration of tannic acid and are toxic to humans if consumed in large quantities. The acid can be removed by repeatedly soaking the acorns in water (until the water is no longer brown) after which they can be roasted, ground into flour or pulverised to extract their oil.

BARK dark grey-brown with ridged, vertical cracks; deeper than those on Sessile Oak

A young English Oak with the characteristic domed canopy (*left*); the acorn cupules on thin stalks and smaller, darker buds distinguish Pedunculate from Sessile Oak (*right*).

FAGACEAE | OAKS

PEDUNCULATE OAK

LEAVES smooth, lobed; LEAF-STALK L <10mm

LEAF-BASE with 'ear'-like auricles

ACORNS on long (L 20–90 mm) stalks from twig

TWIG red-brown to greenish or grey-brown

LEAF-BUD L 3–8 mm; reddish brown; usually hairless; BUD-SCALES <20

SESSILE OAK

LEAVES smooth, lobed; LEAF-STALK L >10mm

LEAF-BASE wedge-shaped; tapering

FLOWERS similar in both species – ♂ hang down in catkins; ♂ tiny in leaf axils (see p. 68)

TWIG red- to green-brown

LEAF-BUD L 5–15 mm; pale brown; usually hairy; BUD-SCALES >20

ACORNS stalks from twig absent or short (L < 20mm)

■ Hybrid (Rose) Oak *Q. ×rosacea* (*Quercus petraea × robur*)

Sessile and Pedunculate Oaks readily hybridize where both species are present. The resulting hybrid is known as a Rose Oak. They are tall trees, like the parents, and in some areas, they can form complete woodland stands. Rose Oaks can be fertile and be pollinated by either parent or by other Rose Oaks, leading to a complex range of mixed and intermediate characteristics in any resulting offspring.

For example a Rose Oak can have both acorns and leaves without stalks, or both acorns and leaves with stalks. This makes identification complicated, especially as the parents are themselves variable. However, some Rose Oaks are obvious and always worth looking out for.

This Rose Oak has a combination of a) long-stalked 'SESSILE' leaves and b) long-stalked 'PEDUNCULATE' acorns

LEAVES highly variable in shape; but can also look exactly like one of the parents

219

Associated species | PEDUNCULATE OAK can support 2,300+ other species comprising 1,178 invertebrates, 716 lichens, 229 bryophytes, 108 fungi, 38 bird species and 31 mammals (excluding fungi, bacteria and other microorganisms that can grow on oaks). Of these species, 320 are found exclusively on oaks – such as the **Purple Hairstreak** butterfly. A further 229 species are almost oak-dependent and are rarely found on non-oak species. Although this is more than any other British or Irish tree species, the number of additional species an individual oak can support depends on factors such as the tree's location, the quality of the surrounding habitat, adjacent land use and local pollution levels.

One important element of oak ecology is that **Jays** eat acorns, and also create winter caches by burying them in the shelter of trees or bushes, often under dead leaves. Jays have been recorded carrying acorns for up to 3 miles (5 km). Individual Jays often forget the locations of caches, and these forgotten acorns have the potential to sprout into new trees far from the parent.

Veteran and ancient oaks support a greater diversity of species compared to younger trees. Ancient oaks in wood pasture, hedges and parklands can support rare invertebrate species such as **Variable Chafer, Moccas Beetle** and **False Mocha Moth**, meaning even a single old oak tree can be a vital habitat in itself.

One of the most important factors for this increased biodiversity in older oaks is the dead wood component – 40% of the total list of species found in oak are associated with dead wood. For example, the decaying wood of old, rotting oaks supports more than 200 fungus species, including the legally protected **Oak Polypore**, protected under Schedule 8 of the Wildlife and Countryside Act, 1981.

Several woodland bat specialist species (*e.g.* **Bechstein's Bat**) also favour oaks for roosting over other tree species, which may be partly related to the high numbers of associated insects on oak trees.

Acorns can be used by *Andricus quercuscalicis*, a tiny gall wasp (), the larva of which twists and distorts an acorn into a **Knopper gall**. Other galls, such as oak apple, are created by other gall wasp species.

Most of the species which live on Pedunculate Oak also live on Sessile Oak. However, SESSILE OAK differs from Pedunculate Oak in having a more open canopy structure which provides nest sites for **Pied Flycatcher** and also favours the development of bryophyte and lichen communities, particularly in western Britain and Ireland. These Sessile Oak communities have been dubbed the 'Celtic oak rainforests' and are of international importance for their assemblages of bryophytes (mosses and liverworts), lichens and ferns. Oak trees in these woods can be covered in communities that include the common **Slender Mouse-tail Moss** (var. *myosuroides*) and liverworts such as **Western Earwort, Spotty Featherwort** and **Prickly Featherwort**, together with lichens such as **Tree Lungwort**, and ferns such as the small **Wilson's Filmy-fern**. The composition and type of species growing on these oaks change with the tree's age.

FAGACEAE | OAKS

Sessile Oak *Quercus petraea*

A tall tree (H to 37 m; trunk girth to 12 m), usually with a narrower crown (W to 18 m) and straighter trunk than Pedunculate Oak (*p. 218*). Like Pedunculate Oak, some of the largest trees may be more than 1,000 years old. It is widespread in Britain & Ireland in hedgerows and woodlands (and often a key component of upland woodlands), growing best in deep, well-drained clays and loams with a preference for areas of high rainfall.
Features: BARK greyish brown; smooth when young, developing cracks and fissures (shallower than those of Pedunculate Oak) with age. TWIGS reddish brown to greenish brown. BUDS egg-shaped (L 5–15 mm); pale brown, typically hairy; usually longer, paler and with more (usually >20) bud scales than those of Pedunculate Oak. LEAVES **on long stalks** (L 10–25 mm); outline oval (L 7–12 cm; widest near the middle) with 3–6 pairs of shallow rounded lobes divided less than halfway to the midrib; leaf-base wedge-shaped, without auricles; underside usually hairless, although can have hairs on the main vein and in the vein axils. FLOWERS ♂ flowers tiny, greenish yellow, in drooping clusters; ♀ flowers small, inconspicuous pinkish, in the leaf-axils. FRUIT acorn (L 20–30 mm) in a cupule; **stalk absent or very short** (L >20 mm); up to 4 acorns in a cluster; green when young, ripening dark brown prior to falling from the cupule.

J F M A M J J A S O N D

BARK greyish brown; fissures shallower than those on Sessile Oak

Identification – *p. 215*

LEAVES long-stalked, base wedge-shaped

ACORNS typically unstalked; very short-stalked at most

Coastal Sessile Oak woodland (*left*) is often stunted and covered in lichens; the acorn cupules that sit directly on the twig and larger, paler, hairy buds are quite noticeable once known (*right*).

FAGACEAE | OAKS Leaves *p.63* | Flowers *p.69* | Fruit *p.74* | Twigs *p.86*

Turkey Oak *Quercus cerris*

A tall tree (H to 30m) with a domed crown, native to southern Europe. It appears to have been brought to Britain & Ireland in the early 18th century, by a Devon nurseryman called Lucombe. The specific date is uncertain but by 1740 Turkey Oak was in cultivation, and by the late 18th century it had been planted in the south and west of England. It grows on light soils and is often planted in parks and gardens. The trees seed freely and, as a result, have become widely naturalized. **Features:** BARK dark grey with orange-tinted cracks creating ± rectangular blocks. TWIGS greyish brown to greenish brown covered in tiny star-shaped greyish hairs. BUDS egg-shaped (L 2–6mm) with **distinctive persistent, long, hair-like stipules** that can cover the bud. LEAVES long oval (L 6–14cm) with 7–8 pairs of irregular lobes divided ±halfway to the midrib; surfaces hairy, rough to the touch. FLOWERS as other oaks. FRUIT acorn (L 25–40mm) in a cupule with **distinctive curved scales** and a short, stout, hairy stalk.

J F M A M J J A S O N D

LEAVES rough; variable, unequal lobes divided ± halfway to midrib; STIPULES hair-like

FLOWERS ♂ catkins; ♀ tiny and inconspicuous (as other oaks)

LEAF-BUD with characteristic long, hair-like stipules

TWIG greenish brown with minute greyish hairs

ACORNS cupule with distinctive curved scales; on short hairy stalk

BARK dark grey, with cracks creating ± rectangular blocks; cracks usually orange tinted

Similar taxa | Lucombe Oak (*p. 225*)

Associated species | In a 14-year study 150 species of beetles were recorded from a single veteran Turkey Oak in Worcestershire. One of these was the **False Click Beetle** (*below*). *Andricus quercuscalicis* (a cynipid wasp) lays eggs in the acorns of Turkey, Pedunculate and Sessile Oaks, forming the distinctive **Knopper Gall** (see *p. 220*). These galls dramatically reduce the fertility of oaks, and therefore Turkey Oaks should only be planted very selectively to avoid the galls affecting the two native oaks.

FALSE CLICK BEETLE

FAGACEAE | OAKS

LC Evergreen (Holm) Oak *Quercus ilex*

A large (H to 28 m), non-native, evergreen tree native to the Mediterranean. The date of its introduction into Britain & Ireland is unknown but it was known to be growing in London by 1581. One tree was described by John Evelyn in 1666 as growing in '*his Majesty's privy garden at Whitehall*'. Widely planted in parks and large gardens; the largest wooded area in Britain & Ireland is at St Boniface Down on the Isle of Wight. It is regarded as an invasive species, as it can spread naturally by seed, and also hybridize with Pedunculate Oak – the hybrid known as Turner's Oak, which John Loudon reported in 1838 as found '*by Mr Spencer Turner, in the Holloway Down Nursery, Essex*' in 1787. **Features:** BARK brown and scaled. LEAVES variable; **from those with spiky margins (resembling Holly (*p. 302*)), to those with margins smooth and untoothed**; very fresh leaves have both surfaces with white or grey hairs which soon fall to reveal the dark green, glossy mature leaf. FLOWERS as other oaks. FRUIT acorn in **a distinctive, deep cup** with grey, downy scales.

J F M A M J J A S O N D

FRESH LEAVES greyish white hairy

ACORN distinctive deep cups with downy grey scales

YOUNGER LEAF

♂

LEAVES dark green; margin untoothed or 'holly-like'; upperside glossy; underside grey woolly

MATURE LEAF

BARK brown, scaled

Similar species | **Cork Oak** (*p. 225*) – thicker, corkier bark; leaves that are less spiny; and acorn cups with hair-like scales. Possibly **Holly** (*p. 302*) – smooth, grey bark; leaves usually stiffly spined.

Associated species | Various micro-moths feed on Evergreen Oak; adults are usually difficult to find but the larvae of some produce obvious leaf mines. First reported in 1996, but not identified until 2001, *Ectoedemia heringella* mines are distinctive and contorted. **Holm-oak Pigmy** mines are almost filled with dark frass.

HOLM-OAK PIGMY MOTH ADULT (left) AND LARVAL MINES (right)

FAGACEAE | INTRODUCED OAKS Leaves *p. 63* | Flowers *p. 69* | Fruit *p. 74* | Twigs *p. 86*

Other oaks

LC Red Oak *Quercus rubra*

A large (H to 28 m) deciduous tree, native to eastern North America. It has been planted in parks and garden since it was first grown by Philip Miller in England in 1739. It is now the most successful American oak in Britain & Ireland. **Features:** BARK silver-grey when young, developing distinctive ridges and grooves with age. LEAVES L 12–20 cm; 7–11 triangular lobes each with 1–3 bristle-pointed tips; lobes are less deeply cut towards the midrib than in most other American Oaks; upperside dark green; underside paler, even whitish, with hairs in the vein axils. FRUIT large acorn (L 15–30 mm); shiny chestnut brown when ripe.

LEAVES turn bright scarlet-red in autumn

J F M A M J J A S O N D

ACORN broad flat base; cupule thick; saucer-shaped

LC Scarlet Oak *Quercus coccinea*

A large (H to 30 m) deciduous tree native to the eastern United States. It was introduced to Britain & Ireland as early as 1691, when it was said to be growing in the garden of Bishop Compton in Fulham. **Features:** BARK brown to dark grey with fine ridges and furrows. LEAVES L 7–16 cm; 5–7 lobes each with 3–7 bristle-tipped teeth; lobes have a deep C-shaped gap between them, extending to near the midrib; upperside shiny green; underside paler and slightly shiny, with hairs in the vein axils. FRUIT large acorn (L 15–20 mm) with rings near the tip; both acorn and cup with fine hairs.

LEAVES turn red in autumn

J F M A M J J A S O N D

ACORN sits in a deep cupule which covers 33–50% of the acorn

LC Pin Oak *Quercus palustris*

A large (H to 25 m) deciduous tree native to the eastern United States. It was introduced c. 1800 by Messrs. Fraser. One of the distinctive features of this species is the drooping habit of the branches. **Features:** BARK grey when young, developing shallow furrows and ridges with age. LEAVES L 5–16 cm; 5–9 lobes each with several bristle points; lobes have a wide, U-shaped gap between them, extending at least halfway to the midrib; yellowish green; underside hairless except for brown tufts of hair in the vein axils. FRUIT small acorn (L to 16 mm); greenish brown and usually obviously striped when ripe.

LEAVES turn scarlet in autumn

J F M A M J J A S O N D

ACORN hemispherical; cupule thin, saucer-shaped

224

FAGACEAE | INTRODUCED + HYBRID OAKS

Cork Oak *Quercus suber*   LC

Medium-sized (H to 15 m) evergreen oak, native to the south-west of Europe and the north-west of Africa. Said to have been introduced to Britain & Ireland in 1699 by the Duchess of Beaufort, although little evidence exists for this; however, they were relatively common by the 18th century. **Features: BARK thick, corky**; grey to red-brown with deep ridges of cork in older trees. **LEAVES** oval (L 3–8 cm); **leathery; shiny dark green**; margin untoothed or with up to 5 pairs of small sharp-pointed teeth. **FRUIT** acorns; in clusters of 2–8; upper scales of cupule grey and hairy.

Similar species | Evergreen Oak (*p. 223*) has thinner corky bark and leaves that typically have, if present, larger points.

J F M A M J J A S O N D

LEAVES leathery; shiny; with or without small teeth

ACORN cupule upper scales grey and hairy

BARK distinctive; thick and corky

Lucombe Oak *Quercus ×crenata*

A large semi-evergreen oak (H to 35 m). A natural hybrid between Cork Oak (*above*) and Turkey Oak (*p. 222*) also known as Spanish Oak. Trees are variable and exhibit both mixed and intermediate characteristics of the parents. The cross that led to the Lucombe Oak occurred naturally at Lucombe's Exeter nursery *c.* 1762 and was first described in a letter written on February 24, 1772. To be a 'true' Lucombe Oak, a tree must be a grafted clone of the original hybrid. Acorns propagated from the original grafted tree in 1792 resulted in trees with a wide range of characteristics. Three of these were grafted and can be distinguished by their much corkier bark compared to the original tree. Now planted widely, particularly in the south-west, the trees grow well in fertile loamy soils and can tolerate shade.

The original tree was felled by Mr Lucombe to make his own coffin. He stored the timber under his bed until it was needed, but then lived much longer than he had thought (102 years old). So, he had another larger tree cut down and stored these new boards under his bed until they were eventually needed.

Features: The original Lucombe Oak hybrid is similar in appearance to Turkey Oak and differs as follows: **LEAVES** smaller (L 10–12 cm), **sub-evergreen** (falling early in their second year), with lobes cut less than halfway to the midrib; **BUD** stipules which are shorter; **FRUIT** acorn cup smaller (L 20–25 mm) and with shorter scales.

COMPARED TO TURKEY OAK:
STIPULES shorter

J F M A M J J A S O N D

LEAVES lobes less deeply cut

ACORN cupule with shorter scales than in Turkey Oak

225

SALICACEAE (WILLOWS + POPLARS) | INTRODUCTION   Leaves *pp. 65–67, 226* | Twigs *pp. 79, 81, 82*

Introduction to willows & poplars

WILLOWS
♂ anthers yellow to purplish red
♀ greyish green to reddish purple
IN FRUIT
FLOWERS in pendent to erect catkins

POPLARS
♀ stigmas yellowish green to pinkish red
♂ anthers purplish red
IN FRUIT
FLOWERS in long, pendent catkins

The Salicaceae is a family of trees and plants (approximately 1,200 species worldwide) which includes approximately 350 species of willows (also called sallows (broad-leaved willows)) and osiers (narrow-leaved willows) and 25–30 species of poplars, aspens, and cottonwoods. Some willows which grow in Arctic and alpine regions *e.g.* Polar Willow *Salix polaris* (N/I) and Net-leaved Willow (*p. 255*), are low-growing creeping shrubs whilst others in lowland habitats can develop into large trees when mature. All poplars are trees (H 15–50 m when mature). Willows have a rich watery sap containing salicylic acid that, as a compound, is used as a wart-treatment, food preservative and antiseptic as well as base material in aspirin production.

Collectively, willows and poplars have simple leaves in a range of shapes ranging from long and narrow to oval, triangular or lobed – usually with toothed or undulating margins; temperate species are typically deciduous. Flowers are tiny and inconspicuous (lacking petals); ♂s and ♀s in unisexual catkins on separate trees. Fruit is a capsule containing seeds with a long plume of hairs arising from the base. Botanically, willows have untoothed bracts and winter buds with 1 scale and poplars have toothed bracts and winter buds, but in Britain & Ireland poplars are easily distinguished by their longer, thinner pendent catkins and triangular or lobed leaf shapes.

Crack-willow, an example of a 'tree' (rather than shrubby) willow (*left*) and the oddly distorted, leaning form of a native Black Poplar (*right*).

Identification of willows & poplars

Within Britain & Ireland both groups are difficult to identify for a number of reasons, primarily:
- their inherent similarity to one another within their species group
- their ability to hybridize freely
- their ability to reproduce identical clones vegetatively.

In practice, it may be necessary to visit an individual willow or poplar a number of times in a season to check buds, flowers and leaves. Even then, some may not be easy to identify without the help of an expert.

Poplar identification – see p. 229

Although the Aspen–Grey Poplar–White Poplar group, the balsam-poplars and the black poplars are easily distinguished as groups, within each of these groups there are identification challenges:
- As Grey Poplar is a hybrid between Aspen and White Poplar, a full range of both intermediate and mixed features can be encountered, making some individual trees hard to identify with confidence.
- Both Black Poplar and Hybrid Black Poplar are represented by a number of cultivars. In these the shape of the crown and the sex of the tree can be important in identification.
- Balsam-poplars are generally similar to each other and details of the tree shape, suckering, sex, leaf-shape on main branches (not suckers) and leaf-stalk hairs are important.

Poplar identification is best attempted between late April and June, when a tree's sex and shape are most apparent, and leaves are mature enough to show the features required.

Willow identification – see pp. 228, 230

Generally, it is not possible to identify willow from the catkins alone. In most cases catkins and both young and mature leaves (from main branches and not suckers or regenerative growth) are the optimal combination. For the best chance of obtaining critical identification features, mature leaves should be examined July–September when still green. Twigs and buds need to be 1–2 year-olds, and not the newest growth. The key features for willow identification are:

- the height of the tree – willow species grow in dwarf forms, short shrubs and tall trees
- the length and shape of the leaves
- whether the leaf-margin is straight or wavy and, if present, the shape and size of the teeth
- the colour of hairs, if present, on the leaves and twigs
- the shape, colour and density of the buds
- the presence/absence of leaf stipules, and whether they are present in late summer.

Striae | Striae are ridges on the surface of the wood of 2nd-year twigs that can be helpful in the identification of some willows, and their hybrids. The optimal 2nd-year twigs are 10–15 mm in diameter and the striae can be revealed by carefully peeling off the outer bark.

Hybridization in willows | Willows hybridize freely, with over 70 combinations recorded, over 20 of which are formed from 3 species. Hybrids can be more common than either parent where both parents occur; or can be found in areas where neither parent is present. Hybrids can be difficult to identify due to inheriting mixed and/or intermediate features from the parents. For pragmatic purposes this book presents identification features that define 'good' individuals of a species and covers just four commonly occurring hybrids. An individual tree that does not fully match the description of a 'good' species (e.g. with a range of differently shaped leaves on the same tree, or having a presence/absence of hairs that is different from a species description) is highly likely to be a hybrid of some sort and will need reference to specialist identification information.

WILLOW + POPLAR LEAF IDENTIFICATION | ID

BALSAM-POPLARS | buds with balsam aroma; ± sticky

leaves ± triangular | underside paler than upperside ▶ p. 260

Other than Balm-of-Gilead, **Balsam-poplars** can be difficult to separate. A combination of leaf-shape, growth form and the extent of suckering (if present) is needed for confident identification.

UNDERSIDE paler than in black poplars

LEAF-MARGIN WITH HOOKED TEETH — LEAF-STALK with long hairs — **Balm-of-Gilead**

LEAF-MARGIN WITH SHORT ROUNDED TEETH
- LEAF-STALK with short hairs — **Eastern Balsam-poplar**
- LEAF-STALK a few long hairs mixed with short — **Hybrid Balsam-poplar**
- the narrowest leaf of the group — **Western Balsam-poplar**

BLACK POPLARS | buds lacking aroma; barely sticky at most

leaves ± triangular to diamond-shaped | leaf underside ± same colour as upperside

Both **Black Poplar** and **Hybrid Black Poplar** have many varieties with variable leaf-shapes. The two taxa are best separated by the features indicated below; differentiating the various varieties is based largely on growth form.

UNDERSIDE ± same colour as upperside

LEAF-BASE GLANDS PRESENT — **Hybrid Black Poplar** p. 262

LEAF-BASE GLANDS ABSENT — spiral galls in leaf-stalk — **Black Poplar** p. 261

leaves oval, lobed or palmately lobed | underside slightly or much paler than upperside

Aspen p. 256
UNDERSIDE slightly paler than upperside

LEAF FROM SHORT SHOOT — **Grey Poplar** p. 258 — UNDERSIDE typically grey hairy

LEAF FROM LONG SHOOT — **White Poplar** p. 260 — UNDERSIDE typically thickly white hairy

229

ID WILLOW + POPLAR TWIG IDENTIFICATION

Willow winter twigs (from *p. 81*)

BUDS appear to be covered by a single scale

Identifying willow twigs in winter | Although willow twigs, as a group, are easy to recognize by their two fused modified leaves which give the appearance of a single bud scale, within the group it is not so straightforward. The best approach to identification is as follows:
- whether the plant is a tree, tall shrub or very low-growing shrub
- whether the 1st-year twigs are hairy or ± hairless and the colour of the twig
- the spacing and colour of the buds and whether they are held away from, or pressed to, the twig
- whether the buds noticeably differ in size, or are all ± the same.

NOTE the key allows for those instances where plants may fall into more than one category (*e.g.* species in which young/1st-year twigs may be hairy or hairless, especially near the tip).

LOW-GROWING SHRUBS (H < 1·5 m; rarely more) – mostly in montane habitats

1ST-YEAR TWIGS typically very hairy, at least near tip and/or BUDS hairy

Eared Willow *p. 236*   HEIGHT to 5 m; but often found as a low sprawling bush

LEAF-BUD L 1·5–2·5 mm
FLOWER-BUD L 4–7 mm

TWIGS typically dark red; very rarely hairless; **BUDS** ± globular; different sizes; ± same colour as twig

Creeping Willow *p. 248*

BUDS all L 2–4 mm

TWIGS yellowish brown to reddish brown; hairy at least near tip; **BUDS** ± pointed egg-shape; ± the same size; yellowish green to red-brown

1ST-YEAR TWIGS minutely or sparsely hairy and/or BUDS hairy

Woolly Willow *p. 251*

FLOWER-BUD broadly egg-shaped; L 6–10 mm; **towards tip of twig**
LEAF-BUD egg-shaped; L ± 5 mm; **towards base of twig**

TWIGS yellowish brown to red-brown; quite hairy; **BUDS** orange-brown dark red-brown; **held away from twig; different shapes and sizes**

Downy Willow *p. 250*

FLOWER-BUD (N/I) egg-shaped; L 6–10 mm; mostly **towards base of twig**
LEAF-BUD narrow; L ± 5 mm; mostly **towards tip of twig**

TWIGS yellowish brown to dark red-brown; sparsely hairy; **BUDS** dark brown; ± pressed to twig; different shapes and sizes; grey hairy

Mountain Willow *p. 252*

BUDS all L 2–4 mm

TWIGS glossy reddish brown; ± hairless; **BUDS** ± pointed egg-shape; sparsely hairy; ± the same size; ± same colour as twig; held away from twig

1ST-YEAR TWIGS + BUDS hairy when young becoming ± hairless; FORM typically < 40 cm

Dwarf Willow *p. 254*	**Net-leaved Willow** *p. 255*	**Whortle-leaved Willow** *p. 253*

TWIGS greenish brown to dark red-brown; **BUDS** all ± narrowly egg-shaped (L to 2 mm); yellow-green; a few sparse hairs at most	**TWIGS** shiny; greenish brown to dark red-brown; **BUDS** all ± egg-shaped (L to 2 mm); shiny; yellowish brown to reddish brown; hairy when young	**TWIGS** shiny; greenish brown to dark red-brown; **BUDS** all ± egg-shaped (L 3–8 mm); shiny; yellowish brown to reddish brown

WILLOW + POPLAR TWIG IDENTIFICATION **ID**

TREES OR TALL SHRUBS (H > 1·5 m; usually over 3 m) — BUDS obviously differently sized

Goat Willow *p. 234* — LEAF-BUD L 3–5 mm
LF-BUD
HAIRLESS / hairy — FLOWER-BUD L 8–12 mm
pointed egg-shape
TWIGS typically green to red-brown; BUDS broad/narrow pointed egg-shape; yellowish green to red-brown

Grey Willow *p. 232* — FLOWER-BUD L 5–10 mm
LF-BUD
DENSELY HAIRY / hairless — LEAF-BUD L 3–5 mm
egg-shaped
TWIGS dull brown-grey to red-brown; BUDS ± broadly egg-shaped; orange to red-brown

Eared Willow *p. 236* — LEAF-BUD L 1·5–2·5 mm
LF-BUD
HAIRY / hairless — FLOWER-BUD L 4–7 mm
± globular
TWIGS typically dark red; usually hairy; BUDS ± globular; yellowish green to red-brown

Tea-leaved Willow *p. 246* — FLOWER-BUD L 7–13 mm
LF-BUD
HAIRLESS / hairy — LEAF-BUD L 5–9 mm
egg-shaped
TWIGS reddish brown; BUDS ± broadly (flower) or narrowly (leaf) egg-shaped; yellow-grey or ± as twig

Purple Willow *p. 238* — BUDS at least some arranged ± oppositely
HAIRLESS — FLOWER-BUD L 10–15 mm — LEAF-BUD L 4–7 mm
± cylindrical
TWIGS greenish to purple-red; BUDS ± cylindrical, usually with pointed tip; yellowish, red-brown to black

TREES OR TALL SHRUBS (H > 1·5 m; usually over 3 m) — BUDS ± the same size

Osier *p. 240* — BUDS overlapping/very small gaps
DENSELY HAIRY — BUDS L 5–9 mm
narrow egg; flat
TWIGS greenish to olive-brown; BUDS narrow egg-shape, tip can be pointed; pale yellowish grey; pressed to twigs

Olive Willow *p. 244*
DENSELY HAIRY — BUDS L 4–6 mm
± cylindrical; flat
TWIGS reddish above, yellow-green below; BUDS flattened cylinder, tip rounded; pale yellow-green with reddish tip

White Willow *p. 241* — BUDS not overlapping
DENSELY HAIRY — BUDS L 5–8 mm
narrow egg; flat
TWIGS olive-green to reddish brown; BUDS narrow egg-shape; pale yellow to reddish brown; pressed to twigs

Crack-willow *p. 242* — BUDS ♂ L 8–12 mm; > twig width; ♀ L ± 6 mm; < twig width
HAIRLESS / HAIRY
± narrow egg; keeled
TWIGS shiny, yellowish brown to grey-green; usually **breaks at base when bent**; BUDS ridged; usually yellowish or as twig

Bay Willow *p. 239*
HAIRLESS — BUDS L 4–8 mm
± pointed egg; keeled
TWIGS shiny; yellow-green to dark red-brown; BUDS shiny; green- to red-brown; pressed to or slightly away from stem

Almond Willow *p. 243* — TWIG aroma / taste of rosewater
HAIRLESS — BUDS L 4–8 mm
± narrow egg; keeled
TWIGS olive- to red-brown; **ridged**; BUDS narrow egg-shape, tip somewhat pointed; greenish brown

Dark-leaved Willow *p. 247*
VELVETY / HAIRLESS — BUDS L 4–7 mm
narrow egg; flat
TWIGS yellowish brown to reddish brown; BUDS narrow egg-shape; colour usually ± as twig

231

SALICACEAE | ROUNDER-LEAVED WILLOWS Leaves *pp. 228, 65–67* | Flowers *p. 68* | Twigs *pp. 230, 81*

LC LC Grey Willow (Sallow) *Salix cinerea*

A many-branched shrub or small tree (H typically 6–10 m; a few to 15 m) that is one of the commonest native British willows, particularly at lower elevations (below 400 m) and occurs as two subspecies (see box *below*). **Features of Rusty Sallow ssp.** *oleifolia*: BARK dull grey-brown; becoming cracked and fissured with age. TWIGS usually dull grey to reddish brown; **densely hairy, especially when young** (NOTE hairs can wear off) and particularly at the junction between a branch and a young twig; 2nd-year twig **with visible raised ridges** (see *striae p. 227*) on wood if bark is peeled back – **usually apparent as raised ridges near the tip of the twig**. BUDS pointed egg-shape; leaf-buds L 3–5 mm; flower buds larger (L 5–10 mm; smaller than Goat Willow's (*p. 234*)). LEAVES broadly oval (L 6–9 cm; W 1–3 cm; 2–4× as long as broad); 10–15 pairs of veins; **leaf-margin ± straight, not wavy**; underside with grey hairs mixed with some stiff rusty/reddish hairs; stipules not persistent. FLOWERS ♂ and ♀ catkins on separate trees; appearing before the leaves; egg-shaped (L 2–3 cm); mature ♂ catkins yellow (as the pollen develops); ♀ catkins greenish grey. FRUIT small seeds with long cottony white hairs, similar to all willows and poplars.

J F M A M J J A S O N D

Grey Willow subspecies

ssp. *oleifolia* (**Rusty Sallow**) is common, found throughout Britain & Ireland. Larger, on average, than Grey Sallow (H to 15 m). LEAVES underside with grey, and at least **some red-brown**, hairs; STIPULES small; BARK **strongly furrowed**.

ssp. *cinerea* (**Grey Sallow**) – is uncommon, occurring mainly in base-rich fens and marshes with a predominantly eastern distribution in East Anglia, Lincolnshire, and the East Midlands but also scattered throughout. Smaller, on average, than Rusty Sallow (H to 6 m). LEAVES underside with **dense pale yellowish grey hairs**; STIPULES large, persistent; BARK smooth.

BARK furrowed

BARK smooth

'Fluffy' seeds of Grey Willow

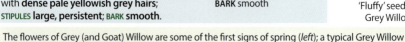
The flowers of Grey (and Goat) Willow are some of the first signs of spring (*left*); a typical Grey Willow growing in a heathland pool (*right*).

Striae explained *p.227* SALICACEAE | LARGER, ROUNDER-LEAVED WILLOWS

IN LEAF

LEAVES oval to broadly oval or pointed-oval; L 2–4× W; 10–15 pairs of veins (*cf.* Eared Willow); **LEAF-MARGIN not wavy**

LEAF-MARGIN not wavy

RUSTY SALLOW

LEAF underside with a mix of rusty and grey hairs

GREY SALLOW

LEAF underside with grey hairs only

FLOWERS
♂ catkins vibrant yellow;
♀ catkins greyish green

TWIGS dull grey- to reddish brown (rarely green); **usually densely hairy**

BUDS oval to pointed-oval; usually brown, red or dark orange; tip angled slightly away from the twig; smaller than those of Goat Willow

2ND-YEAR TWIG longitudinal ridges under bark; absent in Goat Willow

Associated species | There are a wide range of other species which depend on Grey Willow, with 160 species of **epiphytic lichens** having been recorded on Grey and Goat Willow, making the species pair important for this group. They are also very important for providing pollen and nectar to spring insects, particularly bees, and for their leaves, which are fed upon by a wide range of insects.

One of the more important species associated with Grey Willow is the **Willow Tit**, which has the dubious honour of being Britain's fastest declining resident bird species; in 2020 it was estimated that the population had declined by 88–94% since the 1970s. Hole-nesting Willow Tits excavate their own nest chamber but are not as powerful as other excavating birds (such as woodpeckers). Therefore, they need softer woods susceptible to rot and prefer standing rotting deadwood. Willow Tit nests have been found in a range of tree species, but the most widely used are Silver Birch, Elder and Grey Willow. Mostly ♀ birds create new holes every year in April and May, as a pair will use a new cavity for each brood, seeking out short, standing, rotting tree stumps – with most nests located within 1·5m of the ground.

WILLOW TIT

CLIFF MINING BEE

SALLOW KITTEN

SALICACEAE | ROUNDER-LEAVED WILLOWS   Leaves pp. 228, 65–67 | Flowers p. 68 | Twigs pp. 230, 81

LC LC Goat Willow (Sallow)   *Salix caprea*

A small tree or large shrub (H to 10 m, rarely to 15 m), the familiar 'Pussy Willow' is common throughout Britain & Ireland. It tolerates drier conditions than many other willows, so can be found in hedgerows, scrubby areas and woodlands as well as in damper areas and adjacent to water.
Features of Goat Sallow ssp. *caprea*: BARK pale grey; smooth; developing diamond-shaped and irregular cracks with age. TWIGS **upperside green to red-brown**; underside generally greenish; **hairless** or with dense tiny hairs; 2nd-year twig **smooth, lacking any raised ridges** (see *striae p. 227*) on the wood if bark is peeled back. BUDS pointed egg-shape; yellowish green to orange or red-brown; leaf-buds L 3–5 mm; flower buds (L 6–12 mm); **larger than Grey Willow's** (*p. 232*). LEAVES **broadly oval** (L 5–12 cm; W 3–8 cm; 1–2× as long as broad); leaf-margin slightly wavy; underside with greyish white hairs. FLOWERS ♂ and ♀ catkins on separate trees; appearing before the leaves; egg-shaped (L 2–3 cm); mature ♂ anthers yellow; ♀ catkins greenish grey. FRUIT small seeds with long cottony white hairs, similar to all willows and poplars.

J F M A M J J A S O N D

Goat Willow subspecies

ssp. *caprea* – TWIGS typically hairless; sparsely hairy at most (when young); LEAVES average larger (L 5–12 cm; W 3–8 cm); upperside with sparse hairs at most; underside densely hairy; margin wavy with irregular shallow teeth.

ssp. *sphacelata* [N/I] – TWIGS hairy when young; LEAVES average smaller (L 3–7 cm; W 1·5–4·5 cm); upperside typically softly hairy even with age; underside dense hairs pressed to the surface; margin wavy; untoothed or with very small teeth at most.
NOTE *northern distribution but almost certainly under-recorded.*

BARK smooth when young, developing irregular cracks

Did you know? Goat Willow is prone to hybridization with other willows, and known crosses have been found – see *p. 245*.
Historically, the male catkins were frequently used to decorate churches on Palm Sunday.

The vibrant yellow of male Goat Willow catkins is one of the first signs of spring (*left*); when they ripen they produce the 'fluffy' seeds characteristic of all willows (*right*).

Striae explained *p.227* | SALICACEAE | LARGER, ROUNDER-LEAVED WILLOWS

IN LEAF

LEAVES oval to broadly oval or pointed-oval; L 1–2× W – the **broadest of the native willows**

SSP. CAPREA
UPPERSIDE typically hairless; LEAF-MARGIN **slightly wavy**; shallow-teeth

SSP. SPHACELATA
UPPERSIDE typically **hairy**; LEAF-MARGIN small teeth at most

SSP. CAPREA

LEAF underside with soft, greyish downy hairs

FLOWERS ♂ catkins vibrant yellow; ♀ catkins greyish green

IN FRUIT prior to seed release

LEAF-BUD

FLOWER-BUD

TWIGS typically appear hairless; upperside usually red-brown to green; underside greenish

BUDS pointed-oval; variably yellow, green, orange, red or brown; tip angled slightly away from the twig

2ND-YEAR TWIG smooth under bark (*cf.* Grey and Eared Willows)

Associated species | Goat Willow has a mycorrhizal partnership with various fungi, including **Willow Milkcap** and **Willow Brittlegill**. Other fungi found using the tree's deadwood include **Blushing Bracket** and **Willow Bracket**.

Willows host a large number of insects (450), second only to oaks. Many species of moths use the tree, including **Puss Moth**, **Sallow Kitten** and **Poplar Hawkmoth**. Also 24 species of aphids have been recorded on Goat Willow, including common species such as the **Pale Sallow Leaf Aphid**.

WILLOW BRITTLEGILL

PUSS MOTH CATERPILLAR

PURPLE EMPEROR CATERPILLAR

235

SALICACEAE | ROUNDER-LEAVED WILLOWS — Leaves *pp. 228, 65–67* | Flowers *p. 68* | Twigs *pp. 230, 81*

LC LC **Eared Willow** (Sallow) — *Salix aurita*

Deciduous shrub (H up to 5 m), but usually growing as a sprawling, rounded bush. Eared Willow is a colonising species, found in a range of habitats – from wet meadows, willow-carr and wet woodland edges to riversides, stream edges, hedges and dunes in the lowlands to heathland, moorland and rocky hillsides in the uplands – and can be often be the most abundant willow in montane scrub towards the upper part of the treeline. At higher elevations, the leaves may be notably small.

Features: BARK dull grey; smooth; developing vertical cracks with age. TWIGS usually **dark red; hairy**, especially when young, rarely hairless; 2nd-year twig with visible raised ridges (see *striae p. 227*) on wood if bark is peeled back – usually apparent as raised ridges near the tip of the twig. BUDS ± globular to broadly pointed egg-shape; reddish brown to purplish brown; leaf-buds L 1·5–2·5 mm; flower buds larger (L 4–7 mm); **usually obviously smaller than those of either Grey** (*p. 232*) **or Goat Willows** (*p. 234*)). LEAVES broadly to narrowly oval (L 3–8 cm; W 1·5–3·5 cm; 1·5–2·5× as long as broad); usually dull green to greyish green; **leaf-margin usually obviously wavy** (*cf.* Grey and Eared Willows); underside with greyish white hairs; usually **7–9 pairs of veins deeply impressed on the upperside and raised on the underside** (10–15 pairs in Grey Willow); stipules conspicuous, kidney-shaped with a toothed margin, persistent. FLOWERS ♂ and ♀ catkins on separate trees; appearing before the leaves; egg-shaped (L 1–2 cm, smaller than those of Grey and Goat Willows); mature ♂ anthers yellow; ♀ catkins greenish grey. FRUIT small seeds with long cottony white hairs, similar to all willows and poplars.

J F M A M J J A S O N D

BARK dark grey; vertical cracks with age

Did you know? Eared Willow gets its name from the conspicuous, persistent 'ear'-shaped stipules growing at the base of the leaf-stalk. This species, together with Goat and Grey Willows are often known as sallows, derived from the Old English for 'shrubby willow' *salh* or *salch*.

In the lowlands, Eared Willow is often found as a tall shrub (*left*); in the uplands, it is more typically low-growing and bush-like (*right*), and is often the dominant willow species.

Striae explained *p. 227* SALICACEAE | LARGER, ROUNDER-LEAVED WILLOWS

IN LEAF

LEAVES dull green; tip can be twisted; 7–9 pairs of strong deep veins (*cf.* Grey Willow); underside with greyish hairs

UPPER

LEAF-MARGIN wavy

FLOWERS ♀ catkins greyish green and larger than the yellow-anthered ♂ catkins (N/I)

UNDER

STIPULES conspicuous; kidney-shaped; persistent

FLOWER-BUD

LEAF underside with **prominent veins** and **soft, greyish hairs**

TWIGS typically hairy, especially near the tip; **usually dark red**

BUDS globular to oval; reddish; tips close to the twig

2ND-YEAR TWIG longitudinal ridges under bark

Associated species | Beetle and sawfly larvae feed on Eared Willow, as do moth caterpillars, including those of the **Ruddy Highflyer** and the **Cousin German** – a rare species listed in the UK Biodiversity Action Plan.

Pustules on the upper leaf surface are caused by *Aculus laevis* (a mite), and galls have been found of two midges, *Iteomyia capreae* and *I. major*. Another midge, *Rabdophaga cinerearum*, causes the formation of a tiny rosette of leaves at the tip of the twig.

Adults of the **10-spotted Pot Beetle**, a Priority Species, feed on the leaves and the larvae feed on fallen leaves. This beetle has been found in Staffordshire, Cheshire and Rannoch, and Braemar, Scotland.

It supports a range of other species, including *Rabdophaga salicis*, a gall midge which creates galls in the leaf-stalks and veins, and *Melampsora capraearum* (a rust fungus) which causes leaf spots.

Mosses, including the **Common Tamarisk-moss**, and **lichens** grow on the trunk and branches. Common lichens recorded on Eared Willow include *Platismatia glauca* and *Hypogymnia physodes*, and *Pseudocyphellaria crocata*, a rare species – has been found in Scotland.

Eared willow is also an important food winter foodplant for the **European Beaver**, as it thrives in wet sites and coppices well, allowing the species to cope with beaver damage.

10-SPOTTED POT BEETLE

EUROPEAN BEAVER

237

SALICACEAE | NARROWER-LEAVED WILLOWS   Leaves pp. 228, 65–67 | Flowers p. 68 | Twigs pp. 230, 79

LC LC Purple Willow   *Salix purpurea*

An erect shrub (usually H to 3 m, but up to 5 m) widely distributed and found in marshy and boggy areas, and wet woodland. **Features:** BARK grey; smooth; cracking with age. TWIGS hairless; greenish to purple/red (most noticeable in winter); **wood and underside of bark a bright, characteristic yellow.**
BUDS usually of two sizes – leaf-buds (L 4–7 mm); flower-buds (L 10–15 mm); at least **some leaf-buds arranged oppositely, or nearly so**; long; ± flattened cylindrical (usually with pointed tip); yellowish brown to red-brown or purplish black; pressed to the twig. LEAVES hairless, bluish green; long tapering oval (L 2–8 cm), usually widest near the tip. FLOWERS ♂ and ♀ catkins on separate trees; appearing before the leaves; cylindrical (L to 4 cm); mature ♂ anthers dark reddish purple with fused filaments; ♀ greenish grey. FRUIT small seeds with long cottony white hairs, similar to all willows and poplars.

J F M A M J J A S O N D

At least some oppositely, or near-oppositely arranged buds and/or leaves are highly indicative of Purple Willow.

IN LEAF

LEAVES hairless; bluish green; long, tapering oval usually widest near tip

TWIGS hairless; greenish to purple-red

BUDS long; flat; pressed against stem

FLOWERS
♂ stamens with fused filaments and dark reddish purple anthers

BARK grey, cracking as the tree ages; **inside of bark and wood yellow**

Associated species | A range of insects and mites cause galls on the leaves of Purple Willow. *Euura viminalis* (a sawfly) produces a small pea-shaped gall that is warty and green (often with a hint of red), which hangs from the midrib of the leaf underside. Other rare and local sawflies recorded breeding on Purple Willow in Britain & Ireland include *Pristiphora luteipes*, *Arge enodis*, *Euura purpureae*, *E. salicispurpureae*, *E. weiffenbachiella*, *E. polita* and *E. vesicator* which produces a gall along the midrib of the leaf which projects equally above and below the leaf-blade, with sometimes 4 galls per leaf.

EUURA VIMINALIS GALL

SALICACEAE | NARROWER-LEAVED WILLOWS

Bay Willow — *Salix pentandra*
LC LC

Small tree or large shrub (typically H to 10 m, but up to 17 m) of wet ground, pond margins and river edges. Commonest in the north and west of Britain & Ireland and planted mostly as an ornamental in parks and gardens in the southern half. **Features:** BARK smooth and greyish brown but can be heavily cracked and flaky with age. TWIGS shiny greenish yellow to dark reddish brown; pith 5-angled in cross-section. BUDS pointed egg-shape (L 4–7 mm); greenish brown to reddish brown (similar to twig); pressed flat to, or slightly angled away from, the twig which **can be sticky** (slight **aroma of balsam** when fresh). LEAVES resemble those of Bay (*p. 331*); tapered oval (L 5–12 cm); **widest near tip**; leaf-margin with small fine teeth; leaf-stalk with small nectar glands at the base. FLOWERS ♂ and ♀ catkins on separate trees, appearing with the leaves; cylindrical; ♂ (L 2–5 cm), 5+ stamens per flower, mature anthers greenish yellow; ♀ greenish grey (L 1·5–3·0 cm). FRUIT small seeds with long cottony white hairs, similar to all willows and poplars.

J F M A M J J A S O N D

IN LEAF

LEAVES ± shiny; bluish green; long, **tapering oval widest near tip**;

FLOWERS
♂ greenish yellow;
5 or more stamens per flower

BARK greyish brown; cracking with age

BUDS narrowly oval; shiny; green- to reddish brown; **flat to or slightly angled away from stem**

TWIGS hairless, shiny; yellowish green to dark reddish brown

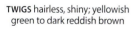

Associated species | Bay Willow contains high concentrations of salicylates which act as a chemical defence and seem to reduce the number of invertebrates using the species in comparison to other willows. However, one species that does feed on the leaves is **Black-spot Sallow Pigmy moth**, the caterpillars of which create distinctive leaf mines. The larvae of *Euura amerinae* (a sawfly) feed on the shoots, causing a large walnut-shaped gall to form on one side of the shoot.

SALICACEAE | NARROWER LEAVED WILLOWS Leaves *pp. 228, 65–67* | Flowers *p. 68* | Fruit *p. 74* | Twigs *pp. 230, 81*

LC LC Osier
Salix viminalis

A tree or large shrub (H to 7 m) that is an ancient introduction, grown for its wide range of uses. It is widespread, in wet ground and alongside rivers and streams. **Features:** BARK greyish brown; developing distinct vertical cracks with age. TWIGS greyish olive to reddish brown; usually softly hairy (especially near tip). BUDS narrowly egg-shaped (L 5–9 mm); pale yellowish grey; those near tip usually very softly hairy; tightly spaced; pressed to the twig. LEAVES **long, very narrow** (L to 18 cm; W 1·5–3·5 cm; 5–14× as long as broad); usually dark green; **underside with silky white hairs**; **leaf-margin downrolled**. FLOWERS ♂ and ♀ catkins on separate trees; appearing before the leaves; cylindrical; ♂ (L to 2 cm), mature anthers yellow; ♀ greenish grey (L to 3 cm); bract dark at apex; ovary hairy. FRUIT as other willows.

J F M A M J J A S O N D

IN LEAF

FLOWERS crowded on twig; ♂ vibrant yellow; ♀ greyish green

♂

♀

♀ FLOWER bract black-tipped; ovary hairy

BUDS closely spaced, even overlapping; pale yellowish grey; pressed to twig

LEAVES very long; margin downrolled; underside with silky white hairs

TWIGS usually greenish to olive-brown

BARK greyish brown, gaining distinct vertical cracks with age

RED-TIPPED CLEARWING

Similar species | The widely planted **Olive Willow** (*p. 244*) is superficially similar.

Did you know? Because of its very flexible twigs (called withies), Osier is a traditional part of the basket-making industry and has been widely used across Europe for thousands of years.
It is also widely used for sculptures and screens, and its rapid growth has made it a valuable part of the short-rotation coppice industry for the wood-fuel market. This range of uses has led to the development of a wide range of hybrids and cultivars, with more than 60 forms being grown.

Associated species | Many moths use Osier, including the **Lackey** and the striking but rarely seen **Red-tipped Clearwing** (*inset right*).

SALICACEAE | NARROWER-LEAVED WILLOWS

LC LC White Willow   *Salix alba*

Tree (H to 30 m), an archaeophyte in Britain but appears to have been introduced to Ireland much later. It has a predominantly lowland distribution (infrequent at elevations above 300 m), with a preference for sunny positions on wet soils, particularly riversides and lake edges. It tolerates coppicing and pollarding well and can be found as a range of cultivars (see box *below*). **Features:** BARK dark grey, cracking with age. TWIGS **olive green to reddish brown**; usually with some hairs at least towards the tip of the branch, although these can be lost with age; **usually flexible** (rarely brittle *cf.* Crack-willow). BUDS narrowly egg-shaped (L 5–8 mm); variable – reddish brown, olive-green to pale yellow; spaced (**not overlapping** *cf.* Osier); pressed to the twig. LEAVES **'spear'-shaped** (L to 10 cm; W 1–3 cm; 6–8× as long as broad); upperside hairy when young; **underside with downy white hairs**; leaf-margin ± flat, finely toothed. FLOWERS ♂ and ♀ catkins on separate trees; appearing with the leaves; cylindrical; ♂ (L to 5·5 cm), mature anthers yellow; ♀ greenish grey (L to 5 cm); bract pale at apex; ovary hairless. FRUIT as other willows.

J F M A M J J A S O N D

FLOWERS spaced out on twig;
♂ **vibrant yellow**; ♀ greyish green

BUDS **spaced,
not overlapping**;
variable from
pale yellow to
reddish brown;
pressed to twig

TWIGS variable; olive-green to reddish brown

♀ FLOWER bract
pale-tipped;
ovary hairless

LEAVES long; margin flat,
finely toothed; underside
with downy white hairs

BARK dark grey,
becoming cracked
with corky ridges

Similar species | Crack-willow (p. 242)

Did you know? Perhaps the best known of the many types of White Willow is 'Cricket Bat Willow' (var. *caerulea* (N/I)). Discovered in Norfolk *c.*1700, it is considered the perfect tree from which to make cricket bats. It is common over much of lowland Britain and is like a typical White Willow (var. *alba*), but usually with larger, broader leaves (L 5–11 cm; W 1·5–2·5 cm), which are densely hairy when young, becoming sparsely so with age. Another common variety is 'Golden Willow' (var. *vitellina*) which has bright orange-yellow young twigs that stand out, especially during winter. Leaf shape is similar to 'ordinary' White Willow (var. *alba*) but with fewer hairs and an upperside which becomes bright green (see p. 244).

SALICACEAE | NARROWER LEAVED WILLOWS **Leaves** pp. 228, 65–67 | **Flowers** p. 68 | **Fruit** p. 74 | **Twigs** pp. 230, 81

LC Crack-willow   *Salix × fragilis*

A medium-sized to tall tree (H to 30 m), many of which have a leaning crown. It occurs in wet places and is the (often pollarded) willow frequently seen on riverbanks. **Features:** BARK grey-brown, becoming coarsely fissured with age. TWIGS yellowish orange-brown to grey-green; shiny, sparsely hairy at most; brittle at the base (*cf.* White Willow). BUDS narrowly egg-shaped (L 6–12 mm); keeled; usually a very similar colour to the twig; pressed to the twig; ♂s usually curved to one side. LEAVES **'spear'-shaped** (L 10–18 cm; W 2–5 cm; 5–9× as long as broad); usually hairy during the early part of the season but upperside ± **hairless when mature**; underside may retain some hairs; leaf-margin leaf-margin ± flat, with irregular coarse teeth. FLOWERS ♂ and ♀ catkins on separate trees; appearing with the leaves; cylindrical; ♂ (L to 6 cm), mature anthers yellow; ♀ greenish grey (L to 8 cm). FRUIT as other willows.

J F M A M J J A S O N D

IN LEAF

LEAVES L 5–9× W; underside sparsely hairy at most

FLOWERS ♂ yellow; ♀ (N/I) greyish green

LF-MARGIN ± flat; **coarsely toothed; 6–10 teeth per cm**

BUDS usually a similar colour to the twig; ♂s wider than twig; ♀s smaller, narrower than twig

TWIG shiny, yellowish brown to grey green; usually **breaks at base when bent**

BARK grey-brown; becoming coarsely fissured with age

Similar species | White Willow (*p. 241*); Eastern Crack-willow (*p. 244*) leaves hairless; larger teeth.

Did you know? The origin of Crack-willow is somewhat uncertain, but it is thought that it is a hybrid between White Willow (*p. 241*) and Eastern Crack-willow (*p. 244*). Crack-willow *Salix fragilis* (a species) is now *Salix ×fragilis* (a hybrid).
Some forms of Hybrid Crack-willow have been propagated in cultivation, including:
• var. *russelliana* (the Bedford Willow) a ♀ form with long narrow leaves;
• var. *furcata*, a ♂ tree with some forked catkins and wide leaves;
• var. *fragilis*, with leaves at first slightly hairy;
• var. *decipiens*, with leaves that are entirely hairless.
These trees are coppiced/pollarded to produce poles used for fencing and basket-making.

SALICACEAE | NARROWER-LEAVED WILLOWS

Almond Willow *Salix triandra*

LC LC

A large shrub or small tree (H to 10 m), form usually leaning and sprawling. It is possibly native, most likely an archaeophyte, and occurs on the edges of lakes and rivers; spreading by seed and broken branches which float and then root. In the past it has been widely planted and cultivated for basket-making materials. **Features:** BARK distinctive; **dull greenish brown; readily flaking to reveal reddish/orange-brown patches**; inner bark with aroma/taste of rosewater. TWIGS olive- to red-brown; ridged and/or angled; hairless, but can be hairy when young. BUDS narrowly egg-shaped (L 5–8 mm) tip somewhat pointed; greenish brown. LEAVES **'spear'-shaped** (L to 10 cm; W ±2 cm); upperside dark green, ± shiny; underside pale green-grey; leaf-margin toothed; **stipules large, ± rounded, toothed.** FLOWERS ♂ and ♀ catkins on separate trees; appearing with, or just prior to, the leaves; cylindrical; ♂ (L 3–9 cm), 3 stamens, mature anthers pale yellow; ♀ greenish grey (L 2–4 cm); bract yellowish green; ovary hairless. FRUIT as other willows.

J F M A M J J A S O N D

♀

FLOWERS ♂ yellow, 3 stamens; ♀ greyish green

♀ **FLOWER** bract yellowish green; ovary hairless

♂

TWIG olive to red-brown; usually ridged

BUDS usually a similar colour to the twig

LEAVES dark green; ± shiny; underside grey-green; conspicuous, persistent stipules

BARK dull greenish brown; with distinctive dark orange underbark

Associated species | Species using Almond Willow have not been widely studied, but there two species known to be unique to Almond Willow: a rust fungus *Melampsora amygdalinae* which distorts the leaf blade and veins and are seen as orange-yellow clusters or spots and a gall-producing sawfly *Euura triandrae* which produces a gall that starts as a smooth, green, bean shape, becoming red with maturity and protruding from both the upper and lower leaf surfaces.

MELAMPSORA AMYGDALINAE

243

SALICACEAE | NARROWER LEAVED WILLOWS Leaves *pp. 228, 65–67* | Flowers *p. 68* | Fruit *p. 74* | Twigs *pp. 230, 81*

Other large willows

LC Eastern Crack-willow
Salix euxina

A tall tree (H to ±18 m) originally from Turkey and the eastern Black Sea. Found along rivers, it appears to be spreading clonally, via broken twigs and limbs drifting downriver and rooting. In Britain & Ireland only males have been recorded. It was originally regarded as a variety of Crack-willow (*S. fragilis* var. *decipiens*) but afforded species status in 2009. **Features:** similar to other crack-willows; differences as per annotations (*right*).

LEAVES similar to those of Crack-willow but always hairless and **leaf-margins coarser** – usually 3–6 teeth per cm

TWIGS pale yellowish brown; brittle at base

CRACK-WILLOW (*p. 242*)

LC Olive Willow *Salix elaeagnos*

A shrub (H to 3 m). Native to southern and central Europe, it was introduced by 1820 and has been planted by riversides and ditches, becoming naturalized in some places. **Features: TWIGS** brownish; densely hairy. **LEAVES** very long and narrow (L to 20 cm; 17–40× as long as broad); grey when fresh; dark green above and whitish below with age. **FLOWERS** ♂ and ♀ catkins on separate trees; appearing with the leaves; cylindrical; ♂ mature anthers yellow.

♀ **FLOWER** bract pale-tipped; ovary hairless

LEAVES very narrow; almost linear; underside whitish matted hairs

Weeping Willow
Salix ×sepulcralis

A tall tree (H to ±22 m) with **distinctive downswept branches**. A hybrid between White Willow (*p. 241*) and *Salix babylonica* (N/I) that is found in gardens, parks and riversides and first known in Britain from *c.* 1869. **Features: TWIGS** brown or yellowish. **LEAVES** 'spear'-shaped (L 7–12 cm) sparsely hairy at most.

LEAVES similar to those of White Willow but usually longer

FORM distinctive downswept branches

Golden Willow
Salix alba var. *vitellina*

Golden Willow is a variety of White Willow with bright, **deep yellow young twigs**, which distinguish it from the '*alba*' variety of White Willow, which has brown twigs. Often planted as a screen or for basket-making.

Similar species | Crack-willow (*p. 242*) twigs can be yellowish, but have ridged buds.

VAR. VITELLINA   VAR. ALBA

YOUNG TWIGS bright yellow   YOUNG TWIGS brown

Common hybrid willows

Salix ×reichardtii
(Goat Willow × Grey Willow)

A shrub or a small tree (H 6–12 m) that is very common (in some areas more so than the parents) at woodland edges and in wet areas such as riversides, bogs and lakesides. Highly variable; with a complete range of mixed and/or intermediate characteristics of the parents. **Features:** typically, individuals usually have striae on the wood of 2nd-year twigs (like Grey Willow) but with leaves that are large, broad, and with pointed tips (more like Goat Willow).

LEAVES highly variable in shape and size, even on the same tree or shrub

2ND-YEAR WOOD striae present

Salix ×multinervis
(Eared Willow × Grey Willow)

A much-branched shrub (H to 5 m) that is common and widespread on infertile soil in a wide range of both dry and wet habitats. A highly variable hybrid with a full range of mixed and/or intermediate characteristics of the parents. **Features:** typically larger and thick-twigged with larger smooth-edged leaves (like Grey Willow) that are dark green and wrinkled, with prominent, persistent stipules, and softly hairy below (like Eared Willow).

LEAVES variably shaped; wrinkled; stipules **conspicuous**; underside softly hairy

Broad-leaved Osier S. ×smithiana
(Osier × Goat Willow)

A tall shrub (to 6 m tall) found in fertile soils in wet places by lakes on riversides and other boggy places. One of the commonest willow hybrids, it has also been planted as a short rotation biomass crop. **Features: LEAVES** Osier-like or 'spear'-shaped (L 6–12 cm; W 1·5–3·0 cm); narrower than any broad-leaved willow, but broader than those of pure Osier; underside densely hairy. **TWIGS** 2nd-year wood smooth, lacking any striae.

LEAVES spear-shaped; underside **dense hairs**

2ND-YEAR WOOD smooth (**striae absent**)

Silky-leaved Osier S. ×holoserica
(Osier × Grey Willow)

A small tree or large shrub (H to 9 m) found in wet lowland sites, often with Broad-leaved Osier. **Features:** it can distinguished fairly readily by the following: **LEAVES** narrow, 'spear'-shaped (intermediate between Osier and Grey Willow), which have short, soft hairs on the underside (more 'velvety' than other similar-looking hybrids; plus **TWIGS** persistently hairy; 2nd-year wood with striae (like Grey Willow).

LEAVES spear-shaped; underside with **short soft hairs**

2ND-YEAR WOOD striae present

SALICACEAE | SMALLER WILLOWS | Leaves *pp. 228, 65–67* | Flowers *p. 68* | Twigs *pp. 230, 81*

LC LC EN Tea-leaved Willow *Salix phylicifolia*

A small shrubby willow (H typically <4 m) found near water, with a preference for alkaline-rich soils and limestone from sea level to just over 1,000 m elevation in Scotland. **Features:** TWIGS reddish brown; sparsely hairy at first, becoming hairless. BUDS yellowish grey or similar to twig; buds differ – leaf-buds narrowly egg-shaped (L 5–9 mm), pressed to twig; flower-buds egg-shaped (L 7–13 mm), usually held slightly away from twig. LEAVES **dark green; leathery; oval** (L to 6 cm; W to 2·5 cm); upperside shiny; underside of all leaves pale greenish grey; leaf-margins with blunt, irregular teeth; stipules usually absent; **no particular taste if chewed**; remains largely green if damaged (*cf.* Dark-leaved Willow). FLOWERS ♂ and ♀ catkins on separate trees; appearing prior to the leaves; egg-shaped (L to 4 cm); bracts black at the tip at least; ♂ mature anthers yellow or orange; ♀ green; ovaries usually hairy. FRUIT as other willows.

J F M A M J J A S O N D

IN LEAF

FLOWER bracts dark or dark-tipped

LEAVES dark green; ± **shiny**; underside greenish grey; stipules usually absent

BUDS flower- and leaf-buds different sizes; ± similar to twig

TWIGS reddish brown; ± **hairless**

DRIED/PRESSED LEAF remains largely green

Associated species | Tea-leaved Willow is an important source of nectar as flowers appear early in the season, becoming a resource for many pollinators, including beetles, moths and butterflies. *Euura plicaphylicifolia*, a sawfly, is dependent on Tea-leaved Willow and creates a gall on one side of the leaf-margin, causing the leaf to sharply fold downward.

On new shoots, the leaf undersides are all greenish grey (*left*); a shrubby Tea-leaved Willow (*right*).

BASAL LEAF

SALICACEAE | SMALLER WILLOWS

Dark-leaved Willow *Salix myrsinifolia*

A small tree or shrub (H to 4 m) of moist habitats such as wet woodlands, lake edges and wet dune slacks but also on rocky cliffs and scree. It is generally a lowland species (although to nearly 1,000 m elevation in Scotland).
Features: TWIGS yellowish brown to reddish brown; usually velvety hairy. BUDS yellowish or similar to twig; flower- and leaf-buds similar; narrowly egg-shaped (L 4–7 mm), pressed to twig. LEAVES **dark green; matt or slightly shiny; thin; broadly oval** (L 5–16 cm; W 2–7 cm); underside of lowest leaves on new shoots green; others greenish grey; leaf-margins toothed; stipules conspicuous, persistent; **lasting foul, bitter taste if chewed; blackens if damaged** (*cf.* Tea-leaved Willow). FLOWERS ♂ and ♀ catkins on separate trees; appearing with the leaves; egg-shaped (L to 4 cm); bracts brown; ♂ mature anthers yellow; ♀ green; ovaries usually hairless. FRUIT as other willows.

J F M A M J J A S O N D

Did you know? Dark-leaved Willow has many chemicals, including high levels of salicylates, to defend against insects, although it is still attacked by sawflies, beetles and moths.

On new shoots, the basal leaves have a green underside; the others on the shoot are pale grey-green (*left*) with a darker, more matt texture than Tea-leaved Willow. Dark-leaved Willows are shrubby trees (*right*) – the name derives from the fact the leaves turn dark if damaged, rather than it having a particularly dark foliage.

SALICACEAE | LOW-GROWING WILLOWS   Leaves *pp. 228, 65–67* | Flowers *p. 68* | Twigs *pp. 230, 81*

Low-growing willows

These willows can be only a few centimetres tall and typically never exceed 50 cm in height. They are long-lived and can be hundreds of years old, usually spreading via underground rhizomes. Apart from Creeping Willow, which occurs in suitable conditions throughout Britain & Ireland, the low-growing willows inhabit the Arctic and alpine zones especially in Scotland. They are important plants in these habitats, providing early-season nectar for insects. Species identification is best achieved by checking leaf shape and twig, bud and flower details.

LC Creeping Willow   *Salix repens*

A small, often creeping, shrub (H to 1·5 m; usually much smaller) of heathland, bogs, dune slacks and fens. Common from sea level to elevations above 900 m in Scotland, although uncommon to rare, and generally confined to wetter habitats, in central, southern and eastern Britain. Some plants can be just a few centimetres tall, especially in heavily grazed areas such as the New Forest. There are three varieties in Britain & Ireland (see box *below*) although intermediates occur. **Features:** TWIGS new-growth **shoots greyish, yellowish or dark red-brown** and typically hairy, especially near the tip when young, but these can be lost. BUDS narrowly egg-shaped (L 2–4 mm); leaf- and flower-buds are similar; tip slightly pointed; typically very similar in colour to the stem; tiny white stomata (use hand lens); held away from the twig. LEAVES small; narrowly oval to oval (L 20–35 mm; W ± 10 mm); **leaf-margins ± smooth and untoothed**; all three varieties have long, silvery, silky hairs on the underside; upperside varies (see box *below* and *opposite*). FLOWERS ♂ and ♀ catkins on separate trees; appearing just before, or with, the leaves; egg-shaped (L 2–3 cm); ♂ mature anthers yellow; ♀ green or reddish; hairy or hairless depending on variety. FRUIT small seeds with long cottony white hairs, similar to all willows and poplars.

J F M A M J J A S O N D

Creeping Willow varieties

var. *repens* – the commonest; a **low creeping shrub** of heaths and moors; LEAF UPPERSIDE ± hairless when mature.
var. *fusca* – **erect shrub** of East Anglian fens; LEAF UPPERSIDE ± hairless when mature.
var. *argentea* – taller, erect shrub found mostly in dune-slacks; LEAF UPPERSIDE **hairy when mature**.

Did you know? Creeping Willow's soft, fluffy seed hairs can also provide nest material for birds. At least 74 species of mycorrhizal fungi have been seen growing around the trees (in studies in Holland). This diversity of fungi help this species survive in some tough environments.

Associated species | A wide range of pollinators – butterflies, moths (including the rare **Dark Bordered Beauty**), bees and beetles (such as the striking bright red **Red Poplar** (or **Creeping Willow**) **Beetle**. In spring, the adults eat large round holes in developing leaves. They lay about 20 eggs on the underside where the larvae feed in a group, often stripping the leaves completely.

RED POPLAR BEETLE   DARK BORDERED BEAUTY

SALICACEAE | LOW-GROWING WILLOWS

♂ FLOWER yellow
♀ FLOWER reddish or greenish

IN LEAF

BUDS similar colour to twig
TWIGS variable; yellowish grey to dark reddish brown; ±hairless

YOUNG LEAF
MATURE LEAF
VAR *ARGENTEA*
VAR *REPENS*

LEAVES small; upperside hairy when young becoming hairless;
LEAVES upperside hairy when mature

The heights of the three varieties: var. *argentea* (top right) is the tallest, with leaves that are very hairy on both surfaces; var. *fusca* (top left) is slightly smaller and has upright stems; var. *repens* (bottom) is low and creeping, and can be just a few centimetres tall.

VAR. *fusca*
VAR. *argentea*

♂ IN FLOWER
VAR. *repens*
♀ IN FRUIT

249

SALICACEAE | LOW-GROWING WILLOWS | Leaves *pp. 228, 65–67* | Flowers *p. 68* | Twigs *pp. 230, 81*

Downy Willow *Salix lapponum*

A small to medium-sized shrub (H to 1·5 m, rarely taller). Widespread, but uncommon, occurring on wet, rocky mountain slopes and cliffs up to ± 1,100 m elevation. **Features:** TWIGS variable, yellowish brown to dark red-brown; hairy when young, becoming hairless. BUDS dark brown; usually held away from the twig; buds differ – leaf-buds narrowly egg-shaped (L 4–5 mm), normally located towards the tip of the twig; flower-buds egg-shaped (L 6–10 mm), normally located towards the base of the twig. LEAVES 'spear'-shaped (L to 7 cm; W to 2·5 cm); dull grey-green; **upperside silky hairy**, veins indented; **underside densely silky hairy**, veins standing out; leaf-margin untoothed and usually curled downwards. FLOWERS ♂ and ♀ catkins on separate trees; appearing before, or with, the leaves; egg-shaped; ♂ (L to 5 cm), mature anthers yellow; ♀ (L to 4 cm) greenish grey. FRUIT as other willows.

J F M A M J J A S O N D

LEAVES 'spear'-shaped; silky hairy on both sides

IN LEAF

BUDS usually dark brown; leaf-buds near tip of twig

TWIGS variable; yellowish brown to dark red-brown; usually hairless

♀ FLOWER greenish

♂ FLOWER yellow

Associated species | Downy Willow is one of the early-flowering willows which provide food for pollinators, including **sawflies**, **bees**, **beetles**, and **moths**. The large, nutrient-rich buds provide a good food source for mountain birds like **Ptarmigan** and **Red Grouse**.

The silver-grey hairy leaves of Downy Willow are distinctive even on wet dull days (*left*), and their pointed shape (*right*) distinguishes them from the rounder Woolly Willow (*facing page*).

SALICACEAE | LOW-GROWING WILLOWS

VU Woolly Willow — *Salix lanata*

A low-growing, branching shrub (H typically <1 m) that is a rare plant of damp rock ledges, crags and cliffs, on base-rich soils usually at 600–1,000 m elevation and in locations that are out of reach of grazing animals. **Features:** TWIGS yellowish brown to greenish brown; variable hairy, from sparse woolly to densely long. BUDS usually held away from the twig; buds differ – leaf-buds narrowly egg-shaped (L 4–5 mm), usually located towards the base of the twig, orange- to red-brown, quite hairy; flower-buds egg-shaped (L 6–10 mm), usually located towards the tip of the twig, orange-brown to dark brown, hairy but less so than the leaf-buds. LEAVES **broadly oval** (L to 7 cm; W to 6·5 cm); dull grey-green; **both upper and underside dense silky hairy, although hairs can rub off with age**; leaf-margin untoothed. FLOWERS ♂ and ♀ catkins on separate trees; narrowly egg-shaped to cylindrical; ♂ (L to 5 cm), appearing before, or with, the leaves, mature anthers yellow; ♀ (L to 15 cm) greenish, densely hairy, appearing with the leaves. FRUIT as other willows.

J F M A M J J A S O N D

IN LEAF

LEAVES broadly oval; usually silky hairy on both sides

LEAF-BUDS

TWIGS yellowish brown to greenish brown; variably hairy

FLOWER-BUDS

♀ FLOWER greenish; hairy

♂ FLOWER yellow

Associated species | An early bloomer of the mountains, the species provides food for many early pollinating insects. **Deer** will also feed on the leaves if they find it, meaning Woolly Willow is rarely able to survive in places that are accessible to grazing animals.

Difficult to find, as mostly located on high inaccessible ledges in the hills of central Scotland, although there are a few at lower elevations (*left*), their rounded silver, hairy leaves being highly distinctive (*right*).

SALICACEAE | LOW-GROWING WILLOWS Leaves *pp. 228, 65–67* | Flowers *p. 68* | Twigs *pp. 230, 81*

LC LC Mountain Willow *Salix arbuscula*

A low-growing shrub (H typically <1 m), usually growing in knee-high clumps. It is uncommon and found mostly in the central Highlands of Scotland in calcium-rich bogs, by small streams or on rocky slopes and mountain ledges, generally above 600 m elevation on base-rich soils.
Features: TWIGS **glossy, reddish brown**; slightly hairy when young but soon becoming hairless. BUDS narrowly egg-shaped (L 2–4 mm); same colour as the twigs; sparsely hairy; held slightly away from the twig. LEAVES **small, oval** (L to 5 cm; W to 3 cm); upperside **dark shiny green**; underside pale, greenish grey, may be weakly hairy; **leaf-margins with small, regular, rounded teeth**.
FLOWERS ♂ and ♀ catkins on separate trees; appearing with the leaves; purplish (L to 3 cm); ♂ mature anthers yellow or reddish purple; ♀ (L to 5 cm) green. FRUIT as other willows.

J F M A M J J A S O N D

IN LEAF

LEAVES dark shiny green; underside pale greenish grey; leaf-margin with rounded teeth

BUDS reddish brown, as twig

TWIGS glossy reddish brown

♀ FLOWER greenish

Associated species | Very little is known about any associated species, but *Euura arbusculae*, a sawfly, is unique to Mountain Willow. Found only on Creag an Lochain, Perthshire, Scotland it is thought to be a very rare British endemic. It was discovered in 1941 but then thought to have become extinct until rediscovered in 2018. ♀ *Euura* sawflies lay their eggs within the leaves of willows, causing galls within which the sawfly larvae develop. In *Euura arbusculae*, these are a small red 'bean gall'.

Often growing as a low knee-high shrubby bush (*left*), Mountain Willow is quite distinctive as its small shiny oval leaves (*right*) have very regular teeth making it 'feel' very different from other dwarf willows.

SALICACEAE | LOW-GROWING WILLOWS

Whortle-leaved Willow *Salix myrsinites*

A short spreading shrub (H typically <40cm) found only in Scotland. Although widely distributed, it is uncommon and often grows in inaccessible places, making it perhaps one of the hardest native trees to find. Whortle-leaved Willow has a preference for wet or moist, often base-rich, soils between 180m and ±900m elevation. **Features:** TWIGS (N/I) shiny; greenish brown to dark reddish brown; older branches dark grey to brown. BUDS broadly egg-shaped (L 3–8mm), bluntly pointed; shiny; yellowish brown to reddish brown; held slightly away from the twig. LEAVES **thick; variably sized; oval to narrowly oval** (L to 7·0cm; W to 2·5cm); tip usually pointed; **bright glossy green on both sides**; leaf-margin with small regular teeth; underside usually long hairy when young, becoming hairless with age; prominent veins; narrow stipules can be present next to the buds on young shoots; **bitter taste when chewed**; withered leaves can persist on the branches. FLOWERS ♂ and ♀ catkins on separate trees; appearing with the leaves; disproportionately large compared to the leaves; narrowly egg-shaped; ♂ (L to 3cm) mature anthers reddish purple; ♀ (L to 5cm) greenish, stigmas purplish. FRUIT as other willows.

J F M A M J J A S O N D

IN LEAF

LEAVES thick; bright glossy green on both sides; leaf-margin with small regular teeth

LEAVES narrow stipules may be present

BUDS yellowish brown to reddish brown

Did you know? The English names Whortle-leaved or Whortle-berry Willow were both used to describe this shrub, after its close resemblance to another mountain species, the Bilberry (or Whortleberry). The leaves have a very strong taste, due to high levels of salicylates.

Whortle-leaved Willow nestling amongst surrounding vegetation (*left*); ♂ catkins are large compared to the leaves and have distinctive reddish purple anthers (*right*).

SALICACEAE | LOW-GROWING WILLOWS   Leaves *pp. 228, 65–67* | Flowers *p. 68* | Twigs *pp. 230, 81*

LC NT Dwarf Willow   *Salix herbacea*

A **low, prostrate, many-branched shrub** (H 5–10 cm) which is the **smallest of the native willows**. It has an extensive system of rhizomes (modified plant stems) which grow underground and can develop into either roots and/or shoots, allowing the plant to spread vegetatively. It is adapted to Arctic/alpine conditions and occurs in snowbeds, stony areas and on open, eroded substrates, as well as in montane heath growing amongst sedges and grasses, from sea-level up to elevations of 1,300 m. **Features:** TWIGS greenish brown, becoming dark reddish brown; can be silky hairy when young. BUDS egg-shaped (L to 2 mm); yellowish green to reddish; sparsely hairy at most; either pressed to or held away from the twig. LEAVES **shiny; broadly oval to ±circular** (L to 3 cm; W to 2 cm); hairy when young, becoming hairless; upperside green; underside paler green; **prominent network of veins on both surfaces**; leaf-margin toothed or scalloped. FLOWERS ♂ and ♀ catkins on separate trees; appearing with the leaves; egg-shaped to globular; ♂ (L to 0·75 cm) mature anthers red, maturing yellow; ♀ (L to 1·3 cm) pinkish red. FRUIT as other willows.

J F M A M J J A S O N D

IN LEAF

LEAVES upperside shiny, dark green; underside paler green

TWIGS greenish brown to dark reddish brown

BUDS tiny; usually yellowish green

LEAF at life-size

Did you know? Dwarf Willow rarely reproduces from seed, the plant mostly spreading vegetatively. Young plants develop lateral branches which grow along the soil surface or just below it. These creeping branches develop roots, allowing the plant to spread effectively.

The low, creeping Dwarf Willow (*left*) is a small plant, although plants can spread to be 50 cm across; nonetheless, even in flower or fruit (*right*) it can be overlooked.

SALICACEAE | LOW-GROWING WILLOWS

Net-leaved Willow *Salix reticulata*

A **small, low, creeping shrub** (H typically <20 cm) that branches extensively both above and below ground. Above-ground stems are often found covered in lichens. It is rare, occurring primarily in northern and central Scotland on wet rocky ledges in limestone and schist mountains, mainly 600–1,100 m elevation. **Features:** TWIGS greenish to dark reddish brown; usually hairy at first, and becoming hairless with age. BUDS egg-shaped (L to 2 mm); yellowish to reddish brown; sparsely hairy at most; held away from the twig. LEAVES **distinctive;** oval to ± circular (L to 6 cm; W to 5 cm); **upperside dull green with deeply impressed veins;** underside grey-green with prominent veins; leaf-margin untoothed but with a distinctive 'knobbly' appearance; both surfaces with silky hairs; although these can wear off. FLOWERS ♂ and ♀ catkins on separate trees; appearing with the leaves; narrowly cylindrical (L to 0·35 cm); upright on hairy red stalks; ♂ filaments yellow, anthers red, maturing yellow; ♀ (L to 1·3 cm) pinkish red, hairy. FRUIT as other willows.

J F M A M J J A S O N D

IN LEAF

LEAVES **distinctive, prominent veins on both surfaces;** 'knobbly' margin

TWIGS greenish brown to dark reddish brown

BUDS tiny; usually yellowish brown

♀ FLOWER pinkish red

Associated species | Net-leaved Willow can often be found growing with **Mountain Avens**, which likes the same wet, lime-rich soils. *Melampsora epitea* (an orange-brown rust) has been reported on Net-leaved Willow. However, *M. epitea* may actually be a complex of species, one of which, *M. reticulatae*, seems to be found only on Net-leaved Willow.

A tiny willow that is easy to overlook in the vegetation of its preferred craggy habitat (*left*); however once the 'net' of leaf-veins (*right*) are seen, this willow is unmistakable.

SALICACEAE | POPLARS   Leaves *pp. 229, 64* | Flowers *p. 68* | Fruit *p. 74* | Twigs *p. 82*

LC LC Aspen   *Populus tremula*

A tall tree (H to 30 m) found throughout Britain & Ireland with a preference for deep, rich, well-drained soil, although it can cope with heavy, cold and damp soils. It is most common in Scotland as it has a preference for cooler climes such as that of the Scottish Highlands, where it is often found growing with Scots Pine and Birch. **Features:** BARK greenish or whitish; smooth when young, developing **small diamond fissures**, and eventually becoming deeply furrowed with age. TWIGS shiny, olive- to dark red-brown with orange lenticels; typically hairless. BUDS egg-shaped with a pointed tip; shiny dark brown; usually hairless; scales can have a pale margin; end-buds (L 6–12 mm); side-buds (L 5–8 mm) held slightly away from the twig. LEAVES **very thin, trembling even in the slightest breeze**; broadly oval (L to 8 cm), tip usually pointed; hairy when young but rapidly become hairless; upperside mid-green; underside paler green; turn bright butter-yellow shades in the autumn; leaf-margin with a few irregular rounded teeth; **leaf-stalks flattened**, and can have 2 cup-shaped glands at the base of the leaf. FLOWERS ♂ and ♀ catkins on separate trees; appearing before the leaves; long, pendent, densely grey hairy, cylindrical; ♂s (L to 10 cm), anthers purplish red; ♀s (L to 12·5 cm), stigmas green to reddish. FRUIT small seeds with long cottony white hairs, similar to all poplars and willows.

J F M A M J J A S O N D

UPPER

UNDER

LEAVES very thin with rounded teeth; hairless when mature; underside much paler

BARK small diamond-shaped fissures; deeply furrowed when mature

Large Aspen trees forming an open woodland in the Scottish Highlands (*left*); the leaves tremble, producing a sound like waves crashing on a beach, in even the slightest breeze (*right*) – from which the tree's scientific name is derived.

SALICACEAE | POPLARS

♂ FLOWERS catkins slightly 'thinner' than the females; anthers purplish red

♀ FLOWERS catkins slightly 'fatter' than the males; stigmas pinkish to red

TWIGS rounded; hairless; shiny; orange lenticels

BUDS pointed egg-shaped; shining brown; not sticky or with strong aroma

IN FRUIT similar to all poplars and willows, the seeds have fluffy white hair-like fibres which help the wind-dispersal of the seeds

Associated species | Aspen supports a wide range of species, including the caterpillars of the **Light Orange Underwing** – an uncommon day-flying moth mostly found in southern England.

The **Leaf-rolling Weevil** is a small striking coppery-green beetle associated with Aspen growing in open scrub and open woodland. The female rolls Aspen leaves into a thin tube, where it then lays its eggs. Largely restricted to a few woods in central and southern England, it is now a rare Red Data Book species.

Galls are also a feature of Aspen. On the leaves can be found the small bright red galls of the gall midge *Harmandiola globuli*, which each contain a solitary orange larva; **Aspen Petiole Gall Midge** galls develop on the leaf-stalk; and the **Small Poplar Borer Beetle** forms woody galls (L 20 mm), most commonly found on young twigs. The beetles themselves can be seen from May to July.

A striking feature of Aspen is an infection caused by the **Aspen Tongue** fungus which impacts ripening ♀ flowers, causing them to swell and turn bright yellow – creating an effect like a mini bunch of bananas. Rarely recorded and seen mostly in Scotland, it was found recently in Hampshire so it may be overlooked.

LIGHT ORANGE UNDERWING

ASPEN LEAF-ROLLING WEEVIL

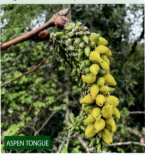

ASPEN TONGUE

Did you know? Aspen, like most poplars, are not long-lived, with individual trees seldom surviving for more than a century. However, it has a remarkable and highly successful method of regeneration through suckers which emerge from the extensive root system, surrounding the 'mother' tree in an ever-extending ring of clonal shoots. It is thought that the Aspen may reach ages of up to 10,000 years and may be one of the oldest living organisms on Earth. In Britain & Ireland, one of the oldest Aspens grows at a golf course in Monmouthshire, where what appears to be a small Aspen woodland is in fact a single tree which occupies more than an acre and comprises several thousand clonally grown trees.

SALICACEAE | POPLARS   Leaves pp. 228, 62, 64 | Flowers p. 68 | Fruit p. 74 | Twigs p. 82

■ Grey Poplar   *Populus ×canescens*

A tall tree (H to 30 m) – a natural hybrid between White Poplar (*p. 260*) and Aspen (*p. 256*). It is planted widely and flourishes in damp locations such as in water meadows and river valleys. Most trees from cultivation are ♂, although ♀ trees occur naturally (as well as being propagated) but appear to be uncommon. Historically, it was considered native (Loudon 1838), although due to the high occurrence of ♂ trees it's spread is most likely to be the result of human intervention. **Features:** Highly variable; individual trees have a range of both mixed and intermediate characters between the parents. The features that follow are from the middle of this range – it may not be possible to confidently identify some trees that are close in appearance to one or other of the parents. BARK greyish or whitish; smooth when young; breaking into small, dark, diamond-shaped lenticels which unite into deep furrows at the base of the trunk. TWIGS greenish brown to shiny grey; ±hairless or with a covering of dense whitish grey downy hairs at least near the tip when young, becoming hairless with age. BUDS ±conical; end-bud (L 3–8 mm), ±hairy; usually 7–12 bud scales (*cf.* White Poplar which has 3–5); side-buds (L 3–6 mm); held slightly away from the twig; 3–5 bud scales; leaf-scars pale brown. LEAVES of two types:
• ON LONG SHOOTS AND SUCKERS (**White Poplar-like**) – arranged alternately; **more 'triangular'** (L 6–8 cm); upperside shiny dark green, slightly hairy; underside with **persistent thick greyish hairs**; leaf-margin with blunted, rounded, teeth; leaf-stalk rounded.
• ON SHORT SHOOTS (Aspen-like) – in clusters; more 'rounded' (L 4–6 cm); **upperside shiny dark green, hairless; underside light green** with only a few scattered grey hairs which are soon lost; leaf-stalk flattened. FLOWERS ♂ and ♀ catkins on separate trees; appearing before the leaves; pendent, cylindrical; ♂'s densely grey hairy (Aspen-like); thicker and longer (L to 10 cm) than ♀s; anthers reddish purple. ♀s green (L to 4 cm); stigmas yellow. FRUIT similar to all poplars.

J F M A M J J A S O N D

BARK dark diamond-shaped marks; like the parents

Grey Poplar (like White Poplar) stands out in the landscape (*left*) due to its white leaves, which flash in the wind; the leaves (*right*) are usually more rounded than the '5-fingered' leaves of White Poplar.

SALICACEAE | POPLARS

UPPER / UNDER — **ASPEN** oval; **UPPERSIDE** mid-green; **UNDERSIDE** pale green; ± hairless

UPPER / UNDER — **GREY POPLAR long-shoot leaves**, (intermediate but more similar to White Poplar) upperside dark green; underside pale grey-green with persistent hairs

UPPER / UNDER — **WHITE POPLAR** 5-lobed; **UPPERSIDE** dark green; **UNDERSIDE** grey; densely hairy; ± hairless

SHORT SHOOT

SHORT SHOOT LEAVES more 'rounded'; in clusters; underside ± hairless

LONG SHOOT

LONG SHOOT LEAVES more 'triangular'; alternately arranged; underside with thick greyish hairs

♂ **FLOWERS** anthers dark red; ♀ **FLOWERS** (n/i) green; stigmas yellow (rarely found)

BUDS dark red-brown hairy (can be hairless); end-buds with 7–12 bud scales

TWIGS hairless or hairy (cf. White Poplar)

LEAF-SCAR pale brown

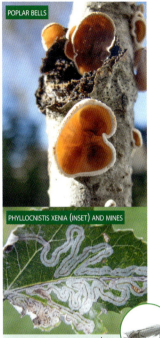

POPLAR BELLS

PHYLLOCNISTIS XENIA (INSET) AND MINES

Associated species

Phyllocnistis xenia, a tiny leaf-mining moth, was first discovered on the leaves of Grey Poplar on the 9th September 1974 in scrubby woodland near Dover, Kent. The caterpillar feeds within the leaf, creating a thin silvery trace on the upperside as it does so, only emerging from the leaf to create a cocoon in a fold of the leaf's edge. The moths fly from July to September, and the mines can be found from June to September. It has spread from Kent and can now be found in East Anglia and Hampshire. Poplars, like willows, live in wet places in conditions that are ideal for a wide range of micro-fungi, many of which occur on both types of tree. One of these is the small upside-down cup-shaped **Poplar Bells**, which can occasionally be found on the broken branches and dead wood of Grey Poplar.

SALICACEAE | POPLARS Leaves *pp. 228, 62, 64* | Flowers *p. 68* | Fruit *p. 74* | Twigs *pp. 82, 85*

LC LC White Poplar *Populus alba*

Medium-sized to tall tree (H to 30m) with a broad, rounded crown. Introduced from southern Europe sometime before 1500, it is often planted as shelter belts, on riverbanks, and for stabilising sand dunes. **Features:** BARK white or grey; smooth when young, developing characteristic dark diamond-shaped marks which become fissures at the base with age. TWIGS **greenish brown; covered in dense whitish grey downy hairs**, at least when young. BUDS egg-shaped, pointed (end-bud L 3–8mm); usually softly hairy; 3–5 bud scales; leaf-scars dark brown (*cf.* Grey Poplar, *p. 258*); side-buds (L 5–8mm) held slightly away from the twig. LEAVES **Maple-like, 5-lobed** (L to 10cm); **both sides thickly covered in white hairs** when young; upperside hairs can wear off with age. FLOWERS ♂ and ♀ catkins on separate trees; appearing before the leaves; pendent, cylindrical; ♂s (L to 7·5cm), anthers dark red. ♀s (L to 5cm) greyish green, stigmas pale yellow. FRUIT similar to all poplars.

J F M A M J J A S O N D

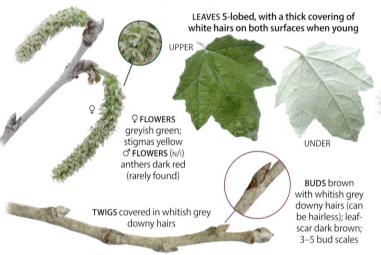

LEAVES 5-lobed, with a thick covering of white hairs on both surfaces when young

UPPER

♀

♀ **FLOWERS** greyish green; stigmas yellow
♂ **FLOWERS** (N/I) anthers dark red (rarely found)

UNDER

TWIGS covered in whitish grey downy hairs

BUDS brown with whitish grey downy hairs (can be hairless); leaf-scar dark brown; 3–5 bud scales

BARK characteristic dark diamond-shaped marks

The white hairy leaves of White Poplar stand out at a distance (*left*); closer up, the distinctive '5-fingered' leaves (*right*) are quite different from the rounded leaves of a typical Grey Poplar.

SALICACEAE | POPLARS

Black Poplar   *Populus nigra*

A tall tree (H to 30 m; trunk girth to 12 m) widespread across lowland Britain & Ireland but scattered in the north and west. The native subspecies (ssp. *betulifolia*) has a preference for deep, rich, well-drained soils but can thrive in heavy cold damp soils, such as wet woodland margins, stream and river boundaries, and boggy field corners. There are other subspecies in Europe, and other varieties and hybrids of Black Poplar are widely planted in Britain & Ireland (see *p. 262*). **Features:** FORM native Black Poplar trees can develop a characteristic lean, usually at ± 60°, and have branches that arch downward with twigs that sweep up at the tips; **rarely infested with Mistletoe** (*cf.* Hybrid Black Poplar *p. 262*). BARK dark grey to deep brown; **mature trees deeply furrowed and frequently with large burrs**. TWIGS cylindrical; shiny; yellowish brown; often hairy when young. BUDS ± conical (L end-bud 8–12 mm; side-bud to 10 mm); shiny; yellowish brown to dark brown; end-bud L 8–12 mm. LEAVES ssp. *betulifolia* variable; triangular to ± diamond-shaped (L 5–10 cm); tip usually drawn-out and sharp-pointed; underside green (*cf.* balsam-poplars *p. 264*); leaf-margin with rounded teeth; leaf-stalks finely hairy; **many leaf-stalks have spiral galls** caused by the aphid *Pemphigus spyrothecae* (diagnostic of Black Poplar ssp. *betulifolia*); **no glands on leaf-stalk near the leaf-blade**; turn dull olive-yellow in autumn (*cf.* Hybrid Black Poplar). FLOWERS ♂ and ♀ catkins on separate trees (vast majority ♂); appearing before the leaves; pendent, cylindrical (L to 6 cm); ♂s, anthers dark red, ♀s green, stigmas yellow. FRUIT similar to all poplars.

J F M A M J J A S O N D

> **Did you know?** Considered to be the rarest of the large native trees, native Black Poplar has an estimated population of ± 8,000 trees. Of these, only ± 500 are female, with the parish of Roydon, Essex, having 30, making it probably the finest collection of female trees in England. Unfortunately, native Black Poplar is a species which does not germinate well from seed, which is viable for only a few weeks and needs to fall on bare, competition-free ground that remains wet. It has typically been propagated from cuttings, with the entire population comprising clones of just 150 individuals. Black Poplar has a variety of distinct forms – see *p. 262*.

BARK dark grey to deep brown and deeply furrowed on old trees; **many have large burrs**

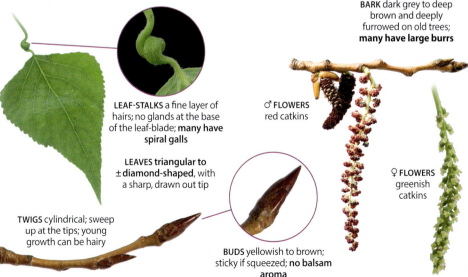

LEAF-STALKS a fine layer of hairs; no glands at the base of the leaf-blade; **many have spiral galls**

LEAVES triangular to ± diamond-shaped, with a sharp, drawn out tip

TWIGS cylindrical; sweep up at the tips; young growth can be hairy

BUDS yellowish to brown; sticky if squeezed; **no balsam aroma**

♂ FLOWERS red catkins

♀ FLOWERS greenish catkins

261

SALICACEAE | POPLARS   Leaves *pp. 228, 62, 64* | Flowers *p. 68* | Fruit *p. 74* | Twigs *p. 82*

Hybrid Black Poplar
Populus ×canadensis

A fast growing tall tree (H to 25 m) – the hybrid between European Black Poplar and the American Cottonwood *Populus deltoides*. Appearance varies; similar to native Black Poplar but differs as follows:
Features: FORM crown generally neater and more symmetrical; side branches arching down less strongly, and twigs rarely swept up. **Can be heavily infested with Mistletoe.** LEAVES typically triangular; underside green (*cf.* balsam-poplars *p. 264*); leaf-stalks **never with spiral galls** and **may have a pair of reddish glands on the leaf-stalk** near the leaf-blade.

Some Black Poplar and Hybrid Black Poplar varieties

HYBRID BLACK POPLAR
'**Marilandica**' | the 'Railway Poplar', a ♀ clone with silvery foliage.
'**Robusta**' | a narrow symmetrical tree with a spring flush of new orange foliage.
'**Serotina**' | a large ♂ clone with larger branches arching out and down, and a deeply fissured trunk coming into leaf very late (*serotinus* is Latin for late). Commonly planted tree of the chalky south of Britain which can look very similar to native Black Poplar because of its large downward-arching branches.
'**Serotina Aurea**' | the 'Golden Poplar' – a tall tree with bright yellow summer foliage
BLACK POPLAR
Lombardy Poplar | a pure form of *Populus nigra* with a strongly upright growth habit and more triangular leaves than ssp. *betulifolia*, without bosses or burrs on the trunk. It can however support a high density of *Pemphigus* spiral galls on its leaf-stalks.

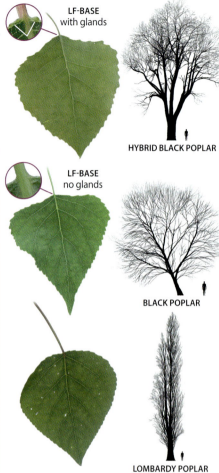

LF-BASE with glands

LF-BASE no glands

HYBRID BLACK POPLAR

BLACK POPLAR

LOMBARDY POPLAR

The somewhat distinctive shapes of (*left to right*) Black Poplar, Hybrid Black Poplar and Lombardy Poplar.

SALICACEAE | POPLARS

HYBRID BLACK POPLAR
('Serotina')
♂ **FLOWERS** red stamens;
♀ **FLOWERS** (N/I) greenish

UPPER

BARK dark, with regular fissures; **burrs absent** (*cf.* Black Poplar)

LEAVES ± triangular with a ± drawn out tip; underside only slightly paler green (*cf.* balsam-poplars)

UNDER

J F M A M J J A S O N D

Similar species | Balsam-poplars (*p. 264*) buds with balsam aroma; leaf underside pale.

TWIGS cylindrical; less upswept than Black Poplar;
BUDS yellowish to brown; sticky; **no balsam aroma**

Associated species | One study found 97 species of invertebrates from 12 families on poplars as a group. Several scarce and uncommon insects, including the bark beetles *Dorytomus filirostris*, *Dorytomus ictor*, *Dorytomus longimanus* and *Byctiscus populi*; a longhorn beetle *Saperda carcharias*; and a leaf miner *Rhynchaenus populi*, are associated with Black Poplar.

Other species include the distinctive **Poplar Hawkmoth** and **Hornet Clearwing**, an unusual moth whose hornet-like yellow protective coloration is an example of Batesian mimicry, in which the harmless moth mimics the look of a dangerous hornet. The larvae of this moth spend two or three years within the stem until they emerge as adults, via a distinctive exit hole at the base of the tree.

POPLAR HAWK-MOTH

HORNET CLEARWING

SAPERDA CARCHARIAS

♂ Black Poplar coming into leaf (*left*); the distinctive red catkins of ♂ Hybrid Black Poplar 'Serotina' (*right*).

263

SALICACEAE | POPLARS   Leaves pp. 228, 62, 64 | Flowers p. 68 | Fruit p. 74 | Twigs p. 82

Balsam-poplars

As a group, the four balsam-poplars differ from poplars in having long, conical buds that have a **sticky surface** with a **strong balsam aroma**, and rounded-triangular leaves that are very pale below. However, differentiating the four species themselves is difficult. Overall shape and the presence/absence of suckers can be helpful but they are perhaps best separated by details of leaves from the main branch (not from suckers or epicormic growth).

WESTERN BALSAM-POPLAR

LEAVES the much paler underside with darker veins is distinctive of balsam-poplars

LC ■ Western Balsam-poplar
Populus trichocarpa

A tall tree (H to 50 m) that is native to the western side of North America. Introduced by 1892, it has a preference for wet conditions and has been widely planted, often adjacent to rivers, lakes and ponds – most trees ♂, few ♀.

TWIGS ridged and/or angled – *other three taxa rounded*

LEAVES narrow triangle; BASE usually **straight**; (some slightly indented); POINT AT TIP abrupt

CROWN usually narrowly conical; lower branches drooping; **BASAL SUCKERS** few at most

■ Balm-of-Gilead
Populus ×jackii

A tall tree (H to 30 m) that is a naturally occurring hybrid between Eastern Balsam-poplar and Eastern Cottonwood (N/I), which was introduced by 1773 and is found in parks and along roadsides – ♀ trees only.

LEAF-MARGIN **obvious hooked teeth** – *other three taxa with short, rounded teeth*

LEAF-STALK **long hairs**

LEAVES rounded triangle; BASE usually **indented**; POINT AT TIP abrupt

CROWN very narrow; **BASAL SUCKERS many**

LC ■ Eastern Balsam-poplar
Populus balsamifera

A tall tree (H to 35 m) that is native to northern North America. Introduced by 1692, and found widely planted in parks, gardens and in roadsides – ♂ and ♀ trees.

LEAF-MARGIN short rounded teeth

LEAVES rounded triangle; BASE usually **rounded**; POINT AT TIP **tapering**

CROWN narrowly pyramidal; some lower branches drooping; **BASAL SUCKERS extensive**

■ Hybrid Balsam-poplar
Populus ×hastata

A tall tree (H to 30 m) that is a hybrid between Western and Eastern Balsam-poplars. Imported by the Forestry Commission in 1948 and widely planted in parks, public spaces and occasionally for timber production – ♀ trees only.

LEAF-STALK **a few long hairs mixed with short hairs** – *Western and Eastern with very short hairs at most*

LEAVES rounded triangle; BASE **rounded or indented**; POINT AT TIP abrupt;

CROWN very narrow; most branches ascending at 45°; **BASAL SUCKERS** absent

BETULACEAE (BIRCH FAMILY) | INTRODUCTION

Introduction to birches, alders, hazels & Hornbeam

Worldwide the Betulaceae are a family of deciduous trees which includes the birches, alders, hazels, hornbeams, hazel hornbeams and hop hornbeams. In total there are ± 167 species worldwide. They are found in the Northern Hemisphere and in the Andes mountains in South America. Their flowers are usually catkins which appear before the leaves.

British & Irish Betulaceae all have ♂ flowers in catkins and fruit that is a winged, or unwinged, nut.

BIRCHES

♀ FLOWERS in smaller catkins

LEAVES broadly triangular

♂ FLOWERS in long catkins

young   mature

FRUIT winged nuts (*right*) held in catkin-like structure

ALDERS

♀ FLOWERS tiny; in small clusters

LEAVES broadly oval to heart-shaped

♂ FLOWERS in long catkins

young   mature

FRUIT winged nuts (*right*) held in cone-like structure

HAZELS

BUD

♀ FLOWERS in unstalked bud-like group

LEAVES broadly oval to heart-shaped with roughly toothed margins

♂ FLOWERS in long catkins

ripe nut

young

FRUIT egg-shaped nut (*right*) surrounded by bracts when young

HORNBEAM

♂ FLOWERS in long catkins

♀ FLOWERS in shorter catkins

LEAVES narrowly oval to oval with roughly toothed margins and strongly pleated veins

FRUIT nuts (*right*) with a 3-lobed bract held in a pendent cluster

BETULACEAE (BIRCHES) | INTRODUCTION   Leaves *pp. 64–65* | Flowers *p. 68* | Fruit *p. 74* | Twigs *p. 84*

Introduction to birches

In Britain & Ireland the group of species called birches have presented identification problems since at least 1753 when Carl Linnaeus, in his *Species Plantarum*, treated Birch as a single, highly variable, species – *Betula alba*. Nowadays it is regarded as two species (**Silver Birch** and **Downy Birch**), with some contentious subspecies and variations within Downy Birch. The third native species, **Dwarf Birch**, looks very different to the other two.

Birch hybrids

Trees which appear to be hybrids between Silver and Downy Birch are known as *Betula ×aurata*. The true extent of this hybridization is poorly studied, but is probably widespread and further complicated by the presence of Dwarf Birch genes. Downy Birch has 4 sets of chromosomes (tetraploid), and both Silver and Dwarf Birches have 2 sets (diploid).

This complex mix, combined with back crosses, results in a wide range of hybrids with variably mixed and intermediate leaf, bark and twig characters, and hairiness which can make identification of some trees very tricky if not impossible, especially in areas where both parents grow. There are also distinct local forms, particularly of Downy Birch, which add to the challenge of identification, and the introduction of non-native species, particularly in urban areas, has introduced further identification pitfalls.

Although it can be difficult to distinguish some individual trees in the hybrid zone between the parents, it is usually possible to identify 'pure' individuals (see table *p. 269*).

Downy Birch subspecies and forms – see *p. 270*
There are two recognized subspecies and one form:

ssp. *pubescens* is widespread throughout Britain & Ireland.

ssp. *celtiberica* ❶ has a western bias but as it has only recently been recognized by some authorities it may be under-recorded due to its, at most, subtle differences.

var. *fragrans* ❷ is a small-leaved variety which seems to have a patchy northern and western distribution. NOTE some small-leaved trees may be due to Dwarf Birch genetic introgression.

top to bottom:
Silver Birch has drooping branches; this **Hybrid Birch** (*middle*) has a mix of drooping and erect branches; **Downy Birch** (*bottom*) has erect branches.

Silver Birches stand out in this heathland setting in the New Forest.

BETULACEAE (BIRCHES) | INTRODUCTION

Did you know? The rising sap of the birch in the spring is sweet like that of the maple. This sweet sap can be used to make birch wine, which goes well with cheeses or cream-based puddings. Birch sap is collected for about a month at the beginning of spring, before the leaves are on the trees, by drilling a hole in the trunk and then using a tube to tap off the sap into a collecting bottle. Although widely undertaken in some countries, particularly around the Baltic, there is little evidence about whether the practice does damage to the trees themselves.

Downy Birch has been used to make canoe skins, rope, drinking cups, roof tiles and a high-quality charcoal. The flexible twigs have also been made into brooms.

Associated species – Silver and Downy Birches | Both SILVER BIRCH and DOWNY BIRCH are excellent for wildlife and have a high conservation value as they provide food and shelter for a wide range of birds including **Redpoll**, **Siskin**, **Goldfinch** and **Greenfinch**.

Birch leaves are eaten by the caterpillars of the **Large Tortoiseshell** butterfly and many moths, including **Buff Ermine** and **Light Emerald**, making areas of birch with egg-laying moths a valuable hunting ground for bats.

Approximately 330 **invertebrate** and more than 230 **lichen** species have been recorded using birches, approximately 60 of which are more or less confined to Scotland as birch epiphytes, with many others more prevalent on birch in Scotland than elsewhere in Britain & Ireland. Approximately 30 **mosses** and 28 **liverworts** have also been recorded on British & Irish birch trees.

Birches support a wide range of **fungi**, including **Chaga** which appears like a large black charcoal-like growth from stems or branches of the tree. In some parts of the world Chaga has been used to produce a fine powder to brew a tea-like drink.

Some associated fungi, like **Birch Polypore**, are parasites, causing brown rot and eventually killing the tree. Others have a symbiotic relationship. Perhaps the most striking of these partnerships is with **Fly Agaric**, the stunning red and white toadstools of which can be found around the roots of birch trees.

Another fungus, *Taphrina betulina*, causes the growth comprising tangled masses of twigs in birches known as 'Witches' Broom' and occurs on all three native species and their hybrids.

In northern areas DOWNY BIRCH provides food for birds, including **Black Grouse**, which feed on the young catkins, and has many plant species associated with it such as **Chickweed Wintergreen**.

267

BETULACEAE | BIRCHES   Leaves *pp. 64–65* | Flowers *p. 68* | Fruit *p. 74* | Twigs *p. 84*

LC LC Silver Birch   *Betula pendula*

A medium-sized, fast-growing tree (H to 30 m), with drooping branch tips giving a 'wispy' appearance at a distance. It has a preference for drier conditions than Downy Birch and is most widespread in the south and east of Britain & Ireland. It occurs naturally on heathlands and in woodlands and is also widely planted in parks and gardens. Mature at approximately 40 years, but can live for up to 70 years. **Features:** BARK light pink-brown when young; gaining the **distinctive white-and-black appearance** with age; the black areas are lenticels – usually diamond-shaped although can be vertical splits or gashes; on older trees the base of the trunk becomes darkened (lacking any white) and splits into vertical cracks and lumpy, corky areas. TWIGS **hairless** (can be hairy on saplings and shoots); **shiny**; red-brown; **usually with white warty resin glands**. BUDS small; narrowly egg-shaped (L 5–7 mm), pointed; usually bicoloured reddish brown and green. LEAVES **roughly triangular** (L to 60 mm) with **drawn-out pointed tip**; hairless; leaf-margin irregularly double-toothed; leaf-stalks hairless. FLOWERS appear with emerging leaves; ♂ and ♀ flowers separate but on the same plant; ♂s in long, pendent, yellowish green catkins (L to 5 cm); ♀ in smaller (L to 3 cm), ± erect catkins. FRUIT in pendent catkins; seeds with papery wings ≥2× seed width and usually extending beyond stigma; fruiting bracts 3-lobed, lateral lobes point towards basal join – 'bird of prey' appearance.

J F M A M J J A S O N D

> **Did you know?** The strong white wood was used for bobbins, flooring and schoolmasters' canes. The bark was used for paper, shoes and roofing. Current uses include parquet floors, furniture and broom handles.
> A single large birch can produce more than 1 million seeds in an autumn.

LC LC Downy Birch   *Betula pubescens*

A medium-sized, fast-growing tree (H to 30 m), with horizontal to erect branch tips giving a 'feathery' appearance at a distance. It has a preference for wetter conditions than Silver Birch and is most widespread in the north and west of Britain & Ireland. It occurs naturally on heathlands and in mixed woodlands but will form single species woodland in the absence of Silver Birch. **Features:** BARK dark reddish brown when young; becoming grey-white with age; usually smooth; although can peel in horizontal strips; horizontal lenticels usually present, especially towards the base of the trunk. TWIGS **hairy**; dull; dark red-brown; **usually lacking resin glands**. BUDS small; egg-shaped (L 4–7 mm), ± rounded tip; usually reddish brown. LEAVES **roughly triangular** (L to 50 mm) with **short pointed tip**; usually hairy when young, becoming hairless although hairs can be retained on the underside along the major veins and in vein axils; leaf-margin coarsely toothed; leaf-stalks can be hairy. FLOWERS as Silver Birch. FRUIT in pendent catkins; seeds with papery wings ± equal to seed width and usually not extending beyond stigma; fruiting bracts 3-lobed, lateral lobes point forwards from basal join.

SSP. CELTIBERICA

VAR. FRAGRANS   J F M A M J J A S O N D

> **Did you know?** The wood was used to make furniture and veneers, while the branches were used to make besom brooms. The sugary sap (also found in Silver Birch) is used to produce birch wine, and oils from the bark can be used as an insect repellent.

Downy Birch forms | see *p. 270* for identification details

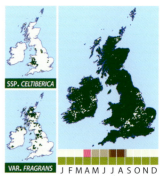

268

BETULACEAE | BIRCHES

Identification of 'pure' Silver and Downy Birches

	Silver Birch	**Downy Birch** (ssp. *pubescens*)
HABITAT	drier and/or warmer areas	wetter and/or colder areas
TWIGS	all downward-pointing	all erect or upward-pointing
BARK AT BASE OF TRUNK	breaks into large dark blocks	thin parallel lines or bands
TWIGS	± shiny; hairless (can have fine hairs on very young twigs)	dull; usually with fine hairs (can be hairless in some forms)
LEAF-TIP	narrower, drawn-out	broader, shorter
LEAF-TEETH	smaller	larger
LF-STALK TO 1ST TOOTH	larger distance between than in Downy	shorter distance between than in Silver
WINGS OF SEED	overtop stigma	do not reach top of stigma

CATKIN IN FRUIT

LF typically with **longer point than in Downy Birch**

BUDS typically more pointed and greener than in Downy Birch

BARK red- to pinkish brown when young; upper parts usually whiter than Downy Birch when mature

vertical to 'diamond' fissures at base

YOUNG TWIGS shiny; hairless; white-dotted

FRUIT BRACT 'bird of prey' – lobes curve backwards

SEED wings extending above stigma; ≥2× W seed

FLOWERS AND FRUITING CATKINS are very similar in both species

ssp. *pubescens*

CATKIN IN FRUIT

LF typically with **shorter point than in Silver Birch**

BUDS typically more rounded and reddish brown than in Silver Birch

BARK red-brown when young; upper parts often brownish or greyish white (can be white in a few)

horizontal fissures at base

YOUNG TWIGS dull; hairy

FRUIT BRACT lobes at least directed slightly forwards

SEED wings typically not extending above stigma; ±W seed

269

BETULACEAE | BIRCHES   Leaves *pp. 64–65* | Flowers *p. 68* | Fruit *p. 74* | Twigs *pp. 84, 85*

Downy Birch subspecies and forms ssp. *pubescens* shown for comparison

var. *fragrans* (ssp. *tortuosa*) – map *p. 268*
Tree (typically H to 5 m; a few up to 15 m) usually with a **short multi-stemmed trunk and contorted branches**. **Features: BARK** usually greyish white but obscured by lichens and bryophytes in many individuals. **TWIGS** ±hairless; erect, spreading or pendulous. **BUDS** sticky; resinous aroma in spring. **LEAVES** most L ≤ 30 mm; sparsely hairy at most.

ssp. *celtiberica* – map *p. 268*
Tree (form as ssp. *pubescens*). **Features: BARK** white to the base (may have dark splits but these are never vertical. **TWIGS** ±hairless. **LEAVES** most L > 30 mm; hairs on underside confined to vein axils. **FRUIT** bracts with downward pointing lateral lobes; width of seed wings **noticeably less than seed width**.

LEAF (actual size)
±hairless

BUDS
sticky

TWIGS sparsely
hairy at most

SSP. *PUBESCENS*

TWIGS sparsely
hairy at most

SEED as ssp. *pubescens* (*box, right*) but **wing width much less than width of seed**

SSP. *PUBESCENS*

FORM
branches
contorted

FRUIT BRACT
lobes
downcurved;
more like
Silver Birch

LEAF hairs confined to axils

Introduced birches

LC ■ Himalayan Birch
Betula utilis

Tree (H to 20 m) native to the Himalayas, and introduced to Britain & Ireland by 1849.
Features: BARK distinctive; varies in colour depending on subspecies – very white (ssp. *jacquemontii*), red (ssp. *albosinensis*), almost black (ssp. *utilis*); peels off in large, horizontal sheets.

LEAF
dull;
7–12 pairs
of veins;
L to 9 cm

IN FRUIT

FRUIT BRACT
wings point
forward

2ND-YR
TWIG
hairy

LC ■ Paper Birch
Betula papyrifera

Tree (H to 30 m) native to the USA and introduced into Britain & Ireland in 1750. **Features: BARK** usually reddish with white lenticels when young; when mature, typically white (usually with black spots and cracks) and flaking off to reveal a salmon-pink inside.

LEAF
±shiny;
5–8 pairs
of veins;
L to 9 cm

IN FRUIT

FRUIT BRACT
wings point
only slightly
forward at
most

2ND-YR
TWIG
hairless

BETULACEAE | BIRCHES

Dwarf Birch *Betula nana*
LC LC

A **branching dwarf shrub** (H to 1 m; often <30 cm in areas grazed by deer or sheep). Found in upland areas of northern England and Scotland on boggy and peaty soils, it is at the southern edge of its main world range and absent from Ireland. **Features: BARK** non-peeling; shiny coppery-red. **TWIGS** dull dark red-brown; hairy; can be resinous. **BUDS** tiny; egg-shaped to ± globular (L 1–2 mm); greenish brown to reddish brown. **LEAVES** small; broadly oval (L 5–16 mm); round-tipped; dark green; hairless; leathery; underside paler green; leaf-margins scalloped with regular teeth. **FLOWERS** ♂ and ♀ flowers separate but on the same plant; ♂s in erect catkins which droop as they mature; ♀s larger, with cone-like appearance as they ripen and are a collection of small, winged seeds. **FRUIT** small winged seeds similar to other birches.

J F M A M J J A S O N D

FLOWERS ♂ catkins pale yellowish brown; ♀ 'cone-like', greenish brown

TWIGS dull brown

LEAVES small; round-tipped; deeply toothed

BUDS smaller than those of other birches

Associated species | Dwarf Birch has formed symbiotic mycorrhizal partnerships with several fungus species (e.g. ***Lactarius helvus***, ***Lactarius rufus*** and ***Leccinum holopus***) as a survival adaptation to the generally low levels of nitrogen and phosphorus found in the harsh growing conditions it inhabits.
The moth ***Swammerdamia passerella*** is one of a few insects that have been recorded in Scotland as specific to Dwarf Birch. Other associated insects include the larvae of three sawflies. One of these, ***Nematus pravus***, had not been recorded in Britain & Ireland until it was found near Loch Ness in 2011. **Ptarmigan** also eat the buds, catkins and twigs.

Did you know? At higher elevations and in colder climates Dwarf Birch produces fewer seeds, so also reproduces vegetatively by branch layering and sprouting. It has an extensive underground root system which can account for up to 80% of the shrub's total biomass. In Greenland some have been recorded as being approximately 150 years old.
Dwarf Birch hybridizes freely with Downy Birch, and this propensity is thought to have led to the Downy Birch subspecies *B. pubescens* ssp. *tortuosa*.

Often growing in boggy places, such as here in Northumberland, Dwarf Birch forms low bushy scrub (*left*) or grows prostrate amongst the vegetation (*right*).

271

BETULACEAE | ALDERS   Leaves *p.64* | Flowers *p.68* | Fruit *p.74* | Twigs *p.81*

LC LC Common Alder   *Alnus glutinosa*

A medium-sized tree (H to 25 m), usually with quite a triangular crown. It is Britain & Ireland's only native alder, found in wet places such as in marshes and by lakes and fens, particularly on wet clays, and is also widely planted.
Features: BARK brown, smooth; becoming rough and fissured with age. TWIGS greenish to reddish brown; hairless; although usually appearing quite rough due to numerous whitish resin glands; ends of smaller branches can feel sticky. BUDS egg- or 'boxing-glove'-shaped (L 6–10 mm); purplish grey, can have a blue tinge; on short (L 2–8 mm), **hairless** stalk. LEAVES **broadly oval** (L to 10 cm); **leaf-tip blunt or notched**; leaf-base wedge-shaped; 4–8 pairs of veins; leaf-margins coarsely toothed. FLOWERS appear before the leaves; ♂ and ♀ flowers separate but on the same plant; ♂s in long, pendent, yellowish to purplish catkins (L 3–12 cm); ♀ small, cone-like (L 4–6 mm); purplish red. FRUIT **cone-like**, egg-shaped (L 8–12 mm); green, woody when ripe/open.

J F M A M J J A S O N D

IN FRUIT

this year's (unripe) fruit

♀s

FLOWERS ♂ and ♀ both form in autumn and open in spring

♂ catkin

last year's (open) fruit

TWIGS typically reddish brown, usually with a rough white covering of glands

BUDS **purplish**; egg- to 'boxing-glove'-shaped

LEAVES oval; **tip blunt to notched**; base wedge-shaped

BARK brown, rough and fissured with age

Riverine Alder woodland (*left*) is an atmospheric habitat, but less common than it was; the purple cones and twigs make Common Alder distinctive in winter (*right*).

Alders compared – *p. 275*

BETULACEAE | ALDERS

Hybrids | **Common Alder** × **Grey Alder** (*A. ×hybrida*) shows mixed leaf and fruit characteristics of the parents.

Did you know? When the wood of Alder is cut it turns bright red/orange, which led to superstitions that the tree bleeds – resulting in the tree's association with evil spirits. The real reason for the discolouring is not entirely clear but appears to be a form of a 'Maillard Reaction' – a reaction between the sugars and amino acids/proteins in wood on exposure to air, similar to what happens when a cut apple turns brown.

The bark, young shoots, wood and catkins can be used as a dye. The flowers are used to make a green dye, whereas the bark dyes cloth a reddish colour known as Aldine Red.

Alder leaf poultices were used to protect wounds against gangrene and to treat rheumatism. More recently, Alders have played a useful role in reclaiming industrial sites and slag heaps as they can improve soil fertility.

The timber of Alder was also highly valued because it does not rot quickly when exposed to continual wetting and drying. Because of this it was used for sluices and canal lock gates. Alder wood was also made into charcoal and used in the manufacture of gunpowder.

Alders grow in conditions that generally lack the nitrates needed for growth, but it can survive as its roots have nodules which contain nitrogen-fixing bacteria that extract nitrogen from the air.

Associated species (all alders) | COMMON ALDER has a high conservation value as the seeds provide good winter food for **Redpoll**, **Siskin** and other seed-eating birds. It has a symbiotic relationship with *Frankia alni*, a filamentous bacteria which lives on its roots, creating nodules. The tree provides the bacteria with carbon, and the bacteria provides the tree with nitrogen. When alder roots grow into water, they can provide shelter for fish such as **trout**. Their leaves also add nutrients to the water that in turn help feed **caddisflies** and **water beetles**.

There are also approximately 140 invertebrate species which feed on the three alder species combined, including **Swallow-tailed Moth**, **Alder Kitten Moth**, **Green Sawfly** and **Alder Leaf Beetle** – a dark metallic blue, possibly native, beetle which was refound in 2004 in Lancashire and is now widely distributed in the Midlands and south-east – which emerges in spring, sometimes in large numbers and can cause significant defoliation. Similarly, the **Large Alder Sawfly** was widespread in southern England until 1904 and thought to be extinct until it was rediscovered in Salisbury in 1997. It has subsequently become much more frequently recorded. The larvae feed on both COMMON and GREY ALDERS, but many recent records suggest that the spread may be linked to accidental reintroduction of sawflies present in the ball of soil around the roots of amenity-planted ITALIAN ALDER.

BETULACEAE | ALDERS   Leaves *p.64* | Flowers *p.68* | Fruit *p.74* | Twigs *p.81*

LC ■ Grey Alder   *Alnus incana*

A medium-sized, freely suckering tree (H to 25 m) with an open rounded to conical crown that is native to continental Europe and widely distributed throughout, growing in boggy areas and on riverbanks. It is thought to have been introduced to Britain & Ireland *c.* 1780 and is widely planted in parks, along roads and in regeneration schemes, such as reclaimed gravel-pits and collieries, as it is tolerant of a wide range of wet and dry soils. It has become naturalized by suckering and by seed on *e.g.* waste ground and along railway embankments. **Features:** BARK smooth, silvery grey at all ages; cracking into light fissures only at the trunk base of older trees. TWIGS hairy, at least near the tip; light to dark brown. BUDS egg- or 'boxing-glove'-shaped (L 6–10 mm); purplish grey; held close to the twig on short (L 2–4 mm), **hairy** stalk. LEAVES broadly oval (L to 11 cm); **leaf-tip pointed; leaf-base wedge-shaped**; 7–15 pairs of veins; leaf-margins coarsely toothed – can resemble those of Hornbeam. FLOWERS appear before the leaves; ♂ and ♀ flowers separate but on the same plant; ♂s in long, pendent yellowish to purplish catkins (L to 10 cm); ♀ small, cone-like (L 2–5 mm); purplish red. FRUIT **cone-like, egg-shaped (L 10–17 mm)**; green, woody when ripe/open.

J F M A M J J A S O N D

LC ■ Italian Alder   *Alnus cordata*

A medium-sized tree (H to 25 m) with a narrowly conical crown. Its native distribution is restricted to Corsica and southern Italy and it was brought to Britain & Ireland in the 1820s. Unlike other alders, Italian Alder thrives on dry and poor soils, which has made it an ideal tree for planting on difficult sites such as spoil heaps, roadside schemes and in other urban regeneration projects. **Features:** BARK smooth, greyish brown and smooth at all ages, although may have some warts; can be fluted at the trunk base. TWIGS hairy, at least near the tip; light to dark brown. BUDS egg-shaped (L 4–8 mm); green; held away from the stalk on long (L 3–10 mm), **hairless** stalk. LEAVES ± **heart-shaped** (L to 11 cm); **leaf-tip pointed; leaf-base incurved**; 8–9 pairs of veins; leaf-margins with small, regular teeth. FLOWERS appear before the leaves; ♂ and ♀ flowers separate but on the same plant; ♂s in long, pendent yellowish catkins (L to 10 cm); ♀ small, cone-like (L 5–10 mm); purplish red. FRUIT **cone-like, egg-shaped (L 15–40 mm)**; green, woody when ripe/open.

J F M A M J J A S O N D

Similar species | Red Alder *Alnus rubra* (N/I) is a tall tree (H to 30 m) from Pacific coastal North America. It has distinctive, pointed-oval leaves with strongly toothed margins and strong well-defined, ladder-like veins. ♂ catkins as Common Alder (L to 14 cm); mature fruit larger (L to 25 mm).
Green Alder *Alnus viridis* (N/I) is a short shrub from mountainous areas of central Europe. It has oval leaves which have very fine, sharp teeth. ♂ catkins as Common Alder (L to 14 cm); mature fruit larger (L to 34 mm).

BETULACEAE | ALDERS

BETULACEAE | HAZELS   Leaves *p.64* | Flowers *p.68* | Fruit *p.74* | Twigs *p.87*

LC LC Hazel   *Corylus avellana*

A medium-sized deciduous tree (H to 20 m) found throughout Britain & Ireland at elevations from sea level to 650 m. It occurs in woods, including as the understory of ancient and old woodlands, and hedgerows on a wide range of mildly acid to alkaline soils, including chalk, limestone and clays. It is uncommon to see Hazel growing as a full tree as it is often coppiced (see *p. 21*). Hazel sends up new shoots from the base of a tree/coppice stool every spring, which allows the stools to spread over time. These new shoots can gain 40 cm height in 1 year and take 3–5 years to produce flowers and fruit. The average life span of Hazel as a mature tree is approximately 80–90 years, but if coppiced this can be extended, with some Hazel stools estimated to be more than 1,500 years old. New stools can also be formed by layering – in which stems are pinned down and then root, forming new stools. This process has led to some very regular spacing of Hazels in some ancient woodlands. **Features:** BARK smooth, pale greenish brown, with small fissures or strips and prominent lenticels. TWIGS flexible; pale brown to olive brown. BUDS small, green, broadly oval (L 4–6 mm); **looking like tiny 'boxing gloves'**; can have brownish fringes to the bud scales. LEAVES **broadly oval to heart-shaped** (L to 12 cm) with a pointed tip; **softly hairy on both sides**; leaf-margin sharply double-toothed. FLOWERS appear before the leaves; ♂ and ♀ flowers separate but on the same plant; ♂s in long catkins (L to 80 mm) ♀ tiny bud-like clusters with star-like red styles when ripe. FRUIT oval pointed nut (L to 20 mm); surrounded by a ruff of deeply lobed bracts; in clusters of up to 4.

J F M A M J J A S O N D

BARK greenish brown; smooth with small fissures

Did you know? Pollen records indicate that Hazel was one of the first plant species to recolonize Britain & Ireland at the end of the last glacial period which ended *c.* 11,700 BCE, with Hazel scrub dominating large areas of western Scotland for around a thousand years.
Humans have domesticated Hazel since Roman times, and more than 400 cultivars have been described, including large-fruited varieties specifically grown for human consumption.

The distinctive yellow catkins of Hazel (*left*) are the first flowers of the year and bring colour to dark winter woodland. The hazelnuts (*right*) appear in clusters, but are often taken by squirrels before they ripen.

BETULACEAE | HAZELS

IN LEAF

IN FLOWER

♂ catkins

♀ **FLOWER** tiny; bud-like with star-like red styles

LEAF broadly oval to heart-shaped; pointed tip; leaf-margin double-toothed

BUDS 'boxing-glove'

FRUIT bracts ± length of nut; lobed < halfway

TWIGS flexible; pale brown to olive-brown

Similar species | Turkish Hazel *Corylus colurna* (H to 25 m) is native to SE Europe and Asia. Introduced to Britain & Ireland in 1804, it differs in fruit bract and leaf details, and its brown buds. **Filbert** *Corylus maxima* (H to 12 m) is thought by some to be a variant of Hazel and differs in details of the fruit bracts.

TURKISH HAZEL — **LEAF** narrower and more pointed; **FRUIT** bracts 2× length of nut; lobed > halfway

FILBERT — **FRUIT** larger, >2 cm; bracts much longer than nut

Associated species | Hazelnuts are particularly attractive to birds such as **Great Spotted Woodpecker** and **Nuthatch** as well as mammals including **Field Vole**, **Yellow-necked Mouse**, **squirrels** and **Hazel Dormouse** – which gathers and stores the nuts as a significant source of nutrients prior to hibernation.

Hazel has a wide range of mycorrhizal fungal partners, from the commonly recorded **Deceiver**, **Ochre Brittlegill** and **Brown Rollrim** to rare species such as **Hazel Gloves,** an IUCN Near-threatened species found occasionally in Hazel woods along the Atlantic coastal fringe of Britain & Ireland.

More than 106 invertebrate species have been found to feed on Hazel, including the widely distributed **Nut Weevil** and the endangered *Smaragdina affinis, a* short-horned leaf-beetle. There are also plants associated with Hazel, such as **Toothwort** – a chlorophyll-lacking parasite of the roots of Hazel which flowers in March and April and is pollinated by early spring bees.

NUTHATCH

HAZEL DORMOUSE

TOOTHWORT

BETULACEAE | HORNBEAM   Leaves *p.65* | Flowers *p.68* | Fruit *p.75* | Twigs *p.83*

LC LC Hornbeam   *Carpinus betulus*

A medium-sized to tall tree (H to 30m) that, as a native, is largely confined to woodland in southeastern England; widely planted elsewhere. It has a preference for damp, fertile soils, although is tolerant of many others, including sands, gravels and heavy clay. It is somewhat like Beech (*p. 214*) although usually more irregular in shape, with a ribbed, twisted and fluted trunk – features that can be particularly apparent in pollarded trees.
Features: BARK smooth and grey with an 'elephant-skin' appearance (*cf.* Beech); **growth irregularities show in the bark as uneven, almost corrugated patches resembling 'stretchmarks'**. **TWIGS** greyish brown; shiny, smooth and hairless; slightly zigzagging form; the base of the current year's growth is marked by ring-like scars. **BUDS** narrowly egg-shaped (L 6–10mm) with a **slightly curved, pointed tip**; pale red- or greenish brown; **pressed against the twig** (*cf.* Beech). **LEAVES** narrowly oval to oval (L to 100mm); **pleated along veins; surfaces appear slightly 'corrugated'**; upperside dark green; underside light green; **leaves turn golden yellow in autumn** and can remain for much of the winter. **FLOWERS** appear before the leaves; ♂ and ♀ flowers separate but on the same plant; ♂'s in long pendent catkins (L to 40mm); ♀ in shorter (L to 20mm) pendent catkins; styles red when ripe. **FRUIT pendent clusters** (L to 80mm) of **small nuts that have distinctive, usually 3-lobed, leafy bracts** (L 25–40mm); bract lobes of individual nuts differ in length.

J F M A M J J A S O N D

BARK grey; 'elephant-skin' with paler vertical 'stretchmarks'

IN LEAF

Often pollarded, Hornbeam's fluted stems become more pronounced with age (*left*); the leaves often have a corrugated appearance (*right*).

BETULACEAE | HORNBEAM

IN FLOWER
♀ catkins
♂ catkins
LEAVES slightly 'corrugated' appearance; margin toothed
SIDE-BUDS pressed against the twig (cf. Beech)
FRUIT in pendent cluster; individual seeds with 3-lobed bract
TWIG smooth; greyish brown; 'zigzagging'

Associated species | Hornbeam is valuable to wildlife, producing nutlets which are eaten by **small mammals** and also **Hawfinch**. More than 50 species of invertebrates have also been found feeding on Hornbeam, such as the **Nut-tree Tussock Moth**. Various mites make galls on the leaves, including *Aceria tenella*, which creates a shiny, smooth bulge on the leaf's upperside between the veins, with a corresponding indent, with abnormal hairs, on the underside. The fungus *Taphrina carpini* causes a Witches' Broom of dense malformed twigs to develop. Coppicing Hornbeam promotes a dense canopy and so, like beechwoods, the ground flora of a true Hornbeam wood is sparse.

HAWFINCH

TAPHRINA CARPINI

NUT-TREE TUSSOCK MOTH

Did you know? The name Hornbeam reflects its hard timber – 'horn' meaning 'hard', and 'beam' an Old English word for 'tree'. However, it is difficult to get long straight pieces of the extremely hard wood and, even if successful, the wood shrinks considerably as it dries, making it difficult to use for construction. Hence traditional uses for the wood were for small domestic and other implements, such as yokes for oxen, cogs for mills and handles for tools.

Although not good for construction, Hornbeam makes great firewood, as it burns 'like a candle' and so trees were regularly managed as coppice or pollards for firewood, which was used to fuel bread ovens, especially in London. These trees were traditionally cut every 8–15 years, resulting in Hornbeams historically rarely reaching their full height. This is particularly noticeable in locations like Epping Forest, Essex, where today there are more than 50,000 veteran pollards of Beech, Pedunculate Oak and Hornbeam – the majority of which are more than 400 years old. Regular pollarding of the trees here ceased between 150 and 200 years ago. To try to maintain the trees over the last few decades, traditional pollarding management has been reinstated, and many trees are now thriving due to a cutting rotation of 10–15 years. However, some trees have died as a result of the management, due to the length of time since their last regular cut.

CELASTRACEAE | SPINDLES Leaves *p. 66* | Flowers *p. 72* | Fruit *p. 76* | Twigs *p. 79*

LC LC Spindle ☠ *Euonymus europaeus*

A branching, deciduous shrub or small tree (H to 6 m) that, as a highly shade-tolerant native species, occurs particularly in old hedges and wood edges with a preference for lowland chalky or good deep rich soils. It is also a popular shrub planted in parks and gardens. **Features:** BARK smooth, grey and relatively nondescript, becoming ridged with age. TWIGS green, usually 4-ridged, at least at the base, giving a corky appearance. BUDS small, egg-shaped (L 6–10 mm) with a pointed tip; green with dark brown edges to the bud scales. LEAVES **oppositely arranged; broadly oval** (L to 8 cm) with a pointed tip; **turning vivid red in autumn**. FLOWERS greenish white (D 8–10 mm); in loose clusters of up to 10; 4 narrow petals alternating with 4 short stamens. FRUIT **striking pinky-red, 4-lobed capsule** (D to 15 mm) containing **bright orange arils** surrounding a whitish seed. The berries are poisonous and, if eaten, can cause liver and kidney damage, and even death.

J F M A M J J A S O N D

IN FLOWER inconspicuous

BARK grey, becoming ridged with age

Did you know? The berries historically formed part of a treatment for head lice, and the hard wood was used to make the spindles used for spinning wool, as well as for pegs and skewers.

The inconspicuous green flowers of Spindle (*left*) are at great contrast to the showy pink seed cases (*right*) and orange arils (*inset*).

CELASTRACEAE | SPINDLES

FLOWERS 4 narrow petals; 4 stamens

LEAVES oval; pointed; thin; leaf-margin untoothed or slightly toothed

SIDE-BUDS green, with dark edges to the scales

TWIGS green; usually 4-ridged

FRUIT pinky-red capsule; lobes obtusely angled

Associated species | Spindle is fairly resistant to mammal browsing but is eaten by various moth larvae, especially those of **Spindle Ermine**, the caterpillars of which generate a webbing nest which is occasionally so extensive an entire shrub can be defoliated. Another moth which uses Spindle is the **Magpie**. **Black Bean Aphids** gather on Spindle in the autumn, and are the most abundant aphid on Spindle during the winter, which led historically to Spindle being removed from hedges in an effort to contain this crop pest. This aphid abundance may explain the high number of ladybird species found on Spindle. It is the food plant for many ladybird larvae and, of all the native shrubs, has the highest recorded diversity of predatory insects. The gall mite *Eriophyes convolvens* occurs on Spindle and rolls the leaf-margins into green or red galls.

BLACK BEAN APHIDS · CATERPILLARS · SPINDLE ERMINE · ADULT

Similar species | **Evergreen Spindle** *Euonymus japonicus* (H to 8 m) is native to Japan, Korea and China. Introduced in 1804, it is planted for hedging. It also has inconspicuous, greenish white flowers but differs in having leathery, evergreen leaves with finely toothed margins. **Large-leaved Spindle** *Euonymus latifolius* (N/I) has much larger leaves (L 7–16 cm); flowers usually with 5 petals; and much larger fruit (D 15–25 mm).

EVERGREEN SPINDLE

LEAVES oval; pointed; widest above middle; leathery; leaf-margin finely toothed

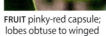

FRUIT pinky-red capsule; lobes obtuse to winged

FLOWERS greenish white; similar to those of Spindle

281

PLATANACEAE | PLANES — Leaves *p.62* | Flowers *p.69* | Fruit *p.74* | Twigs *p.81*

■ London Plane — *Platanus × hispanica*

A tall deciduous tree (H to 40 m) that is probably a hybrid between Oriental Plane and American Plane, although it is possibly a form of Oriental Plane. First noticed about 1663, it might be that the hybrid cross occurred in Tradescant's garden in Lambeth, where both species were growing. London Plane is drought-tolerant and wind-resistant when established and also copes well with pollution. These characteristics mean that it was planted widely in urban streets and squares, particularly by the Victorians, and is a significant feature of some cities. **Features:** BARK grey-green; developing distinctive large plates which flake off to reveal pale cream patches. TWIGS hairless, reddish brown to grey-brown. BUDS shiny; reddish; conical (5–7 mm), curved; hairless. LEAVES arranged alternately (*cf.* maples (*pp. 284–289*); **relatively thick; palmately lobed** (L to 20 cm); lobe cut to ±50% of the way to the midrib; central lobe longer than wide; leaves show characteristics that are intermediate between the putative parent species, being more deeply lobed than those of American Plane and more shallowly lobed than those of Oriental Plane. FLOWERS ♂ and ♀ flowers separate but on the same plant; in pendent clusters of usually 2, but up to 4 ball-like flower-clusters consisting of many tiny flowers; ♀s (D ± 25 mm) with red stigmas; ♂s in smaller green 'balls' with brownish stamens. FRUIT **ball-like clusters** (D 25–35 mm), like large female flowers, which break up over the winter, releasing small seeds that have short, stiff hairs.

J F M A M J J A S O N D

Did you know? One of the oldest London Planes in the country is in the garden of King's Ely School, Cambs. Thought to have been planted *c.*1680 by Peter Gunning, the then Bishop of Ely it is still growing well in 2025. In 2009, a London Plane planted in Berkeley Square in 1789 was valued at £750,000 by the London Tree Officers Association, for its 'amenity value', making it the most valuable tree to have been assessed.

London Plane's large leaves and broad canopy provide valuable shade, but they are not without their difficulties as the hairs on the seeds can cause nasty irritation to human skin and eyes.

Often planted in streets, London Planes form large city trees (*left*); and their distinctively shaped large leaves and fruits are readily recognizable (*right*).

PLATANACEAE | PLANES

IN FLOWER

♂ ♂

BEWARE flower clusters can fall off, leaving just one, *cf.* American Plane

BUDS shiny; conical; characteristically curved

♀ **FLOWER** in bud

TWIGS usually grey-brown; hairless

♂

FLOWERS in 2–4 ball-like clusters; ♂ brownish stamens; ♀ red stigma

♀

LEAVES usually 3–5-lobed; end-lobe less than half leaf length and longer than wide

FRUIT ball-like clusters

BARK grey-green; flaking to reveal creamy patches

VU Oriental Plane
Platanus orientalis

Known to be growing in Britain & Ireland by the end of the 16th century, but not common, it is occasionally found in parks and gardens in southern England. **Features: BARK** grey-brown, **flaking to reveal a mix of brown and green patches. LEAVES** 5–7-lobed; much more deeply cut (>66% of the distance to the midrib) than those of London Plane. **FRUIT** ball-like; **3–6 per stalk**.

LEAVES usually 5-lobed; **more deeply cut than London Plane**; end-lobe at least half leaf length

BARK grey-brown; flaking to reveal brown and green patches

FRUIT

LC American Plane
Platanus occidentalis

Rarely planted in Britain & Ireland, **Features: LEAVES** more shallowly lobed than those of London Plane; central lobe that is generally wider than long. **BARK** grey-green, flaking to reveal a mix of white, cream, green and grey (*cf.* London Plane). **FLOWERS/FRUIT** only **1 flower/fruit ball per stalk**.

LEAVES as London Plane but **less deeply lobed; end lobe wider than long**

BARK grey-green; flaking to reveal white, cream, green and grey patches

FRUIT

283

SAPINDACEAE (MAPLES) | MAPLES   Leaves *p. 62* | Flowers *p. 69* | Fruit *p. 75* | Twigs *p. 79*

LC LC Field Maple   *Acer campestre*

A medium-sized tree (H to 14 m) that is native to southern England (becoming increasingly rare north of Shropshire, Derbyshire and Lincolnshire) and Wales, and introduced in Scotland and Ireland. Generally a lowland species (rare above elevations of 300 m), occurring in hedges and as the understorey in woods and copses, with a preference for lime-rich soils although will tolerate other conditions – *e.g.* in Wales its distribution is largely associated with the limestone and old red sandstone areas of southern Wales. Field Maple is not a pioneer species and usually arises in pre-existing scrub or hedges. It is commonly grown in parks and gardens for its autumn colours and ability to tolerate air pollution.

Features: BARK can be corky on young trees, becoming fissured and scaly with age. TWIGS red-brown; can be covered in minute hairs and can develop distinctive corky ridges. BUDS bud scales typically brown (can be purplish or greenish) fringed with white hairs; end-buds usually comprising one (or two) main egg-shaped bud flanked by two smaller buds, the group (L 3–5 mm); side-buds single, similar; pressed to the twig. LEAVES arranged oppositely; variable; **palmately 3–7-lobed** (L 3–11 cm); **leaf-margin untoothed**; upperside dark green; underside light green, can have scattered hairs that can be dense in the vein junctions. FLOWERS open with the leaves; yellowish green (D ± 6 mm); 5-petalled; in rounded branched clusters of approximately 10 flowers which start erect but droop with age. FRUIT a pair of **winged seeds angled at ± 180°**; wings L 25–30 mm.

J F M A M J J A S O N D

> **Did you know?** Field Maple generally live for about 200 years, but one Field Maple in Boldre, Hampshire, was at least 150 years old when it was recorded by William Gilpin in 1800 and was still alive in 1950.
> Historically, Field Maple was used for topiary, and the soft wood used in wood turning, producing veneers as well as for making furniture and harps. The Anglo-Saxon ship burial found at Sutton Hoo, Suffolk contained a musical instrument believed to be a lyre made from Field Maple. Hair attached to the wood suggests that the lyre was kept in a beaver skin case.

BARK fissured and scaly with age

A fully grown Field Maple is an uncommon sight (*left*), as they more typically grow close to other trees where a full canopy cannot develop although the straight-winged fruit and palmately lobed leaves (*right*) are recognizable even in a managed hedge.

SAPINDACEAE (MAPLES) | MAPLES

IN LEAF

LEAVES variable in shape, even on the same tree; typically with **3 main lobes and 2 smaller basal lobes**

IN FLOWER

FLOWERS greenish; in clusters which are erect at first, but droop as they age

END-BUDS typical maple form; bud scales usually **reddish brown with pale margin**

FRUIT pair of winged seeds; **wings angled at ± 180°**

TWIGS red-brown; buds opposite

An ancient Field Maple growing in Suffolk.

MAPLE PROMINENT

Associated species | Field Maple is an important habitat for up to 51 invertebrate species including the **Maple Prominent** moth, the caterpillars of which feed on the leaves.

The fruit 'keys' are eaten by **small mammals**.

285

SAPINDACEAE (MAPLES) | SYCAMORE   Leaves *p.62* | Flowers *p.69* | Fruit *p.75* | Twigs *p.79*

LC ■ Sycamore   *Acer pseudoplatanus*

A tall tree (H to 35 m) which is probably introduced from Europe but may be native (see below). Sycamore grows in a wide range of habitats and soil types. It is an excellent colonizer and is now one of the most widely distributed trees in Britain & Ireland. **Features:** BARK smooth and greyish when young, fissuring and scaling off in large strips on older trees to reveal a pale reddish colour underneath. TWIGS grey- to greenish brown; shiny; hairless. BUDS hairless; bud scales yellowish green; with a dark brown margin; end-buds usually comprising one (or two) main egg-shaped bud flanked by two smaller buds, the group (L 5–10 mm) looking ± 'cabbage-shaped'; side-buds single, smaller (L 3–8 mm); at 45° to the twigs. LEAVES arranged oppositely; **palmately 5-lobed** (L 10–15 cm) **with irregular teeth around the leaf-margin**; prominent side veins look darker than the rest of the leaf and give the leaf a slightly ridged appearance; upperside dark green, shiny; underside paler, can have reddish hairs on the main vein. Leaves turn brownish in autumn and many have **distinctive black blotches caused by the Tar Spot fungus.** FLOWERS open when the leaves are fully developed; yellowish green (D ± 6 mm); 5-petalled; in pendent ± cylindrical spikes of 25–150 flowers. FRUIT pair of **winged seeds angled at ± 45°**; wings L to 25–45 mm; ripening seeds typically have a reddish pink tinge.

J F M A M J J A S O N D

Did you know? There is a great deal of controversy about the date when Sycamore may have arrived in Britain & Ireland. Some suggest it was introduced by the Romans, others in The Middle Ages, or even to Scotland in the 1550s. The Sycamore was first written about in England in 1578 by the botanist and early herbalist Henry Lyte, although there is a carving at Christ Church, Oxford on a shrine to St Frideswide, the patron saint of Oxford, which includes leaves and seeds of Sycamore which is dated 1289. Over the last decade, it has been suggested that the species may be a native after all and should be called the Celtic Maple.

BARK greyish and fissured with age

Mature Sycamores are impressive, stately trees (*left*); the distinctly angled pairs of seeds are very obvious for most of the summer (*right*).

SAPINDACEAE (MAPLES) | SYCAMORE

IN LEAF

FLOWERS yellowish green; in pendent cylindrical spikes

LEAVES palmately 5-lobed; leaf-margin irregularly toothed

TWIGS grey- to greenish brown; buds opposite

END-BUDS typical maple form; bud scales **green with dark brown margin**

FRUIT pair of winged seeds; **wings angled at ±45°**

Associated species | Sycamore supports only a limited number of insect species, but it can support large numbers of individuals of certain aphids, particularly **Common Sycamore Aphid** and **Common Periphyllus Aphid**. As an example, in one study 2·25 million aphids were recorded on a single mature 20 m tall tree. Consequentially, migrating insectivorous birds such as **warblers** and **flycatchers** can often be found feeding in Sycamores in the autumn.

High numbers of aphids have an impact on the tree's ability to produce new growth, by removing sugars that would otherwise be laid down as wood. Curiously, however, a high number of aphids may have a different advantage for the tree, as high levels of sugary honeydew fall to the ground around the tree, increasing the levels of nitrogen-fixing bacteria, which potentially benefit the tree.

Sycamores also support good **lichen** growth, particularly in the west of Britain & Ireland, with more than 170 species being recorded. The ability of lichen populations to develop is due to the relatively high alkalinity level of the bark (similar to elms). One of the rarer species is the legally protected **Golden-hair Lichen**, which usually has a bright yellow 'brillo-pad' appearance, although when it grows on trees it can have a grey-green colour. One of the best sites for this species in Britain is a line of old Sycamores in coastal Devon.

COMMON SYCAMORE APHID

CHIFFCHAFF

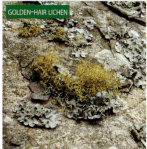
GOLDEN-HAIR LICHEN

SAPINDACEAE (MAPLES) | MAPLES   Leaves *p. 62* | Flowers *p. 69* | Fruit *p. 75* | Twigs *p. 79*

LC Norway Maple   *Acer platanoides*

A medium-sized tree (H to 30 m) with a broad, rounded crown. A widely distributed European native introduced to Britain & Ireland in the late 17th century and now widely naturalized as well as planted in streets, parks and gardens. **Features:** BARK grey-brown, smooth when young; developing small furrows with age. TWIGS hairless and green at first, soon becoming pale brown. BUDS shiny; green at the base and reddish brown at the tip, or wholly reddish brown; end-buds usually comprising one main egg-shaped bud flanked by two smaller buds, the group (L 6–10 mm) looking 'turban-shaped'; side-buds single, smaller (L 3–8 mm); pressed to the twig. LEAVES arranged oppositely; shiny green **palmately 5–7-lobed** (L 8–25 cm) with **bristle points at the lobe tips and a few other pointed teeth around the leaf-margin** (*cf.* Cappadocian Maple). FLOWERS open before the leaves; yellowish green (D ± 8 mm); 5-petalled; in upright clusters of 10–30 flowers. FRUIT pair of **winged seeds widely angled up to 180°** (*cf.* Field Maple (*p. 284*)); wings L 35–70 mm.

J F M A M J J A S O N D

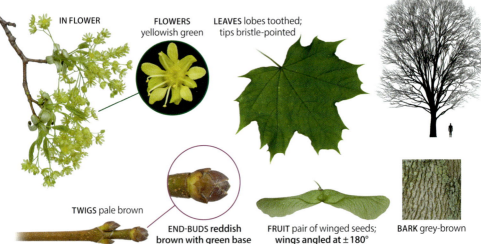

IN FLOWER

FLOWERS yellowish green

LEAVES lobes toothed; tips bristle-pointed

TWIGS pale brown

END-BUDS reddish brown with green base

FRUIT pair of winged seeds; wings angled at ± 180°

BARK grey-brown

Did you know? Norway Maples have been recorded in Europe as being 250 years old, but their lifespan is usually much shorter. It is the occasional host of Mistletoe.

Purple-leaved forms (*e.g.* 'Crimson King') makes a handsome street or garden tree (*left*); the early flowers of Norway Maple add touch of colour to the late winter landscape (*right*).

SAPINDACEAE (MAPLES) | MAPLES

Other planted maples

LC Silver Maple *Acer saccharinum*

A tall tree (H to 40 m), native to North America and introduced into Britain & Ireland in 1725 by Sir Charles Wager. **Features: TWIGS** greenish. **BUDS** end-bud usually angled. **LEAVES deeply 5-lobed**; lobes long with pointed tips and serrated triangular teeth; upperside shining green, **turning pale yellow in autumn**; underside silvery white with scattered hairs; leaf-stalk lacks milky sap. **FRUIT** pair of **winged seeds angled at ± 90°** on slender drooping stalks, ripening early in May or June although rarely seen in Britain & Ireland.

FRUIT wings angled at ± 90°

LEAVES deeply 5-lobed with large triangular teeth; underside silvery-white

LC Cappadocian Maple *Acer cappadocicum*

A medium-sized tree (H to 25 m), native to Asia and introduced in 1840. Widely planted as a tree of gardens and public spaces. **Features: TWIGS** greenish, although can become reddish on the sunlit side. **LEAVES** simple **palmately lobed, with 5–7 pointed lobes and no teeth** (*cf.* Norway Maple). **FLOWERS** ± erect clusters like those of Norway Maple. **FRUIT** resembles that of Norway Maple; wings L 35–70 mm.

FRUIT wings widely angled *cf.* Norway Maple

LEAVES 5–7-lobed; **lobes untoothed** *cf.* Norway Maple

LC Red Maple *Acer rubrum*

A medium-sized to tall tree (H to 40 m), native to North America and the first American maple to be introduced (1656) into Britain & Ireland. Occasionally planted for its spectacular red colour in autumn. **Features: TWIGS** reddish. **LEAVES 3 or 5 shallow palmate lobes**, with a toothed margin. If 5-lobed then the 2 basal lobes are very small. **FRUIT** a pair of light brown to reddish winged seeds acutely angled to 60° on long slender stems ripening from April to June although rarely seen in Britain & Ireland.

FRUIT wings reddish; acutely angled to 60°

LEAVES 3 large (± 2 tiny basal) lobes with **toothed margins**

LC Ashleaf Maple *Acer negundo*

A medium-sized unisexual tree (H to 25 m), native to North America and introduced into England in 1688, cultivated in the Fulham Garden by Bishop Compton. **Features: TWIGS** greenish with slightly 'frosty'-looking buds. **LEAVES** pinnate; 3 or 5 oval, stalked leaflets. The end leaflets can also be 3-lobed and have a sharply pointed tip. Leaves turn yellow in autumn. **FRUIT** pair of **winged seeds angled at < 90°, in pendent cluster** somewhat like those of Ash (*p. 298*).

♀ in fruit

FRUIT wings angled *cf.* Norway Maple

LEAVES pinnate with 3 or 5 oval, stalked leaflets; end leaflet can be 3-lobed

SAPINDACEAE (MAPLES) | HORSE CHESTNUTS   Leaves *p. 60* | Flowers *p. 71* | Fruit *p. 74* | Twigs *p. 79*

VU ■ Horse Chestnut ☠   *Aesculus hippocastanum*

A medium-sized to tall tree (typically H to 25 m; a few reach 40 m; trunk girth to 6 m) with a flat-topped crown (D to 20 m) and upswept branches that dip down to form low, spreading limbs that turn sharply up at the ends. It is native to the Balkans, where it considered globally VULNERABLE. It was introduced into Britain & Ireland in the early 17th century and is extensively planted – commonly found in streets, parks and gardens, even in inner city areas as it is tolerant of atmospheric pollution. It is often self-seeded and has a preference for moist, well-drained soils although it is found across a wide range of soil types and conditions, from chalk and dry sandy soils to wet clays. **Features:** BARK dark grey-brown, smooth when young, but can form long fine strips or scales that can fall from the tree with age. TWIGS stout; with distinctive pale lenticels and conspicuous leaf-scars. BUDS **large, pointed egg-shape** (L 15–30 mm); **red-brown and very sticky.** LEAVES oppositely arranged; **palmate with 5–7 pointed-oval leaflets** (L to 25 cm); leaf-margins irregularly toothed. FLOWERS white (D 25–35 mm) with basal spots that start yellow then turn pink in erect spikes (L 10–30 cm) of 20–50 flowers; ♂s at the top, above bisexual and then ♀s below; petals 4 or 5 (at least some flowers have 5); stamens just exceeding petals (by ≤ 1 cm); 2–5, rarely 8 of the ♀ flowers at the base of the spike develop into fruit. FRUIT **distinctive spiny capsule** (D ± 65 mm) containing 1 (rarely 2 or 3) shiny, red-brown nut ('conkers').

J F M A M J J A S O N D

> **Did you know?** Since 1965 there has been an annual World Conker Championship in Northamptonshire on the second Sunday in October. The main threat to the native populations in the Balkans is the species' limited ability to disperse to new areas – landslides, clear cutting and soil erosion have left the remaining wild population small and declining. Overall, it is estimated that the population is smaller than 10,000 trees with no area containing more than 1,000 trees.

Whether in fruit (*left*) or in flower (*right*), Horse Chestnut is distinctive. From mid-summer the leaves can show signs of damage caused by the caterpillars of a leaf-mining moth (see *opposite*) which hollow out the leaves.

FLOWERS at least some 5-petalled; stamens just exceeding petals *cf.* Indian Horse Chestnut

SAPINDACEAE (MAPLES) | HORSE CHESTNUTS

IN FLOWER

BARK dark grey-brown with long strips/scales

HORSE CHESTNUT LEAF-MINER: ADULT AND LEAF DAMAGE

Associated species | The nuts provide food for **deer** and other mammals and its flowers provide pollen for insects. There are increasing problems caused by the **Horse Chestnut Leaf-miner** moth which appears to decrease the weight of the seeds (conkers), impacting seedling germination and vigour.

LEAVES palmate; 5–7 **unstalked** leaflets; **widest near the abruptly pointed** tip

END-BUDS large; **sticky** *cf.* other horse chestnuts (see *p. 79*)

TWIGS stout with conspicuous leaf-scars; buds

FRUIT spiny capsule that contains conkers

NUT (conker)

LC Indian Horse Chestnut
Aesculus indica

A medium-sized tree (H to 30 m) native to the Himalayas, between Nepal and Kashmir. It was introduced to Britain & Ireland in the mid-19th century and commonly planted in parks and gardens.
Features: END-BUDS not sticky; FLOWERS 4-petalled; white with variable pink, red and yellow markings; stamens exceed petals by ± 2 cm. FRUIT globular; lacking protuberances.

J F M A M J J A S O N D

FLOWERS 4-petalled; stamens much exceeding petals

LEAVES palmate; 5–7 **stalked** leaflets; **narrower than Horse Chestnut**; tip **gradually pointed**

FRUIT smooth

Red Horse Chestnut
Aesculus ×carnea

A small tree (H to ± 12 m) that is a hybrid between Red Buckeye (N/I) and Horse Chestnut. It seems to have first appeared in Germany some time before 1820, arriving in Britain & Ireland soon after. **Features:** END-BUDS not sticky; FLOWERS bright pink to red. FRUIT globular; a few blunt protuberances at most.

J F M A M J J A S O N D

FLOWERS red

LEAVES palmate with 5–7 **very shortly stalked** leaflets; tip **abruptly pointed**

FRUIT ± smooth

MALVACEAE (MALLOW) | LIMES   Leaves p.64 | Flowers p.69 | Fruit p.74 | Twigs p.81

Limes *Tilia* species

The three native limes share the same basic leaf and flower morphology. The buds are a distinctive 'boxing-glove' shape; the leaves are broadly heart-shaped with a pointed tip (the base can look asymmetric, reminiscent of an elm (*p. 197*)); the very long and papery flower bracts are also distinctive and obvious from a distance; flowers and fruits are in clusters (which can be dangling, spreading or erect, depending on the species).

(EUROPEAN) LIME – *pendent clusters with numerous flowers*

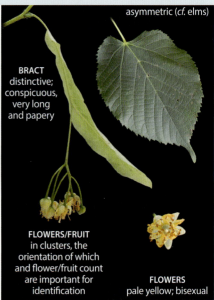

asymmetric (*cf.* elms)

BRACT distinctive; conspicuous, very long and papery

FLOWERS/FRUIT in clusters, the orientation of which and flower/fruit count are important for identification

FLOWERS pale yellow; bisexual

Identifying Limes

Much of the confusion that arises when identifying limes is due to the considerable variation that can be found both within populations of each species but also on individual trees – related to *e.g.* environment and age. **It is important to check leaves, flowers and fruits on well-lit flowering shoots**, as these show a tree's true features. Shaded leaves, such as those under the canopy, can be similar in all three species, making them unhelpful for identification purposes.

		Large-leaved (*p. 292*)	**Common Lime** (*p. 291*)	**Small-leaved** (*p. 290*)
TWIGS		very hairy when young	± hairless	hairless
LEAVES	LENGTH	L 6–12 cm	L 6–9 cm	L **3–6 cm**;
	UNDERSIDE	**hairy**	hairless except for a **few whitish forked hairs** in the vein axils (can be hairy along veins *cf.* Large-leaved Lime)	hairless except **tufts of orange hairs** in the vein axils
BUDS		L 4–10 mm; 3 visible scales		L 3–6 mm; **2 visible scales**
FLOWERS		clusters usually hanging down		clusters usually **± erect**
		3–5 per cluster	5–10 per cluster	5–15 per cluster
FRUITS	RIBS/RIDGES	**prominent**	slight	faint or absent
	WALLS	thick and robust	woody and hard	densely hairy, thin, fragile

Similar species | **Silver Lime** (*p. 297*) has leaves with a densely hairy underside.

MALVACEAE (MALLOW) | LIMES   Leaves *p.64* | Flowers *p.69* | Fruit *p.74* | Twigs *p.81*

LC LC Small-leaved Lime   *Tilia cordata*

A large, tall tree (H to 40 m), with a dome-like crown. An uncommon native of oak and ash woodland as far north as Cumbria, and occurring as lime woodland in Hampshire, Worcestershire, Wales and East Anglia. It is a planted tree in Scotland and Ireland and is also widely planted in parks, woodlands and as a street tree throughout its native range. It has a preference for sunny or semi-shaded conditions on moist loamy alkaline to neutral soils but will also grow on slightly acid soils so long as not too dry or too wet. It can also tolerate exposure to wind. Small-leaved Lime spreads readily via shoots and was historically coppiced for its wood; some trees are thought to be up to 2,000 years old. **Features:** BARK smooth and grey; developing narrow fissures and scaly ridges with age; burrs absent (*cf.* Common Lime). TWIGS green when young becoming dark reddish brown with age. BUDS 'boxing-glove' shaped (L 3–6 mm) – smaller and rounder than other limes; green to rusty red; **2 of the 3 bud scales visible** (the third usually only visible at the tip; the distance between buds (L 0·5–4·0 cm) is similar to the much larger-budded Large-leaved Lime but much less than in Lime. LEAVES **heart-shaped (L 3–6 cm);** dark green and largely hairless, although **tufts of orange hairs are usually present in the vein axils on the underside**; side veins not particularly prominent and more irregularly spaced than in other limes. FLOWERS yellowish green; in clusters of 5–15 **flowers held ± erect above the leaves** above a conspicuous, long, pale green bract; 25–30 stamens per flower. FRUIT ± spherical (D 6–8 mm); densely hairy; faintly ridged at most.

J F M A M J J A S O N D

BARK burrs absent

> **Did you know?** In Roman times Small-leaved Lime was known as 'the tree of a thousand uses'. The wood is soft and was used in carving and wood turning, model making and for making keys for pianos and organs. The best-known carver of lime was Grinling Gibbons (1648–1721), whose work decorates churches and stately homes around the country. Lime honey is highly prized across Europe. A form of woven rope was made from the under or inner bark (bast). Lyndhurst, in the New Forest, is thought to owe its name to the prevalence of this species in ancient times (Lynd = lime; hurst = hill).

Often seen as coppiced trees (*left*), Small-leaved Limes are readily recognized by their conspicuous bracts and clusters of flowers and fruit (*right*) which are held above the leaves, unlike other limes.

Limes compared *p. 293* MALVACEAE (MALLOW) | LIMES

■ Common (European) Lime *Tilia ×europaea*

A large, tall tree (H to 40m), which seems to be a hybrid between Small-leaved and Large-leaved Limes that occurs naturally in limestone woods (where both parents are present) in the Wye Valley, parts of Derbyshire, Staffordshire and Yorkshire. Elsewhere it appears the tree may have first been planted in Britain & Ireland during the early 17th century, probably from European stock. It is widely planted in streets, parks and gardens and commonly found in woods and scrub on a wide range of soils, although it has a preference for those that are lime-rich. It spreads readily via suckers; mature trees arising from seed are very rare. **Features:** BARK smooth and grey with vertical rows of lenticels when young which, with age, develop into shallow fissures and ridges which crack the surface into small rectangular scales; large irregular growths on the trunk (burrs) present on most (*cf.* Common Lime); **characteristic masses of young shoots which develop from the trunk** (epicormic growth) are also present on many. TWIGS usually hairless; green when young becoming dark reddish brown with age; those in sun the most red. BUDS 'boxing-glove' shaped (L 5–10mm); green to reddish brown; all 3 bud scales visible; the distance between buds (L 5–10cm) is much greater than in other limes. LEAVES **heart-shaped (L 6–9cm)**; upperside dark green, hairless; underside lighter green with whitish to buff hairs in the vein axils and along the veins in some (*cf.* Large-leaved Lime). FLOWERS yellowish green; **in pendent clusters of 5–10** below a conspicuous, long, pale green bract. FRUIT ± spherical (D 7–8mm); slightly ridged; seeds not usually viable.

J F M A M J J A S O N D

BARK burrs usually present

A significant number of suckers at the base of a trunk is characteristic of Lime.

Common Lime can have epicormic growth – shoots emerging from the trunk (*left*); and are often recognizable at distance by their conspicuous bracts with pendent flowers or fruit in clusters of 5–10 (*right*).

MALVACEAE (MALLOW) | LIMES | Leaves p.64 | Flowers p.69 | Fruit p.74 | Twigs p.81

LC LC Large-leaved Lime *Tilia platyphyllos*

A large, tall tree (H to 35 m in woodland; to 25 m in hedgerows), with a broad crown; can be multi-stemmed after coppicing or damage. A rare native of woodland in the Wye Valley, Pennines and South Downs and introduced in Scotland and Ireland. Widely planted in parks, woodlands, hedges, along old boundary banks and occasionally as a street tree. It has a preference for sunny or semi-shaded conditions on lime-rich soils but will tolerate sandy areas. Large-leaved Lime spreads readily via shoots from the base of the trunk.
Features: BARK smooth and grey with vertical rows of lenticels when young which, with age, develop into shallow fissures and ridges which crack the surface into small rectangular scales; burrs absent (*cf.* Common Lime). TWIGS greenish brown or reddish brown with long dense hairs near the buds at least when young. BUDS 'boxing-glove' shaped (L 4–10 mm); green to rusty red; all 3 bud scales visible; distance between buds (L 1·5–3·0 cm) similar to the much smaller-budded Small-leaved Lime but much less than in Common Lime. LEAVES heart-shaped (L 6–12 cm); upperside dark green with many short hairs; **underside lighter green with whitish simple hairs along the veins** (*cf.* Common Lime) and also in the vein axils. FLOWERS yellowish green; in pendent clusters of 3–6 below a conspicuous, long, pale green bract; 40–45 stamens per flower. FRUIT ± spherical (D 6–8 cm); with 3–5 prominent ridges.

J F M A M J J A S O N D

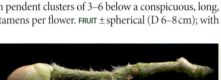

Young twig showing the highly distinctive long hairs and the 'boxing-glove' bud with 3 bud scales.

Did you know? Large-leaved Lime flowers produce considerably more nectar than Small-leaved Lime, and honeybees use the flowers, making the tree important to beekeepers, who can produce a pale and richly flavoured lime-tree honey.
As with the other limes, the wood of this species resists splitting and was used for artificial limbs, wooden clogs and toys.

BARK burrs absent

Large-leaved Lime (like Small-leaved Lime) is often found as old coppice stools (*left*) and is usually recognizable at distance by its conspicuous bracts with pendent flowers or fruit in clusters of 3–6 (*right*).

Associated species | Small-leaved Lime, Large-leaved Lime and Common Lime all have value to wildlife. Lime flowers attract insects, particularly bees. It is one of the main hosts of **Mistletoe**.

LIME is known to host 31 insect species. LARGE-LEAVED LIME flowers produce the most nectar and are a major source of it for honeybees and bumblebees, especially **White-tailed**, **Buff-tailed** and **Early Bumblebees**.

As well as bees, LARGE-LEAVED LIME flowers are used by many moths, including **Heart and Club**, and **Shuttle-shaped Dart**. The foliage is used by the caterpillars of the **Lime Hawkmoth** and is also very palatable to browsing animals, particularly **Fallow Deer**.

Eriophyes tiliae, a tiny mite (L ≤ 0·2 mm), creates a nail gall. As a mite drink saps from the leaf, chemicals are released, causing a red or yellow-green distortion to develop on the upperside of the leaves. The galls do not seem to impact a tree's health. The substance that drips like fine rain in summer from these limes is honeydew – a sweet sticky residue excreted by sap-sucking insects such as scale insects and aphids.

The **Lime Aphid** is particularly noticeable in urban areas, creating large quantities of honeydew which can drop onto cars and streets beneath the trees. This residue is also prone to the development of a black mould which can cause concern. **Ants** 'farm' the aphids to supplement their diet, by collecting the honeydew from them, like milk from a cow.

SMALL-LEAVED LIME and COMMON LIME are used by **Great Spotted Woodpeckers** for 'sap-sucking' in which the birds drill horizontal lines of holes into the bark and feed on the sugary sap that oozes from them. The fruits are a source of food for **Nuthatches** – which have been observed wedging fruit into grooves in the bark, making it easier for the birds to use their strong beak to crack the fruit open and get at the seeds within. Small mammals such as **Bank Vole** and **Wood Mouse** also crack open the fruits and eat many of the seeds. Seed predation was thought to be behind the lack of regeneration of the trees but it is now thought to be due to the grazing of young trees by the increasing deer population.

MISTLETOE

NAIL GALL

GREAT SPOTTED WOODPECKER SAP-SUCKING HOLES

LIME HAWKMOTH

LC ■ Silver Lime *Tilia tomentosa*

A tall tree (H to 35 m) native to southern Europe and the Balkans, where it can form extensive woodlands (*e.g.* in Romania). It was introduced to Britain & Ireland in 1767 and has been planted in parks and gardens. **Features:** LEAVES heart-shaped (L 6–11 cm); upperside dark green; **underside densely whitish hairy**; flashes conspicuously in the wind; unlike other limes in Britain & Ireland, the leaves do not produce honeydew. FLOWERS 6–10 flowers in pendent cluster.

IN LEAF

LEAVES heart-shaped; **underside with dense covering of whitish hairs**

OLEACEAE | ASH   Leaves *p.61* | Flowers *p.69* | Fruit *p.75* | Twigs *p.79*

NT LC Ash   *Fraxinus excelsior*

A tall, spreading tree (H to 30 m) found in woodland, scrub and hedgerows. It is one of the commonest trees in Britain & Ireland, estimated at 2 billion trees (including seedlings and saplings). It is tolerant of shade (saplings growing under cover of other trees) and prefers deep loamy soils, although can thrive in very exposed positions, such as at the coast, where it may become shaped by the wind. It is not widely planted for amenity as it lacks strong autumn colour. **Features:** BARK smooth and grey when young; becoming cracked and fissured with age; many are covered in lichens. TWIGS grey; smooth. BUDS **distinctive; velvety black**; end-buds like a 'bishop's mitre' (L 5–13 mm); side-buds much smaller (L 1–6 mm). LEAVES oppositely arranged; **pinnate** (L 20–35 cm) with **9–13 pairs of pointed-oval leaflets** (L to 7 cm); leaflet margins toothed; stipules absent (*cf.* Rowan *p. 140*). FLOWERS petal-less ♂ and ♀ flowers separate but on the same plant, appearing before the leaves; ♂'s with 2 purple stamens in a spiky-looking purple sphere; ♀s in looser clusters with dark purple stigma. FRUIT **1-winged 'keys'** (L ±4 cm) in large pendent clusters (L to 20 cm); green, ripening dark brown over 2 seasons.

J F M A M J J A S O N D

Did you know? Ash is a valued part of the national treescape, especially in limestone areas such as the Cotswolds, where it has been dominant and historically was managed as old pollards, particularly for wood fuel.

Its strong wood is flexible and was used as an element in wheel making, for skis, oars and tool handles, and is still used for high-quality furniture.

The native population is being impacted by Ash Dieback, caused by *Hymenoscyphus fraxineus*, a fungal pathogen from Asia which first arrived in Europe during the 1990s and then spread rapidly. The first official record in Britain & Ireland was in 2012, but there is retrospective proof that trees were affected back in 2004. If infected, most trees will die, although evidence from Europe suggests that 10% of trees are moderately tolerant, and 1–2% highly tolerant, to infection.

Ash trees have flowers which may be ♂, ♀, or hermaphrodite. In one population studied, 20.5% of trees were ♂, 68.1% hermaphrodite and 11.4% female. Intriguingly, ♂ trees can become female and ♀ trees male. Why and how these changes happen is still poorly understood. However, although Ash gender may have a genetic component, it appears that an increase in the 'femaleness' of the population is linked to higher average minimum daily temperature in the first half of April.

BARK grey; cracked and fissured on older trees

One of the most impressive Ash trees in Britain (*left*), this ancient tree has a huge girth (± 9 m); the dangling clustered 'keys' of fresh, green, winged seeds (*right*).

OLEACEAE | ASH

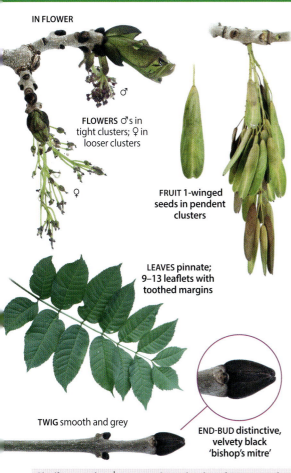

IN FLOWER

FLOWERS ♂'s in tight clusters; ♀ in looser clusters

FRUIT 1-winged seeds in pendent clusters

LEAVES pinnate; 9–13 leaflets with toothed margins

TWIG smooth and grey

END-BUD distinctive, velvety black 'bishop's mitre'

Similar species | **Narrow-leaved Ash** and **Manna Ash** (both *p. 340*); **Rowan** (*p. 140*).

Associated species | Ash is a hugely valuable native tree and supports almost 1,000 species, including a huge variety of **lichens** (more than 780 species), **invertebrates** and other wildlife. It is a species with a high wildlife value, and it is used as a nesting site by **woodpeckers** and other hole-nesters, including **Redstart**. The unwelcome fungus *Hymenoscyphus fraxineus* causes Ash Dieback.

HYMENOSCYPHUS FRAXINEUS

REDSTART

Growing on limestone pavements, these dwarf Ash trees may be several hundred years old (*left*); Ash Dieback is now seriously impacting woodlands, where many of the trees are dying (*right*).

FABACEAE (PEAS) | FALSE-ACACIA   Leaves p.61 | Flowers p.72 | Fruit p.74 | Twigs p.80

LC False Acacia   *Robinia pseudoacacia*

A medium-sized, fast-growing, suckering, spiny, deciduous tree (H to 25m) that is native to the coastal eastern USA and introduced into Britain & Ireland sometime in the 17th century. The first reference is in a catalogue from Tradescant's garden in London in 1634. Originally introduced as a garden and ornamental tree, the species has been widely planted (especially the cultivar 'Frisia' – the Golden Acacia) and has established itself in the south-east. Its nitrogen-fixing ability has enabled its spread along transportation corridors and rivers, in harsh urban habitats, rocky areas, and open woodland.

Features: BARK grey-brown to dark brown, becoming fissured with age. TWIGS quite brittle and liable to wind damage; reddish to purplish brown; spiny; spines (see LEAVES below) (L to 2cm) **flattened with a broad base**; in pairs. BUDS minute, hard to see; naked with minute hairs around the leaf scars. LEAVES **pinnate** (L 15–30 cm) **1 terminal leaflet plus 2–12 pairs of oval leaflets** (L 2–5 cm); stipule downy at first, becoming a woody, persistent spine. FLOWERS white; 'pea-like' (L ±2cm); upper petal (standard) with yellow basal blotch; slight orangey aroma; in pendent clusters (L 8–18 cm). FRUIT **hairless pod** (L 5–9 cm) containing 4–10 seeds.

J F M A M J J A S O N D

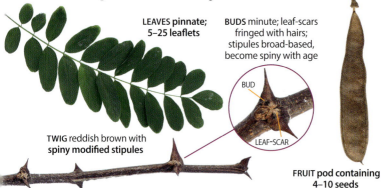

LEAVES pinnate; 5–25 leaflets

BUDS minute; leaf-scars fringed with hairs; stipules broad-based, become spiny with age

BUD

LEAF-SCAR

TWIG reddish brown with spiny modified stipules

FRUIT pod containing 4–10 seeds

BARK grey; fissured on older trees

Similar species | Honey Locust (*p. 343*) which has thorns that have a narrow base.

Associated species | False Acacia hosts **Mistletoe**, and in mainland Europe more than 69 species of **fungi** have been recorded growing on and in the tree. Twelve species of **beetles**, **bugs** and **moths** have been recorded in Britain & Ireland, including **Carnation Tortrix**. Another well-marked micro-moth *Epicallima formosella*, feeds on the tree but has not been recorded in Britain since 1845.

False Acacia's fragrant, creamy white, typical pea flowers (*left*); False Acacia forms dense clumps (*right*).

Leaves *p. 67* | Flowers *p. 72* | Fruit *p. 75* ERICACEAE (HEATHERS) | STRAWBERRY TREE

LC NT Strawberry-tree *Arbutus unedo*

A **small evergreen tree** (H to 10 m; trunk girth to 3 m), generally only reaching its largest size in Ireland. Although generally regarded as a Mediterranean shrub, Strawberry-trees are seemingly wild in parts of western Ireland based upon pollen identified from 2000 BCE, although another explanation is that it may have been brought to Ireland during the Bronze Age. It is difficult to be certain when it was introduced to mainland Britain, but there is some evidence it arrived in the mid-16th century. To thrive, Strawberry-trees need conditions that are warm in winter and damp in summer, with seedlings not being able to withstand cold winds and frost damage. **Features:** BARK grey- to reddish brown; fissured; **peeling off in small flakes to reveal cinnamon-coloured inner bark**. TWIGS reddish brown; hairy. LEAVES laurel-like; narrowly to broadly oval (L 4–11 cm) with a tapering pointed tip; leaf-stalks very short (L >1 cm); upperside glossy dark green; underside dull green; leaf-margins variable, coarsely toothed to untoothed. FLOWERS autumn-flowering; heather-like, **bell-shaped** (L to 9 mm); **white**, can be tinged pink or green; in short drooping clusters. FRUIT **globular berry** (D ±2 cm) **with a rough, warty skin**; green to yellow, ripening one year after flowering; with last year's berries occurring at the same time as the current flowers.

J F M A M J J A S O N D

IN FLOWER/FRUIT

FRUIT 'strawberry-like' when ripe

FLOWERS bell-shaped ('heather-like')

LEAVES glossy dark green; leaf-margins usually coarsely toothed

BARK greyish brown to reddish brown with cinnamon-coloured inner bark

Did you know? The tree's scientific name is *unedo*, from the meaning 'I eat one', because although the berries look like strawberries, you can eat only one as they have a gritty texture, due to the tiny seeds, and a very unpleasant sour taste.

301

AQUIFOLIACEAE | HOLLIES   Leaves *pp. 63, 67* | Flowers *p. 71* | Fruit *p. 77*

LC LC Holly ☠   *Ilex aquifolium*

A medium-sized, evergreen, multi-branched tree (H to 23 m), found in open situations, although more usually encountered as understorey in woodlands or as part of a managed hedge. Holly occurs on any soil and is shade-tolerant. Trees are usually either ♂ or ♀, or at least predominantly so, with the vast majority of flowers being of one sex and a few of the other.
Features: BARK grey and smooth becoming finely fissured with age. TWIGS green, becoming grey with age. LEAVES **highly distinctive**; oval (L 5–10 cm); dark green; **leaf-margin undulating with strong spines**; leaf-stalk unwinged (*cf.* Highclere Holly) younger leaves have fewer spines that are less stiff; some cultivars can be variegated. FLOWERS ♂ and ♀ typically on separate trees; white (D 6 mm); in clusters; petals fused at the extreme base; ♂s with four stamens; ♀s with 4-lobed stigma and 4 infertile stamens. FRUIT ± **globular scarlet-red berry** (D 6–10 mm); in clusters of up to 60 per twig.

J F M A M J J A S O N D

> **Did you know?** Historically, pure Holly woodland was common. Today, this habitat is uncommon, although it still persists in the New Forest, Hampshire, with trees of up to 300 years old being recorded. Holly woodland typically starts to establish when a Holly seed is dropped by a bird flying across heathland. The developing Holly drops its leaves, which rot down and improve the soil around the tree, eventually producing a nutrient-rich soil ideal for other Holly trees to thrive in and eventually forming an often near-circular wood, known locally as a 'Ringwood'.
> Recent research showed changes in the tree populations of Britain's woodlands over the last 50 years. Of all the tree species, Holly had done best, appearing to show increases in all sizes of trees and with rapidly increasing numbers of seedlings. This appears to be a response to climate change, but may also be as a result of the increase in the successful hybrid Highclere Holly (*facing page*).
> The hard, white wood takes stain well and was traditionally known as 'English Ebony'. It was particularly used for carving and inlay, whilst the shoots with berries are used for Christmas decorations.

BARK grey; finely fissured at most

The beginnings of a ring of Holly forming on heathland in the New Forest (*left*) which occurs as a result of birds spreading the bright red berries (*right*).

AQUIFOLIACEAE | HOLLIES

IN FLOWER

♂

IN FRUIT

♂

♀

FRUIT red berry

FLOWERS
♂'s with 4 stamens;
♀ with stigma and
4 sterile stamens

LEAVES glossy dark green; leaf-margins wavy and spiny; cultivars (*below*) can be variegated

young leaves can lack spines but will have a slightly wavy margin

HOLLY BLUE

HOLLY LEAF MINER MINES

LEAF MINE PREDATED BY BLUE TIT

■ Highclere Holly *Ilex ×altaclerensis*

A large evergreen shrub (H to 15 m) that is a fertile hybrid between Holly and Madeiran Holly *Ilex perado* [N/I]. It arose in the 1830s at Highclere Castle, Hampshire, between a greenhouse Madeiran Holly and a native Holly in the grounds. The hybrid and its offspring have now been widely planted in parks, churchyards and gardens, and it has become established on waste ground, along roadsides, in woods and other places. It can be difficult to separate from Holly, due to the range of cultivars and back-crosses.
Features: LEAVES **narrower and more pointed than those of Holly with few, if any, spines;** leaf-stalk can be slightly winged. FLOWERS + FRUIT slightly larger than in Holly.

LEAVES narrower than Holly's; few spines at most; margin flat (*below*)

HIGHCLERE HOLLY

HOLLY

Associated species | The berries are eaten by **birds** and the foliage by **deer** and **Rabbits**. Invertebrates using Holly are scarce, with only 10 species having been recorded feeding on it. The leaves are the food plant of the **Holly Blue** butterfly, but perhaps the most obvious sign of use is provided by **Holly Leaf Miner**, a leaf-mining fly. The larvae burrow in the leaves, leaving distinctive pale blotches, and are often predated by **Blue Tits**, who leave characteristic V-shaped tears in the leaf as they dig out the larvae.

303

CORNACEAE (DOGWOODS) — Leaves *p.66* | Flowers *p.71* | Fruit *p.76* | Twigs *p.78*

LC LC Dogwood ☠ *Cornus sanguinea*

A deciduous large shrub (H to 2 m) or small tree (H to 6 m), native or planted throughout much of Britain & Ireland, except for northernmost Scotland – the distribution seemingly linked to summer temperatures. Dogwood can be fast-growing (up to 1 m per year when young), especially from root suckers or after damage (*e.g.* from coppicing or browsing). It spreads freely by seed, but is largely clonal in the north, spreading from root suckers and forming groups which can cover up to 10 m². Trees typically live for 30 years, but have been noted up to 80 years old. It is common in scrub and hedgerows and, along riversides and woodland edges on a wide range of soils but with a preference for well-drained sunny locations. There are two subspecies, which differ in details of the hairs on the leaf underside – the native **ssp. sanguinea** and the widely planted **ssp. australis** (see *facing page*). **Features:** BARK smooth and grey; developing ridges with ages. TWIGS green when young, becoming **purplish red**; 'rotting fish' aroma when crushed. BUDS flattened cone (L 4–8 mm); reddish brown; **bud scales 0** ('naked') such that the detail of the young, curled leaves can be seen. LEAVES broadly oval (L to 8 cm) with an abrupt, sharply pointed tip; arranged oppositely. FLOWERS 4-petalled; creamy-white (D 5–10 mm) in flat-topped clusters. FRUIT Small, globular (D 5–8 mm) 'berry-like' drupe with high levels of vitamin C and a bitter taste.

J F M A M J J A S O N D

Did you know? An experiment on seed viability showed that seed close to the soil surface was ±90% viable for the first year and a half, but burial to a depth of 5 cm reduced its viability to ±30%.

The wood is dense and noticeably hard, elastic and tough, and it is these qualities that give Dogwood its name. Prehistoric Europeans made arrows from viburnum and Dogwood shoots, and some were found in the possession of Ötzi the Iceman, who lived between 3350 and 3105 BCE and whose remains were found on the border between Italy and Austria.

The veins contain a strong, stretchy latex. This can be seen by folding the leaf widthways and gently pulling apart the two halves. The latex is so strong that the two halves will hold together on very thin threads.

BARK smooth and grey, becoming ridged when older

Similar species | **Cornelian Cherry** *C. mas* (N/I) has yellow flowers and small, red, cherry-like fruit; other dogwoods (*facing page*).

When in flower, the dense clusters of white flowers attract many insects (*left*) whilst the red stems make the species very obvious in winter (*right*).

CORNACEAE (DOGWOODS)

LEAVES oval; green above; paler underside; **3–5 pairs** of veins which curve towards the leaf-tip

FRUIT 'berry-like'; purple-black

FLOWER + LEAF-BUDS

BUD 'naked'; reddish brown

LEAF-BUDS

LEAF-HAIRS (UND.)

1ST-YEAR TWIG green when young; turning purplish red

SSP. SANGUINEA unbranched; upstanding

SSP. AUSTRALIS 2-branched; appressed

FLOWERS 4-petalled; white; in flat-topped clusters

Associated species | The leaves provide food for the caterpillars of some moths like **The Engrailed**. The berries are eaten by birds (including **Robin**, **Starling**, **Redwing**, **Blackbird** and **Song Thrush**) and mammals such as **Fox** and **Badger**. The seeds are readily eaten by rodents, particularly by **Yellow-necked Mouse**, **Wood Mouse** and **Bank Vole**.

THE ENGRAILED

Other dogwoods | Red-osier and White Dogwoods are similar to Dogwood in form and the small white flowers in flat-topped clusters but differ in their **larger leaves** with **6 or 7 pairs of veins** and **whitish fruit**. However, as a species pair they are very similar; differences shown below.

LC ■ Red-osier Dogwood
Cornus sericea

Shrub (H to 3 m). Introduced to Britain & Ireland by 1683. **Features:** TWIGS dark red when young. LEAVES L to 10 cm; tip tapers to point. FRUIT berry-like (D 4–7 mm); white.

1ST-YEAR TWIG dark red

FRUIT white

LEAF-TIP tapers to point; 6 or 7 pairs of veins

SEED flattened egg-shape; base rounded; **length ± width**

LC ■ White Dogwood
Cornus alba

Shrub (H to 3 m). Introduced to Britain & Ireland by 1741. **Features:** TWIGS bright red when young. LEAVES L to 10 cm; tip abruptly pointed. FRUIT berry-like (D 4–7 mm); white, tinged blue.

1ST-YEAR TWIG bright red

FRUIT whitish with a pale blue tinge

LEAF-TIP abruptly pointed; 6 or 7 pairs of veins

SEED flattened egg-shape; tapered at top and base; **length > width**

ADOXACEAE (MOSCHATEL) | ELDERS   Leaves *p.61* | Flowers *p.71* | Fruit *p.76* | Twigs *p.78*

LC LC (Common) **Elder**   *Sambucus nigra*

A deciduous non-suckering shrub or small tree (H to 10 m) widespread in Britain & Ireland, but uncommon in northern Scotland, although it does occur in Shetland and Orkney. It has a preference for open areas and woodland edges and is often found in locations with organic rich soil and on disturbed nitrogen-rich soils. Elder can live for up to 400 years.
Features: BARK deeply furrowed and usually corky when mature.
MATURE TWIGS grey with obvious lenticels; pith white. BUDS conical (L 4–15 mm); the tips of new leaves can be visible at the end of the bud. LEAVES **pinnate; 5 or 7 oval or narrowly oval leaflets** (L to 90 mm) with a pointed tip; leaf-stalk lacks glands (*cf.* Red-berried Elder). FLOWERS 5 fused petals; creamy-white (D 5–6 mm) in flat-topped clusters (D 10–20 cm); distinctive sweet, honey-like aroma when fresh, although can smell like urine with age. FRUIT **shiny black;** ± spherical (D 6–8 mm) berries in distinctive drooping heads.

J F M A M J J A S O N D

Did you know? The flowers of Elder have been widely used to produce cordials or 'champagne', whilst wine is made from its berries. Fried in batter, the flowers can make superb fritters. Dyes have been made from parts of the tree, while the leaves can be made into an insecticide.

The loose flower clusters of Elder (*left*) do not usually appear on plants until they are at least three years-old; they have a distinctive aroma and form dense clusters of edible black berries in autumn (*right*).

ADOXACEAE (MOSCHATEL) | ELDERS

Associated species | Birds, especially **Woodpigeon**, **Bullfinch**, **Blue Tit** and **Great Tit** eat the berries. **Jelly-ear Fungus** occurs commonly on dead branches and a crust fungus known as **Elder Whitewash** can often be seen growing on the stems of older trees. A 1970s study found a total of just 19 species of plant-eating insects recorded feeding on Elder – the low number thought to be due to the presence of cyanogenic glycoside toxins. One of these is the dark green to yellowish brown **Elder Aphid** which can form dense colonies in spring on young Elder shoots, overwintering as eggs on the roots. Nineteen is, however, a relatively small number of species. The only trees and shrubs that have fewer insects recorded are Dogwood (18, *p. 304*), Holly (13, *p. 302*), Yew (6, *p. 116*), and Box (4, *p. 313*).

Similar species | **Red-berried Elder** (*below*); **American Elder** *S. canadensis* (N/I) is a suckering shrub (H to 4 m) which has leaves with 5–11 leaflets; purplish black fruit and 2nd year twigs with few lenticels; and **Dwarf Elder** *S. ebulus* (N/I) is herbaceous (H to 1·5 m), and has leaves with 7–13 leaflets.

LC ■ Red-berried Elder
Sambucus racemosa

Non-native shrub (H to 4 m) introduced in the 16th century. It is found in hedges and woodland north from the English Midlands but is uncommon in Ireland. **Features:** TWIGS grey; pith a distinctive orange-brown. LEAVES 5 or 7 leaflets very similar to Elder; unpleasant musky aroma when crushed; leaf-stalk with stalked glands at the base. FLOWERS very similar to Elder in smaller clusters (D 4–12 cm). FRUIT red; ± spherical (D ± 6 mm) berries.

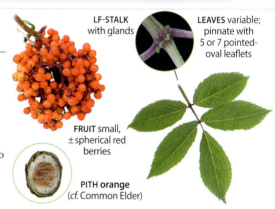

LF-STALK with glands

LEAVES variable; pinnate with 5 or 7 pointed-oval leaflets

FRUIT small, ± spherical red berries

PITH orange (*cf.* Common Elder)

The Elder-like clusters of white flowers (*left*), produce inedible red berries (*right*), hence the English name.

ADOXACEAE (MOSCHATEL) | GUELDER-ROSES   Leaves *p. 62* | Flowers *p. 71* | Fruit *p. 76* | Twigs *p. 78*

LC LC Guelder-rose ☠   *Viburnum opulus*

A deciduous, multi-stemmed shrub (H to 4 m) that can live for up to 50 years. It is the more widespread of the two native *Viburnum* and is found in woodland, scrub and hedgerows throughout England and Wales and Ireland, although less abundant in Scotland. Guelder-rose is not shade-tolerant and has a preference for moist moderately alkaline soils, although it will tolerate moderate acidity. The combination of its colourful berries and vivid red autumn leaves has made Guelder-rose a popular ornamental shrub with a range of cultivars. **Features:** BARK smooth and grey. TWIGS greyish or reddish brown with a few raised lenticels; hairless. BUDS leaf and flower buds noticeably different. Leaf-buds very narrowly oval (L to 8 mm). Flower-buds reddish green to brown; egg-shaped to ± globular (L 5–10 mm), with a pointed tip; bud scales 2, fused into 1. LEAVES arranged oppositely; **palmately 3-lobed** (L to 80 mm); leaf-margin toothed; upperside hairless; underside can have hairs, particularly in main vein axils; leaf-stalk long (L 1–3 cm) with a shallow groove on the upperside and two unstalked glands near the leaf-base. FLOWERS in **flat-topped clusters** (D to 10 cm) consisting of **small (D 6 mm), greenish white, fertile inner flowers and larger (D 5–20 mm), white sterile outer flowers**; with a somewhat unpleasant aroma that, if detected, has been described as 'old socks'. FRUIT globular (D 8–11 mm) bright red berries; orange or yellow in some cultivars.

J F M A M J J A S O N D

Associated species | Both native *Viburnum* support relatively low numbers of plant-eating insects, with 24 species recorded on both and an additional 6 found only on Guelder-rose. Only Dogwood (*p. 304*), Elder (*p. 306*), Holly (*p. 302*), Yew (*p. 116*) and Box (*p. 313*) support fewer. For both native species the **Viburnum Leaf Beetle** is the main source of defoliation. Female beetles burrow into the terminal twigs and create spaces in the central pith in which they lay their eggs. In this space the eggs are protected during the winter prior to hatching and the young beetles emerge in spring to feed on the leaves. The flowers are pollinated by a range of beetles and hoverflies including, *Eristalis arbustorum, E. nemorum, E. tenax* and *Eristalinus sepulcralis*. The berries are eaten by a range of birds which distribute the seeds in their droppings.

VIBURNUM BEETLE

Did you know? In 1597, English herbalist John Gerard named the species the 'Rose Elder'. A sterile variety of the species raised in Gelderland in Holland was named the 'Ghelderscher Roose'. From this mixture of names at some point arose the common name – Guelder-rose.

The distinctive larger sterile flowers of Guelder-rose stand out at a distance (*left*); whilst the red berries and red-orange leaves make identification easy in autumn (*right*).

ADOXACEAE (MOSCHATEL) | GUELDER-ROSES

FLOWERHEAD consists of smaller yellowish central fertile flowers with yellow anthers (INSET) and much larger outer sterile flowers

FRUIT globular red berries

LEAF-STALK with **unstalked** rounded glands

LEAVES 3–5 lobed; margins toothed

BUDS oppositely arranged; flower buds (LEFT) globular; leaf buds (RIGHT) like small leaves

TWIGS hairless; green when young maturing to greyish brown or reddish brown; few lenticels

LEAF-MARGIN with relatively **short**, **unpointed** teeth

Similar species | Two closely related species, which some authorities regard as subspecies of Guelder-rose, are widely planted and may be overlooked. **American Guelder-rose** *V. trilobum* differs in leaves with longer sharper leaf-margin teeth and short-stalked glands on the leaf-stalk.
Asian Guelder-rose *V. sargentii* differs in having purple anthers.
Sycamore (*p. 286*) has similar leaves but has very different flowers and fruit and solid leaf-stalks without glands.

LEAF-STALK with **stalked** rounded glands

AMERICAN GUELDER-ROSE

ASIAN GUELDER-ROSE

LEAF-MARGIN with **relatively long, sharply pointed** teeth

309

ADOXACEAE (MOSCHATEL) | WAYFARING-TREES   Leaves *p.64* | Flowers *p.71* | Fruit *p.76* | Twigs *p.79*

LC LC Wayfaring-tree ☠   *Viburnum lantana*

A deciduous, multi-stemmed shrub or small tree (H to 6 m) that is a widespread native in southern Britain and a neophyte in northern Britain and Ireland. It occurs in hedgerows and scrubby areas, and along woodland edges with a preference for warm lime-rich soils in drier conditions.
The native distribution appears to be determined by the species' need for warm temperatures in summer. **Features:** BARK brown; normally smooth but can be scaly. TWIGS grey-brown becoming reddish brown as they reach their second and third year; somewhat hairy, hairs star-like. BUDS leaf and flower buds noticeably different; both can have short stalks (L to 15 mm). Leaf-buds very narrowly oval (L 10–20 mm); pale greyish yellow to greenish yellow; can have short stalks (L to 15 mm). Flower-buds ± globular (L to 10 mm), with a pointed tip and star-like hairs; usually partly covered by bracts. LEAVES arranged oppositely; **broadly oval** (L to 100 mm); **veins deeply impressed**; leaf-margin finely toothed; upperside typically hairless; underside densely hairy; short-stalked hairs greyish white and star-like. FLOWERS in flat-topped clusters (D to 10 cm) consisting of small (D 5–6 mm), creamy-white flowers with a faint 'floral' aroma that has been described as lily-like. FRUIT egg-shaped (L to 8 mm); green when fresh; ripening red then black.

J F M A M J J A S O N D

> **Did you know?** Although the origins of its name are lost in history, John Gerard in his 1597 'Herball' described it as 'The Wayfaring Man's tree' and 'The Wayfaring-tree'.
> Wayfaring-tree has not been widely used in recent history, but intriguingly, a mummified body of a man was found in the Swiss Alps in September 1991. He lived between 3400 and 3100 BCE and was found to be carrying 14 arrows – two of which were ready to be used. These are the best-preserved Neolithic arrows in Europe and were made from Wayfaring-tree and Dogwood. Three feather halves were attached to the end of the arrows with birch tar glue and bound on with thin nettle fibres.

The tight clusters of white flowers (*left*), the thick hairy leaves, and the berries which ripen from red to black (*right*) mean that Wayfaring-tree is easy to recognize from spring to autumn.

ADOXACEAE (MOSCHATEL) | WAYFARING-TREES

FLOWERHEAD flattish head of small white flowers

LEAF underside (especially younger leaves) with unstalked star-like hairs

FRUIT egg-shaped berries ripening from green to red then black

LEAVES broadly oval; finely toothed; underside very hairy

BANK VOLE

LEOPARD MOTH

BUDS oppositely arranged; leaf buds (LEFT) like small leaves; flower buds (RIGHT) more globular; end-buds can be flower or leaf

TWIGS reddish brown when mature with soft, star-like hairs

Similar species | Chinese Wayfaring-tree *V. veitchii* (N/I) which is planted for hedging and possibly overlooked, which differs in having leaves with stalked star-like hairs; and two evergreens with ± untoothed leaves – **Laurustinus** (*p. 335*) which has blue-black fruit; and **Wrinkled Viburnum** (*p. 335*) which has yellowish flowers and globular red (turning black) fruit.

Associated species | There is data available on the attractiveness of the seed to rodents, the seeds of Wayfaring-tree being the third-most attractive, (behind Blackthorn (*p. 171*) and Wild Cherry (*p. 176*)) particularly to **Yellow-necked Mouse**, **Wood Mouse** and **Bank Vole**.

In central Europe, the caterpillars of the **Green Hairstreak**, **Leopard Moth**, **Emperor Moth**, **Oak Eggar**, **Common Emerald**, **Orange Moth**, and **Dotted Clay** all feed on Wayfaring-tree, but no research has been conducted in Britain & Ireland. However, all these species are present and could therefore be using Wayfaring-tree here too.

OLEACEAE (ASH) | PRIVETS   Leaves *p.66* | Flowers *p.71* | Fruit *p.76* | Twigs *p.79*

LC LC Wild Privet ☠   *Ligustrum vulgare*

A common semi-evergreen small shrub (H to 3 m) of hedgerows, scrubs and woodland edges. **Features:** BARK grey-brown. TWIGS + BUDS hairy. LEAVES oppositely arranged; glossy green; oval (L to 6 cm). FLOWERS in spikes (L 3–6 cm); white; 4-lobed petal-tube (D 4–5 mm) with a pungent aroma described as 'honey with subtle notes of vanilla and mandarin'. FRUIT globular black berry (D 6–8 mm).

LC Garden Privet *Ligustrum ovalifolium*

Differs from Wild Privet in being more evergreen with broader leaves; longer flowers; and hairless twigs and buds. Very widely planted as hedging and found naturalized on waste ground and along roads and railways.

Similar species | **Wild Privet** and **Garden Privet** can be hard to distinguish, the key differences are shown below.

Associated species
Wild Privet is the main food plant of the **Privet Hawkmoth** and also provides cover for small birds and other animals.

The spiked cluster of flowers (*left*) or fruit (*right*) are both distinctive indicators of privet.

312

Leaves *p. 66* | Flowers *p. 69* | Fruit *p. 74*   BUXACEAE | BOX

LC LC Box   *Buxus sempervirens*

A slow-growing, evergreen shrub or small tree (H to 9 m) native to western Europe. It is an introduction into Ireland and most of Britain, although probably native in southern England – the best-known sites are Box Hill, Surrey, where it occurs with Yew and Large-leaved Lime, and in the Chilterns. However, it was also used extensively in Britain by the Romans in their gardens, and it may have spread naturally from these settlements. In more recent times, Box has frequently been used as hedging material or for topiary.
Features: BARK grey and smooth when young, becoming cracked and lumpy with age. TWIGS green; with dense short white hairs when young, becoming hairless. LEAVES arranged oppositely; **simple oval** (L to 25 mm); **dark green and leathery**, persisting for five or six years. FLOWERS tiny (D 2 mm); petal-less; yellow-green; in small clusters in leaf axils; comprising a single female above several males. FRUIT **distinctive '3-horned' oval capsule** (L 7–11 mm); green, ripening brown; containing small black seeds.

J F M A M J J A S O N D

FLOWERS in clusters

♀ FLOWERS with 3 styles

♂ FLOWERS with 4 stamens

FRUIT egg-shaped; with the styles persisting as 'horns'

LEAVES broadly to narrowly oval; leathery and shiny

Did you know? In Britain, Box is usually a shrub, with the largest trees being at Box Hill, Surrey. In Europe some of the largest trees were described in the 20th century from the Caucasus, with trunks 63–96 cm in girth. The largest of these trees is reported to be more than 600 years old. One of the earliest mentions of Box in Britain is in place names – the Saxon name for the present Berkshire village of Boxford was *Boxora* meaning 'Box shore' (*i.e.* Box tree covering the riverbank). Box wood is very dense and fine-grained, and it is so heavy when green that it sinks in water. The wood has an ivory-like texture, and, as a result, it was used extensively for turning and inlay.

Associated species | Box plays host to just four native insects. Unfortunately, the **Box Moth** from south-east Asia, which arrived in 2007–8, is devastating Box populations around the country and may present a long-term threat to the species here. The caterpillars form webs amongst the leaves and can defoliate shrubs completely as they develop.

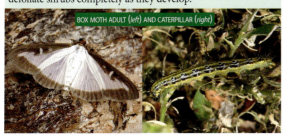
BOX MOTH ADULT (*left*) AND CATERPILLAR (*right*)

313

MYRICACEAE | BOG-MYRTLE Leaves *p.67* | Flowers *p.68* | Fruit *p.74* | Twigs *p.87*

LC LC Bog-myrtle *Myrica gale*

A **small, distinctive, suckering shrub** (H to 1·5 m) of boggy areas, wet heathland and fens. The stem and leaves have a strong, sweet Eucalyptus-like aroma. **Features:** TWIGS hairy when young, becoming hairless; **reddish brown** with a few shiny yellowish glands. BUDS dark reddish brown; clustered near tip of twig; bud scales with pale margins; ♀ (L to 15 mm) with 6–8 scales; ♂ (L 5–8 mm) with 20–30 scales. LEAVES narrowly oval (L to 60 mm); upperside grey-green, hairless; underside downy. FLOWERS ♂ and ♀ in stiff catkins typically on separate trees; appearing before the leaves; ♂ stiff, reddish spike (L to 15 mm); ♀ ± globular (D to 5 mm) with red styles. FRUIT 2-winged seed.

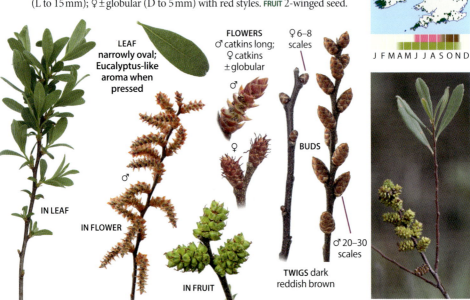

J F M A M J J A S O N D

Bog-myrtle in summer with yelow Bog Asphodel and white Common Cottongrass (*left*); in winter its purplish orange-brown twigs are distinctive even at distance (*right*).

Leaves *p.60* | Flowers *p.72* | Fruit *p.77*   ASPARAGACEAE | BUTCHER'S-BROOM

LC LC Butcher's-broom *Ruscus aculeatus* ☠

A perennial evergreen, slow-growing, much-branched shrub (H to 1 m) that is native to southern England and Wales and introduced to Scotland and Ireland. It is tolerant of dense shade and occurs predominantly in woodlands and hedges, although it can also be found on stony ground at the coast.
Features: STEMS **dark green with stiff leaf-like extensions** (cladodes) which are pointed-oval (L 10–30 mm) with a spiny tip. LEAVES small triangular scales (L <5 mm) at the base of the cladodes and flowers. FLOWERS ♂ and ♀ typically on separate plants, although some plants can be bisexual; similar-looking; small (D ± 3 mm); 6 greenish white 'petals'; solitary and located on the upper surface of the cladode. FRUIT ± spherical red berry (D ± 10 mm).

TWIGS green; the apparent 'leaves' are actually sharp extensions of the stem

FRUIT bright red berry

♀ with 'cauliflower'-like stigma

FLOWERS

♂ with 3 stamens

FLOWERS located on the upperside of the cladodes

IN FRUIT

J F M A M J J A S O N D

Did you know? Known locally as 'Knee Holly', a stem and spiny 'leaves' of Butcher's-broom was traditionally used to sweep small bits of meat from a butcher's block.

The evergreen, short, spiky form of Butcher's-broom is obvious all year (*left*) and the toxic red berries at the centre of the 'leaf' are highly distinctive (*right*).

BERBERIDACEAE | BARBERRIES — Leaves *pp. 63, 65, 67* | Flowers *p. 72* | Fruit *p. 77* | Twigs *p. 80*

Barberry — *Berberis vulgaris*

A deciduous shrub (H to 4 m) that arises from a mass of basal stems. It is either native or an ancient introduction. Historically planted as hedgerows, during the First World War it was removed to eliminate a rust, *Puccinia graminis*, that uses both Barberry and cereals as a host. This risk is no longer present as rust-resistant strains of wheat are now grown and Barberry is now found widely in hedges and woodland edges, on waste ground, and on cliffs. **Features:** BARK greyish and grooved. TWIGS grooved; greyish (can be reddish when young) with **extremely sharp 3-parted spines**. LEAVES small (L to 5 cm); growing in small clusters above the spines. FLOWERS in loose pendent spikes. FRUIT sharply acidic, but edible, red egg-shaped berries although care is needed when picking on account of the spines.

J F M A M J J A S O N D

FRUIT egg-shaped; red when ripe

TWIGS greyish; grooved; with 3-parted spines

LEAVES oval with a finely toothed margin

FLOWERS yellow; in pendent clusters

Did you know? In Britain, Barberry has conservation value as the very rare and legally protected **Barberry Carpet**, a moth that relies on the plant, is only known from a few sites in southern England.
The berries are used for jams and are an important food for many small bird species which disperse the seeds in their droppings.

Similar species | Planted barberries (*facing page*) which differ in leaf, flower and spine details.

Barberry, with its delicate yellow flowers, is obvious in spring (*left*) and can form thick hedgerows (*right*).

BERBERIDACEAE | INTRODUCED BARBERRIES

Introduced barberries *Berberis* species

Of the many species that can be encountered in parks and gardens or naturalized in hedges and scrub, the most regularly seen are **Darwin's Barberry** *B. darwinii* and **Thunberg's Barberry** *B. thunbergii*. Also increasing are **Gagnepain's Barberry** *B. gagnepainii*; **Chinese Barberry** *B. julianae*; and **Mrs Wilson's Barberry** *B. wilsoniae* (N/I – like Thunberg's but with much narrower leaves and 3-parted spines).

Features of introduced barberries: LEAVES evergreen or deciduous; broadly to narrowly oval; untoothed to spiny. FLOWERS yellow to orange; in loose clusters of varying form. FRUIT egg-shaped to ± spherical berry; bluish black covered with a paler bloom or red.

Identification is by a combination of at least some of the following: leaf-shape, flower-cluster structure, spine configuration, and berry shape/colour – with leaf-shape being perhaps the most helpful.

J F M A M J J A S O N D

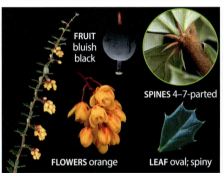

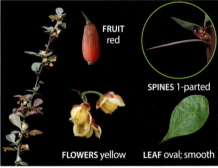

Darwin's Barberry *B. darwinii*

A spiny evergreen shrub (H to 3 m) introduced in 1848. FLOWERS orange; in clustered heads with a common stalk. LEAVES ± oval (L 10–30 mm); spiny teeth near apex. SPINES 4–7-parted (L <1 cm). FRUIT ± spherical (D 4–7 mm); bluish black.

Thunberg's Barberry *B. thunbergii*

A spiny deciduous shrub (H to 2 m) introduced in 1882. FLOWERS yellow. LEAVES oval (L 10–24 mm); untoothed. SPINES usually 1-parted. FRUIT egg-shaped to ± spherical (L 7–10 mm; D 4–7 mm); red.

EVERGREEN BARBERRIES WITH SPINES USUALLY 3-PARTED AND BLUISH BLACK FRUIT

Gagnepain's B. *B. gagnepainii*

Shrub (H to 1·5 m) introduced 1901. FLOWERS yellow. LEAVES narrow oval (L 3–10 cm); margins with spiny teeth. FRUIT egg-shaped (D 4–7 mm); bluish black; style unstalked.

Chinese Barberry *B. julianae*

Shrub (H to 3 m); introduced 1900. FLOWERS yellow, tinged red. LEAVES narrow oval (L 3–10 cm); ± shiny; margins spiny. FRUIT egg-shaped (D 4–7 mm); bluish black; style stalked.

Hedge Barberry *B. ×stenophylla*

Shrub (H to 3 m); hybrid (arose c. 1860). FLOWERS yellow; in clusters. LEAVES narrow oval (L 15–25 mm); margins untoothed, folded. SPINES usually 3-parted. FRUIT ± globular (D 4–7 mm); bluish black.

317

FABACEAE (PEAS) | GORSES   Leaves pp. 60, 67 | Flowers p. 72 | Fruit p. 74

Native woody peas   *Ulex, Cytisus* and *Genista* spp.

Members of the pea family (Fabaceae) all have the same highly distinctive flower shape. Most are herbaceous; some are trees and shrubs, represented in Britain & Ireland by the spiny gorses (*Ulex*), the spiny or spineless greenweeds (*Genista*) and the spineless brooms (*Cytisus*).

Gorse is by far the most widespread; the two smaller gorses (**Western** and **Dwarf**) can be difficult to separate as they show many overlapping characteristics, including spine length, height, flower colour and pod size. **Broom** is also widespread and is tall with trifoliate leaves (beware the planted **Spanish Broom** with oval leaves). **Greenweeds** are much less common and smaller than all but Dwarf Gorse. Two are spineless, one is spiny, but all have narrowly oval leaves.

Did you know? As a species that has evolved to cope with fire, Gorse burns readily, with temperatures of up to 1,500 °C recorded at the fire's centre. This fierce heat means that fire burns quickly through the above-ground vegetation, reducing the chances of a fire burning deep into the roots. Consequently, after a fire Gorse rapidly recolonizes an area both by new shoots arising from the roots and from dormant seeds that have evolved to germinate after being scorched.

LC LC Gorse
Ulex europaeus

An evergreen shrub (H to 3 m), a familiar sight throughout the country. **Features:** BRANCHES + STEMS main branches usually ascending; modified into dense grey-green, strongly furrowed, stiff spines (L to 30 mm). FLOWERS golden yellow (L 10–20 mm); usually with a strong aroma of coconut; flowering all year. SEPALS ± ⅔ length of petals; hairs spreading. BRACTS W 2–5 mm; > 2× flower-stalk width. FRUIT hairy pod (L 12–19 mm); green ripening blackish brown.

LC LC Western Gorse
Ulex gallii

An evergreen shrub (H to 2 m but usually smaller) with a preference for heathy acid soils and coastal and mountain areas. **Features:** BRANCHES + STEMS main branches usually ascending; modified into dense dark green, weakly furrowed, stiff spines (L to 30 mm). FLOWERS bright yellow (L 10–13 mm); may have weak aroma of coconut; flowering July to December. SEPALS ≥ ⅔ length of petals; hairs pressed flat; teeth at calyx tip incurved. BRACTS W < 0·8 mm; ≤ 2× flower-stalk width. FRUIT hairy pod (L 9–13 mm); green ripening blackish brown.

LC LC Dwarf Gorse
Ulex minor

A dwarf evergreen shrub (H to 1·5 m but usually < 50 cm) found mostly in the south and east of Britain, usually on acid or heathland soils. **Features:** BRANCHES + STEMS main branches spreading horizontally to weakly ascending; modified into dense green, weakly furrowed, flexible spines (L to 20 mm). FLOWERS bright yellow (L 8–10 mm); flowering July to September. SEPALS ± length of petals; hairs pressed flat; teeth at calyx teeth spreading. BRACTS W < 0·8 mm; ≤ 2× flower-stalk width. FRUIT hairy pod (L 6–9 mm); green ripening blackish brown.

J F M A M J J A S O N D
J F M A M J J A S O N D   J F M A M J J A S O N D

318

FABACEAE (PEAS) | GORSES

Similar species | Spanish Gorse *Genista hispanica* (right) has branched spines, simple leaves, and flowers in terminal spikes. *U. ×breoganii* – the hybrid between **Gorse** and **Western Gorse** (INSET *below*) flowers July–February and shows features intermediate between the parents. It is perhaps most readily recognized by its mid-green foliage (between the grey-green of Gorse and dark green of Western Gorse); golden-yellow flowers and a mix of spreading and flattened hairs on the sepals.

SPANISH GORSE

FABACEAE (PEAS) | BROOM + PETTY WHIN   Leaves *pp. 60, 67* | Flowers *p. 72* | Fruit *p. 74*

LC Broom *Cytisus scoparius*

A deciduous spineless shrub (H to 3 m), looking like a gorse that lacks spines. Common on heaths, woodlands and hedgerows, and at the coast.
Features: FLOWERS bright yellow (L 16–18 mm); vanilla aroma. LEAVES trifoliate or absent. FRUIT hairy pod (L to 40 mm); black when ripe; splits open with an audible 'pop'.

ssp. *maritimus* grows on cliffs and beaches and differs in its low-growing, almost prostrate, form (H typically ≤ 40 cm) and hairy young shoots.

POD hairy; black when ripe

LEAVES trifoliate, but not always present

STEM ridged; spineless

FLOWER typical pea

Similar species | **Spanish Broom** *Spartium junceum* (widely found in gardens) can become naturalized on light soils, particularly in the lowlands and near the coast. It differs in its smooth stems; simple (not trifoliate) leaves and green pods.

SPANISH BROOM

J F M A M J J A S O N D

NT Petty Whin *Genista anglica*

An uncommon small hairless spiny shrub (H to 1 m but usually much smaller) of heathlands and the dry edges of damp meadows and boggy areas. **Features:** BRANCHES + STEMS with curved spines. FLOWERS bright yellow (L 7–10 mm). LEAVES narrowly oval (L to 8 mm); usually held quite close to the stem. FRUIT inflated pod (L to 20 mm).

POD hairless; inflated; brown when ripe

STEMS with curved spines

LEAVES narrowly oval; typically held quite close to the stem

FLOWER typical pea

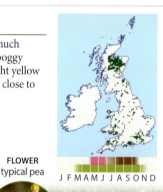

J F M A M J J A S O N D

Did you know? Whin is an old, colloquial name for gorse; hence the name means 'Little Gorse'.

FABACEAE (PEAS) | GREENWEEDS

NT Dyer's Greenweed
Genista tinctoria

A short, spineless, hairless shrub (H to 1 m) of unimproved grassland and heathland. **Features:** FLOWERS yellow (L 10–15 mm); on long branches. LEAVES narrowly oval (L to 30 mm); leaf-margin hairy. FRUIT hairless pod (L 20–30 mm).

Did you know? Dyer's Greenweed supports a range of rare and scarce insects and is the sole food plant for five rare or NATIONALLY SCARCE moths, including the Large Gold Case-bearer.

NT Hairy Greenweed
Genista pilosa

A short, spineless, **hairy** shrub (H to 50 cm) of cliff tops, cliff edges and heathland; rare and declining; **differing from other native** *Genista* **species in its hairy leaves and flowers**. **Features:** FLOWERS yellow (L 7–11 mm); on short branches. LEAVES narrowly oval (L 3–5 mm). FRUIT hairy pod (L to 22 mm).

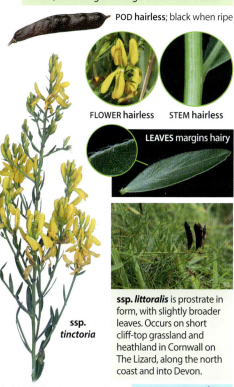

POD **hairless**; black when ripe
FLOWER **hairless**
STEM **hairless**
LEAVES margins hairy

ssp. *tinctoria*

ssp. *littoralis* is prostrate in form, with slightly broader leaves. Occurs on short cliff-top grassland and heathland in Cornwall on The Lizard, along the north coast and into Devon.

LEAVES underside softly hairy
STEM softly hairy
POD softly hairy
FLOWERS hairy

J F M A M J J A S O N D

J F M A M J J A S O N D

GROSSULARIACEAE | CURRANTS   Leaves *p.62* | Flowers *pp.69, 73* | Fruit *p.77* | Twigs *p.83*

Currants   *Ribes* spp.

LC ▢ Flowering Currant *Ribes sanguineum*

Shrub (H to 2·5 m), native to the western United States, often planted in gardens and occasionally spreading into nearby woodlands. LEAVES hairy; sweet 'blackcurrant' smell when crushed. FLOWERS distinctive; pink-red (D 6–10 mm). FRUIT distinctive; grey-blue berry (D 6–10 mm).

IN FLOWER
LEAVES palmately lobed; hairy
FLOWERS distinctive pink-red; in clusters
FRUIT dark blue berry with a grey-blue bloom
BUD pointed-oval (L to 15 mm); green to reddish purple
TWIG reddish brown with tiny hairs

J F M A M J J A S O N D

LC LC Black Currant *Ribes nigrum*

Small shrub (H to 2 m) introduced by the 17th century; now naturalized and widespread in woodlands, hedges and by riversides. It is host to the Currant Clearwing moth, the caterpillars of which feed inside the centre of the stems. LEAVES palmately lobed; distinctive sweet 'blackcurrant' scent when crushed or rubbed.
FLOWERS 'saucer'-shaped; 'closed'-looking (D 6–10 mm).
FRUIT black berries (D 10–15 mm); rich in vitamin C.

LEAVES 3-lobed; underside with glands; **distinctive sweet aroma when crushed**
FLOWERS greenish yellow (can have purplish tinge); 'closed' appearance
FRUIT distinctive black berry when ripe
TWIG usually pale (purplish) brown
BUD yellowish green to purplish; yellow glands present

J F M A M J J A S O N D

322

GROSSULARIACEAE | CURRANTS

Six short shrubs, five of which are usually found wild or naturalized in woodland or hedgerows, and one (Flowering Currant) that is widely planted in gardens although can be found naturalized.

LC LC Red Currant *Ribes rubrum*

Small, probably native, shrub (H to 1·5 m) of woods, hedges and scrubby places. LEAVES downy; unscented. FLOWERS small (D 4–6 mm); saucer-shaped; 'open'-looking; greenish yellow; stigma surrounded by a raised ring, look 'flatter' than in Downy Currant. FRUIT bright red berry (D 6–10 mm); slightly sharp taste (widely used in cooking, especially with game and in puddings, jams and syrups).

LEAVES palmately lobed; (A) narrowly angled between basal lobes; (B) underside sparsely downy

FLOWERS 'saucer'; anthers with 'cross-bar'

FRUIT bright red berry

IN FLOWER

TWIG typically pale grey brown

BUD pointed-oval (L to 7 mm); dark brown

J F M A M J J A S O N D

LC LC Downy Currant *Ribes spicatum*

Uncommon native shrub (H to 1 m) of northern England and Scotland. Usually found in upland areas in river valleys, on shady islands or riverbanks, or in limestone woodland; often in clonal groups; flowers and fruit often absent, especially in deep shade. LEAVES generally quite hairy; unscented. FLOWERS look 'deeper' than in Red Currant; ring around stigma absent. FRUIT as Red Currant.

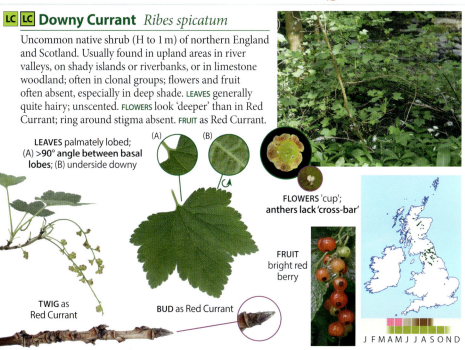

LEAVES palmately lobed; (A) >90° angle between basal lobes; (B) underside downy

FLOWERS 'cup'; anthers lack 'cross-bar'

FRUIT bright red berry

TWIG as Red Currant

BUD as Red Currant

J F M A M J J A S O N D

323

GROSSULARIACEAE | CURRANTS — Leaves p.62 | Flowers pp.69, 73 | Fruit p.77 | Twigs p.83

LC LC Mountain Currant *Ribes alpinum*

Small native shrub (H to 2 m) occurring naturally in limestone woodlands and rocky places (especially in the north of England; particularly the Peak District) but can be found as a garden escape in other areas. FLOWERS ♂ and ♀ on separate shrubs (D 5–10 mm); **in upright clusters unique among native currants**. LEAVES resemble those of Gooseberry; usually 3-lobed (L to 5 cm); leaf-margins with rounded teeth. FRUIT small red berry (D 5–7 mm).

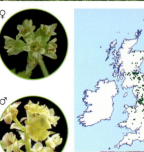

LEAVES 3-lobed

FLOWERS in ± upright clusters; ♂'s containing more flowers than ♀s

FRUIT small red berry

BUDS whitish; tinged green or pinkish (L 4–9 mm)

TWIGS pale grey-brown; spines absent

J F M A M J J A S O N D

Gooseberry *Ribes uva-crispa*

Small **spiny** shrub (H to 1 m) – the only widely occurring spiny *Ribes* in Britain & Ireland. Possibly native but probably introduced in the 13th century. It is widely cultivated but has naturalized in woods, hedges and scrubby areas. FLOWERS usually 'bell'-shaped (D 6–12 mm) greenish yellow (can be red-tinged). LEAVES usually 3-lobed (L to 5 cm); unscented; can be hairy. FRUIT ± globular (L to 20 mm); greenish yellow (can be red-tinged).

IN FLOWER

LEAVES usually 3-lobed

TWIGS pale brownish white; **usually with conspicuous spines**

BUDS whitish or brown (L 4–10 mm)

FRUIT greenish, edible berry

J F M A M J J A S O N D

324

Leaves *p.67* | Flowers *p.72* | Fruit *p.77* | Twigs *p.83*   THYMELAEACEAE | SPURGE-LAUREL + MEZEREON

Other native shrubs

LC Spurge-laurel *Daphne laureola* ☠

A small evergreen shrub (H to 1·5 m) occurring mainly in woodlands and hedges on lime-rich or clay soils. BARK yellow-grey; very thin; strong aroma of diesel/paint thinner if cut; the sap can cause skin rashes. LEAVES oval to spear-shaped (L to 12 cm); dark green. FLOWERS petals absent; 4 sepals (D 8–12 mm); yellow-green; in clusters of up to 10 at the junction of the leaves and stem. FRUIT black; lemon-shaped berry (L 8–11 mm) that is toxic to humans.

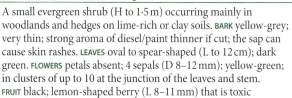
FLOWERS yellowish green; in clusters

IN FRUIT

FRUIT black berry

J F M A M J J A S O N D

LC VU Mezereon *Daphne mezereum* ☠

A small deciduous shrub (H to 1·5 m) occurring in two very different habitats – lime-rich woodlands and calcareous fens. Now rare in the wild but common in gardens. BARK greyish brown, dotted with black; skin irritant if touched. LEAVES long oval (L to 10 cm); light green. FLOWERS appear before leaves; bright pink (D 10–15 mm) on the bare stem in clusters. FRUIT is a red berry (D 7–12 mm) that is highly toxic to humans.

BUDS usually purplish black; terminal (L 7–12 mm) much larger than laterals (L 2–4 mm)

FLOWERS pink; before leaves

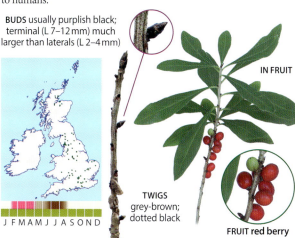
IN FRUIT

TWIGS grey-brown; dotted black

FRUIT red berry

J F M A M J J A S O N D

WIDESPREAD INTRODUCED SHRUBS + TREES | Leaves pp. 58–67 | Flowers pp. 68–73

Widespread introduced shrubs and small trees

In Britain & Ireland there is a long history of importing trees and shrubs from all over the world, particularly for planting in gardens and public spaces. The earliest evidence is from the Bronze Age and the Roman era, with many of the species involved now being part of the British & Irish landscape and ecosystems. Perhaps the most significant period correlates with the rise in expeditionary travel starting in the 18th century. Today the number of plant species and varieties that have been important runs into the thousands.

Rhododendron can take over large areas of woodland, shading out native species

A subset of these imported species are those which, having been planted, have spread from their original sites and become naturalized, though in most cases usually not too far away from either historical or current habitation.

A few of these species, such as **Rhododendron** and **Snowberry**, spread easily and are of little value to wildlife, taking over areas of habitat, ousting other species and having a negative impact on the local environment.

The species in this section are generally not as widespread as the archaeophytes and neophytes included in the main species accounts but are included as they are regularly encountered.

From a conservation perspective it is important to record these species if encountered in a naturalized situation in order to be able to keep track of any changes in their distribution.

LC ◾ Tamarisk *Tamarix gallica*

An evergreen bushy shrub or small tree (H to 5 m) native of the western Mediterranean and southern Europe. It was introduced into British gardens before 1597 and was known from the wild by 1796. It very resistant to salt and so has been planted on the coast, and along water courses and stream sides. Some of the largest trees in Britain are in hedges on the coast of Hampshire. **Features:** LEAVES tiny, almost scale-like; arranged alternately. FLOWERS tiny; 5-petalled; pink in long catkin-like spikes.

J F M A M J J A S O N D

IN FLOWER
distinctive

LEAVES
tiny; almost scale-like

FLOWER
clustered in tight spike

BUDS tiny; ±globular (D to 1 mm); usually partially hidden behind papery scale

TWIGS thin; usually purplish brown

326

Rhododendrons *Rhododendron* spp.

The most familiar and widespread species is *Rhododendron ponticum* – a dense shrub (H to 8 m). Although it appears to have occurred in Britain & Ireland prior to the last ice age (± 10,000 years ago), it did not recolonise naturally afterwards. A remnant species of the original laurissilva (laurel-based) forests that covered much of Europe 15–40 million years ago, its modern native range is disjunct – a pocket in Spain and Portugal, and from Turkey eastwards into Asia. It was found by botanists in Spain c.1750 and introduced to Britain & Ireland in 1763. It has become naturalized and is classed as an invasive species in parts of Britain & Ireland as it colonizes heathland, moorland and woodland, forming patches which shade out other native flora. Honey produced from the flowers can be poisonous to humans and also to some bee species, although Buff-tailed Bumblebee does not seem to be affected. **Features:** LEAVES evergreen; large, long-elliptical or oblong (L to 20 cm). FLOWERS 5-petalled (3 upper, 2 lower); funnel-shaped (D to 5 cm); violet-purple; can have a yellow central blotch (indicating a Turkish origin).

J F M A M J J A S O N D

IN FLOWER distinctive; groups of showy flowers

FLOWER distinctive; funnel-shaped with 5 spreading petals (3 upper, 2 lower)

LEAVES large, similar to those of Cherry Laurel (p. 178)

Yellow Azalea *R. luteum* (left) is a deciduous bushy shrub with very fragrant bright yellow flowers in late spring. **Labrador Tea** *R. groenlandicum* (right) is an evergreen shrub (H to 1 m) that grows naturally in Greenland but has been found in the English Peak District on boggy ground. However, whether it has escaped from cultivation or been dispersed by migrating birds (such as 'Greenland' Wheatear) is uncertain.

YELLOW AZALEA

LABRADOR TEA

WIDESPREAD INTRODUCED SHRUBS + TREES | Leaves pp. 58–67 | Flowers pp. 68–73

'Hedge Hebe' *Veronica ×franciscana*

An evergreen shrub (H to 1·5 m) that is a garden hybrid (first recorded 1859) between two species from New Zealand and now naturalized in seaside locations and on sea-cliffs. **Features:** LEAVES oval (L 3–9 cm); tip rounded; ± leathery. FLOWERS speedwell-like; violet-purple; in dense spikes. Other hebes may be much more rarely encountered as garden escapes.

The gap between the newest leaves at the top of a branch will distinguish 'Hedge Hebe' from other hybrids, which all have a very small gap at the most

LEWIS'S HEBE

LEAVES

FLOWERS speedwell-like

'HEDGE HEBE'

oval; tip rounded — 'HEDGE HEBE'
narrow; tip pointed — LEWIS'S HEBE
very narrow — NARROW-LEAVED HEBE

Similar species | **Narrow-leaved Hebe** *V. salicifolia* LEAVES long, very narrow; FLOWERS whitish or purplish in long spikes. **Lewis's Hebe** *V. × lewisii* LEAVES narrower; FLOWERS purple fading to white.

LC Butterfly-bush (Buddleia) *Buddleja davidii*

A spreading shrub (H to 5 m) native to central China and Japan. Introduced in the 1890s and commonly grown as a garden plant, it is now widely naturalized, especially along railway lines and around derelict buildings, and is now considered an invasive species. **Features:** BARK light brown, becoming deeply fissured with age. TWIGS thin; light brown. BUDS naked; consisting of a pair of tiny densely hairy leaves. LEAVES arranged oppositely; large, 'spear'-shaped (L 7–13 cm); underside densely grey hairy. FLOWERS purple, pink or white (L 9–11 mm) in dense, pyramidal clusters.

IN FLOWER Jul–Sep; distinctive

FLOWERS tubular, with 4 petal-like lobes

LEAVES 'spear'-shaped; underside grey hairy

BUDS 'naked'; a pair of tiny hairy leaves

TWIGS light brown

328

Fruit *pp. 74–77* | Twigs *pp. 78–85* **WIDESPREAD INTRODUCED SHRUBS + TREES**

■ Brideworts *Spiraea* spp. + cultivars

Usually a suckering shrub with multiple, tall erect stems (H to 2 m). Already growing in Britain by 1640, this group of deciduous shrubs are regularly found as garden escapes. **Features:** LEAVES narrowly oval (L 3–7 cm); leaf-margins ± toothed. FLOWERS in branched spike; 5-petalled; pinkish or white with long stamen filaments.

> There are perhaps a dozen species and hybrids that can be encountered; most are varieties or hybrids involving Steeple-bush *S. douglasii*. Identification, if possible, depends upon flower-cluster and leaf details beyond the scope of this book. Two examples of the genus are shown.

J F M A M J J A S O N D

INTERMEDIATE BRIDEWORT *S.* ×*rosalba*

STEEPLE-BUSH ssp. *menziesii*

LEAVES

INTERMEDIATE BRIDEWORT

INTERMEDIATE BRIDEWORT

toothed in apical half; underside sparsely hairy at most

toothed near tip; underside densely grey hairy

LC ■ Lilac *Syringa vulgaris*

A large shrub or small tree (to 6 m) native to the Balkans. It had arrived in Britain & Ireland, in the gardens of John Gerard, by 1596 and is still widely planted, occasionally becoming naturalised. **Features:** BARK grey-brown; furrowing and flaking with age. TWIGS plain brown. BUDS conspicuous; egg-shaped, pointed (L 5–10 mm); greenish or reddish; end-buds usually in pairs. LEAVES arranged oppositely; large, ± heart-shaped (L 7–13 cm); underside densely grey hairy. FLOWERS sweet-scented; purple, pink or white (L 8–12 mm) in dense, pyramidal or conical cluster.

IN FLOWER Apr–May; distinctive

LEAVES ± heart-shaped

FLOWERS tubular, with 4 petal-like lobes

J F M A M J J A S O N D

TWIGS plain brown

BUDS large; those at tip of twig usually in pairs

329

WIDESPREAD INTRODUCED SHRUBS + TREES   Leaves pp. 58–67 | Flowers pp. 68–73

LC Laburnum *Laburnum anagyroides* ☠

A small deciduous tree or large shrub (H to 10 m). A member of the pea family, it was introduced to Britain & Ireland by 1596 by the famous gardener and plant collector Tradescant and can now be found growing widely throughout the country on most soils, provided they are not over dry. The seeds are poisonous and contain cytisine, highly toxic alkaloid – the side-effects of which can include nausea, vomiting, heart pain, headache and in larger doses even death via respiratory failure.

J F M A M J J A S O N D

FLOWERS yellow (L 17–23 mm); hanging in clusters (L 10–20 cm)

LEAVES trifoliate

LABURNUM   SCOTTISH LABURNUM

PODS hang in clusters; most filled with seeds (cf. hybrid)

BUDS green; usually covered in white hairs

TWIGS greenish; can be hairy

Similar species | **Scottish Laburnum** *L. alpinum* has longer flowerheads (L 15–35 cm) with smaller flowers (L 13–18 mm) and leaf-stalks, flower-stalks and leaf underside ± hairless. Hybrids between the two show intermediate features and are best identified by the pods which contain few seeds.

LC Spotted Laurel *Aucuba japonica* ☠

An evergreen shrub (H to 5 m), native to Japan and first introduced in 1783 although it was not until the 1860s that it became naturalized and widespread in parks, gardens and hedges. **Features:** BARK grey-brown; furrowing and flaking with age. TWIGS plain brown. BUDS conspicuous; egg-shaped, pointed (L 5–10 mm); greenish or reddish; end-buds usually in pairs. LEAVES narrowly oval (L 8–20 cm), tip pointed; dark green but usually with variable yellow blotches. FLOWERS ♂ and ♀ flowers on separate trees; 4-petalled; reddish purple to purplish brown (D 4–8 mm); ♂s in branched spikes; ♀s in clusters. FRUIT conspicuous; bright red; berry-like (D 10–15 mm); containing a single stone.

J F M A M J J A S O N D

LEAVES distinctive; dark green; usually blotched yellow

♀ FLOWER

FRUIT conspicuous; bright red; in loose clusters

♂ FLOWER

330

Fruit *pp. 74–77* | Twigs *pp. 78–85* WIDESPREAD INTRODUCED SHRUBS + TREES

■ Bay *Laurus nobilis*

A large shrub (H to 18 m) that is native to the Mediterranean and cultivated in Britain & Ireland since The Middle Ages and first recorded as naturalized in 1924. **Features:** LEAVES oval, pointed-oval (L to 12 cm); hairless; **strongly aromatic**. FLOWERS pale yellow-green (D ± 10 mm); 4-petalled; ♂ and ♀ flowers on separate trees. FRUIT shiny black; berry-like (D ± 10 mm).

IN FLOWER

FRUIT black, in loose clusters

♂ FLOWER

LEAVES margin untoothed, but can be wavy

IN FRUIT

Similar species | Cherry Laurel (*p. 178*) has large spikes of creamy white flowers; larger (L 10–25 cm) leaves and cherry fruit.

■ Oregon-grape *Mahonia aquifolium*

An evergreen shrub with vertical stems (H to 1·5 m) that is native to western North America. Introduced to Britain in 1823 and has become naturalized in gardens, woodlands, hedges and roadsides and spreading via stolons. **Features:** LEAVES pinnate; 5–9 oval leaflets (L 3–7 cm); glossy dark green; holly-like with spiny margins. FLOWERS yellow; ± globular (D 6–8 mm) in dense spikes. FRUIT berry (D ± 10 mm); green ripening blue-black with grey coating.

IN FRUIT

FLOWERS yellow petal-like segments in 2 whorls of 3

FRUIT blue-black with greyish bloom when ripe

LEAVES pinnate with 5–9 holly-like leaflets

331

WIDESPREAD INTRODUCED SHRUBS + TREES | Leaves *pp. 58–67* | Flowers *pp. 68–73*

LC ▪ Snowberry *Symphoricarpos albus* ☠

An upright suckering shrub (H to 2 m) with arching stems that is native to North America. Snowberry was introduced into cultivation in 1817 and has been planted in gardens and parks from where it has spread widely.
Features: LEAVES broadly oval (L 2–5 cm) and untoothed. FLOWERS small; bell-shaped (D 5–8 mm) comprising 4 or 5 erect petals that are pink on the outside and white and hairy on the inside; flowers in dense clusters of up to 16 located at the tip of a twig. FRUIT distinctive, fleshy, white globular berry (D 8–15 mm).

FLOWERS bell-shaped; largely pink

J F M A M J J A S O N D

IN FRUIT

FRUIT all white

LEAVES (L 2–5 cm)

▪ Hybrid Coralberry *Symphoricarpos ×chenaultii* ☠

Shrub (H to 1·5 m) rooting from its arching stems (rather than suckering). A hybrid between two garden escapes (Snowberry and probably Pink Snowberry *S. microphyllus*) both originating from North America.
Features: LEAVES broadly oval (L 2·0–2·5 cm) and untoothed. FLOWERS small; bell-shaped (D 5–8 mm) comprising 4 or 5 erect petals that are pink on the outside and white and hairy on the inside; flowers in dense clusters of up to 16 located at the tip of a twig. FRUIT distinctive, fleshy, globular berry (D 6–10 mm); whitish, flushed pink on the side exposed to the sun; contrasting white spotting in the pink areas and pink spots in the white.

FLOWERS bell-shaped

LEAVES (L 2·0–2·5 cm)

J F M A M J J A S O N D

IN FRUIT

HYBRID CORALBERRY

FRUIT pink and white with contrasting spots

Similar species | **Coralberry** *S. obiculatus* (N/I) is similar but has smaller leaves (L 1–2 cm) and smaller (D 4–6 mm) all-pink berries. Coralberry and Hybrid Coralberry also hybridize as '**Doorenbos Coralberry**' *S. ×doorenbosii* (N/I) – very like Hybrid Coralberry and separated most easily by the fruit, which lacks any white-on-pink and pink-on-white contrasting spots.

WIDESPREAD INTRODUCED SHRUBS + TREES

LC Fuchsia *Fuchsia magellanica*

A spreading shrub (H to 3 m) with curving stems that is native to Chile and Argentina. It was introduced in 1822 and has become established widely in western Britain and Ireland, where it can be found along roadsides and in hedges. The flowers of Fuchsia as a genus are unmistakable. Those found naturalized are predominantly the bright red relatively slight flowers of the cultivated form *F. magellanica* 'Riccartonii' which appears to have arisen in a Scottish nursery in the village of Riccarton in East Ayrshire. Those that are larger flowered or differently coloured are garden escapes.

J F M A M J J A S O N D

'GARDEN' FUCHSIA

FUCHSIA unmistakable; much more delicate than any similar 'garden' cultivar

LEAVES narrowly oval

FRUIT black berry

There are many Fuchsia cultivars which can become naturalized; those 'Garden' Fuchsia (*left*) that have the same sepal and flower-tube colouring as Fuchsia (*right*) typically have a much larger flower-tube.

Duke of Argyll's Teaplant *Lycium barbarum*
Chinese Teaplant *Lycium chinense*

Deciduous, arching, suckering spiny shrubs (H to 2·5 m) both native to China. They are very similar and probably best regarded as examples of a single, highly variable species. Duke of Argyll's Teaplant was first recorded as an established introduction in 1696, and first noted growing wild in 1848. They are usually found planted in boundary hedges or naturalized on waste ground and other areas from garden throw-outs, expanding from there via suckers. **Features:** LEAVES narrowly oval (L to 10 cm); greyish. FLOWERS funnel-shaped (D to 17 mm) with purplish lobes and a variable paler throat with darker veins. FRUIT egg-shaped red berry (L 10–20 mm).

J F M A M J J A S O N D

IN FRUIT

FLOWERS purple; 5-petalled

FRUIT egg-shaped; red (Goji berry)

LEAVES narrowly oval

333

WIDESPREAD INTRODUCED SHRUBS + TREES

Leaves pp. 58–67 | Flowers pp. 68–73

■ Firethorn *Pyracantha coccinea*

A spiny shrub (H to 6 m) which is a European species grown in gardens since the late 16th century. It can be found as a garden escape in hedges and on waste ground and roadsides. **Features:** TWIGS sharply spiny; LEAVES very narrowly oval (L 2–7 cm); leaf-margin finely toothed; leaf-stalk hairy. FLOWERS white (D ± 8 mm); 5-petalled. FRUIT orange to scarlet berry (D 5–6 mm).

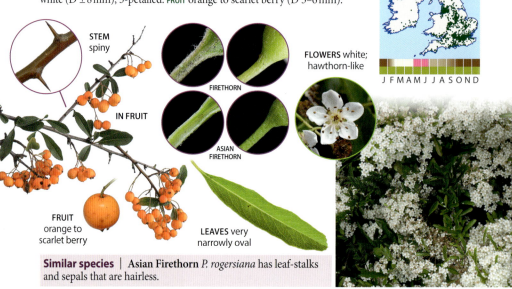

Similar species | **Asian Firethorn** *P. rogersiana* has leaf-stalks and sepals that are hairless.

LC ■ Mock-oranges *Philadelphus* spp.

Shrubs (H to 3 m) that originate from central and southern Europe, North America and China. Mock-orange *P. coronarius* was introduced from Europe by 1596 and escaped from gardens and is now occasionally found in hedges, by roadsides and in scrub-covered wasteland. However, most plants encountered are probably one from a range of hybrids of complex horticultural origin (including hairy and 'double-flowered' groupings). **Features:** LEAVES oppositely arranged; pointed-oval (L 5–10 cm); distinctly toothed towards the tip. FLOWERS white; 4-petalled (D 3–5 cm); sweet orange aroma. FRUIT small capsule (D 15–30 mm).

Note | Images are of a member of the '**Lemoinei Group**' (**Hairy Mock-orange**) that have cup-shaped flowers and usually with sepals, flower-stalk and leaf underside that are hairy. Mock-orange *P. coronarius* has hairless green flower parts and leaf undersides sparsely hairy at most. Identification of these taxa is beyond the scope of this book.

Fruit pp. 74–77 | Twigs pp. 78–85   WIDESPREAD INTRODUCED SHRUBS + TREES

LC Laurustinus *Viburnum tinus*

A large evergreen shrub (H to 6 m) that is native to the Mediterranean and southwest Europe. Introduced by 1596 and originally planted in gardens, it has become established in scrubby areas on sea cliff, roadsides and railways and many other places – especially on chalk and limestone. **Features:** LEAVES pointed-oval (L 4–10 cm); dark green; leathery; leaf-margins untoothed. FLOWERS very similar to Wayfaring Tree (*p. 310*); small (D ±6 mm); white; in slightly domed clusters (D 5–10 cm). FRUIT 'metallic' blue-black berries (D 7–10 mm).

IN FRUIT

FRUIT 'metallic' blue-black

FLOWERS white; in slightly domed cluster

LEAVES pointed-oval; untoothed

J F M A M J J A S O N D

LC Wrinkled Viburnum *Viburnum rhytidophyllum*

A large evergreen shrub (H to 6 m) that is native to central and western China. Introduced *c.*1900 into gardens and now found in hedges, woodlands and along roadsides in England, Wales and Scotland. **Features:** LEAVES long; narrowly oval (L to 20 cm); leaf-margin untoothed; upperside with wrinkled appearance; underside densely hairy. FLOWERS white to yellowish/white (D ±6 mm); in slightly domed clusters (D 8–20 cm). FRUIT egg-shaped (L to 8 mm); ripening red then black; white star-shaped hairs.

IN FRUIT

FRUIT red, ripening black; star-shaped hairs

FLOWERS white; in slightly domed cluster

LEAVES narrowly oval; untoothed; underside densely hairy

J F M A M J J A S O N D

335

WIDESPREAD INTRODUCED SHRUBS + TREES — Leaves pp. 58–67 | Flowers pp. 68–73

Juneberry *Amelanchier lamarckii*

A small tree (H to 10 m) that is native to North America. It was introduced in 1746 and has become naturalized, especially on sandy soils in the New Forest and the heathlands of north Hampshire and Surrey. **Features:** LEAVES oval (L 3–7 cm); leaf-margin with small regular teeth. FLOWERS white (D 2–5 cm); 5-petalled; petals narrow and not overlapping. FRUIT ± spherical berry (D ± 1 cm); ripening red then purplish black (although not often produced); persistent sepals on top of the fruit.

IN FRUIT

FRUIT if formed; red; ripening purple to black

FLOWERS white, petals narrow

J F M A M J J A S O N D

TWIGS reddish brown with grey skin

BUDS pale yellowish green tinged purplish

LEAVES broadly oval; apple- or cherry-like

LC Medlar *Mespilus germanica*

A small tree (H typically 6–8 m) that is native to southwest Asia, and possibly southeast Europe. Planted in Britain for its fruit since medieval times, it can be found growing in hedges and woodlands as a relic of cultivation and does not readily spread by seed. **Features:** LEAVES narrowly oval (L 5–12 cm); hairy. FLOWERS white (L 3–5 cm); 5-petalled with 30–40 stamens; sepals longer than petals. FRUIT distinctive; ± globular (D 2–6 cm); orange; with wide-spreading, persistent sepals on top of the fruit.

FRUIT distinctive; apple-like with conspicuous sepals

FLOWERS white; 5-petalled; apple-like

J F M A M J J A S O N D

BUDS egg-shaped with sharply pointed tip

TWIGS densely hairy

LEAVES narrowly oval

Did you know? The fruit tastes best after being exposed to the first frosts and can be eaten raw or used for jam, jelly and medlar 'cheese' – strained pulp cooked with eggs, sugar and butter.

WIDESPREAD INTRODUCED SHRUBS + TREES

LC Stag's-horn Sumach *Rhus typhina*

A deciduous shrub (H to 5 m) that is native to eastern North America. A widespread garden plant, known in Britain by 1629, and has become established on roadsides and waste grounds, spreading by root suckers. The name comes from the shapes created by the branches. **Features:** TWIGS covered in dense brown hairs. LEAVES large (L 25–55 cm); pinnate; typically 11–21 pointed-oval leaflets (L 5–12 cm); margins toothed. FLOWERS ♂ and ♀ flowers usually on separate trees; tiny; in dense pyramidal spike (L 10–20 cm); ♂'s tend to be in looser arrangement. FRUIT red; hairy (D 3–4 mm).

IN FRUIT
FRUIT red hairy
FLOWERS tiny
LEAVES pinnate; 11–21 leaflets with toothed margins
TWIGS fuzzy brown hairs (like deer antler velvet)
BUDS conical; surrounded by raised leaf-scar

LC Fig *Ficus carica*

A species of small tree (7–10 m) which produces edible fruits. A native of the Mediterranean, it has been cultivated for centuries and was brought here by the Romans. **Features:** LEAVES distinctive, large 3–7-lobed. FLOWERS hidden; formed on the inner surface of the pear-shaped receptacle that is slightly open at the top. FRUIT green-skinned; pear-shaped (L 3–5 cm); usually ripens to brown or purple.

FRUIT distinctive; pear-shaped; ripening brown or purple
LEAVES distinctive; large; 3–7-lobed
TWIGS stout; usually ridged
BUDS naked; conical with a long, drawn-out, pointed tip

PARK, STREET & GARDEN TREES + SHRUBS   Leaves *pp. 58–67* | Flowers *pp. 68–73*

Park, street & garden trees and shrubs

There are over 500 tree and shrub species from all around the world that have been introduced into the British & Irish landscape. The most significant period for this was from the end of the 18th century into the Victorian era. Population growth was the driver for a rapid expansion in urban areas which saw a decline in air quality and population health. The creation of public urban spaces and ancillary areas, such as cemeteries, was a response to this urban growth in an attempt to improve living conditions, and many of these spaces became sites where exotic species were planted. Tree planting became part of urban planning and still continues today, with amenity planting of non-native shrubs for screening and landscape.

In parallel, gardening changed from being restricted to the wealthy to being a pastime available to all. The associated increase in nurseries and garden centres saw a huge increase in obtainable species for deliberate planting. The following are a selection of the most widely encountered species showing the main characteristics useful in their identification.

EN Ginkgo *Ginkgo biloba*

A large, deciduous gymnosperm tree (H to 30 m). Young trees have a pyramidal shape, becoming broader and more spreading with age. ♂ trees are more pyramidal and upright in habit and ♀ trees are more compact in shape. Growing naturally in China although now rare in the wild, the species grows in habitats similar to those thought to have been used by fossil *Ginkgo* species, such as rock crevices, stream banks, and steep rocky slopes. Introduced into Britain & Ireland in 1754, when the first specimen was planted at Kew, it is now widely planted and thrives in deep, well-drained, rich soil. Ginkgo appear largely resistant to attack by insects and fungi, and the leaves resist pollution, making it a regular choice for planting in towns and cities.

TWIGS red-brown to grey; with a distinct zig-zag; often with short woody spurs (L to 10 mm)

BUDS small (L 2–4 mm); brown; hairless; on twig or as end-bud on woody spur

BUD
WOODY SPUR

Did you know? Ginkgo trees are the only survivors of a group of plants (Ginkgoales) which were widespread 190 million years ago. Thought by some scientists to be the first trees to have evolved, similar species even grew in England 60 million years ago, and fossil remains can still be found along the coast at Scarborough. About 30 million years ago, the species formed extensive stands across the London basin but then disappeared from the native flora because of climatic changes.
Ancient Ginkgoes produce strange nodular aerial roots which hang from the underside of branches. In Japan these are known as *chichi*, Japanese for breast.
In Japan, a group of Ginkgo trees stood within 2·2 km of where an atomic bomb was dropped on Hiroshima on August 6, 1945. The bomb created a vast fireball and exceptional temperatures, and little vegetation was left standing alive, except the Ginkgo trees, which are still growing today. These trees have been dubbed *hibakujumoku* in Japanese: 'the trees that survived the bombing'.

LC Tulip-tree *Liriodendron tulipifera*

A tall deciduous tree (H to 50 m) that is native to eastern coastal USA. Probably introduced to Britain & Ireland in the middle of the 17th century. **Features:** TWIGS shiny reddish brown. BUDS egg-shaped, purplish. LEAVES 4-lobed; creating a distinctive truncated look. In autumn, the leaves turn bright yellow-gold. FLOWERS greenish yellow; 'tulip-like'. FRUIT light brown cone-like structure containing winged seeds.

J F M A M J J A S O N D

TWIGS distinctive; shiny, red-brown with large pale leaf-scars

FRUIT

BUDS distinctive; purplish; usually wider than adjacent twig

Cabbage-palm *Cordyline australis*

An evergreen perennial, branched and tree-like when mature (H to 20 m). A member of the Asparagus family introduced from New Zealand in 1823. It has escaped from gardens and can be found naturalized on waste ground and in coastal locations, particularly in western Britain and south-east Ireland. **Features:** LEAVES long; sword-like (L 40–100 cm); in tufts at the tips of branches; upper leaves in tuft upright; lower drooping. FLOWERS creamy white in distinctive ± cylindrical erect to drooping spikes (L 60–120 cm). FRUIT white berry.

J F M A M J J A S O N D

IN FRUIT

Spanish-dagger *Yucca gloriosa*

A striking evergreen perennial plant (H to 5 m), usually with a thick stem from which a crown of leaves arise. A member of the Asparagus family introduced from North America by 1596, it is grown widely as a garden plant and can be found naturalized in sand dunes and coastal areas, particularly in south-west England and south Wales. **Features:** LEAVES long; stiff; sword-like (L 40–100 cm); tip with sharp spine. FLOWERS creamy white (can have red or purple tinge); on an erect, narrowly conical spike (L 90–240 cm). FRUIT oblong capsule.

J F M A M J J A S O N D

FLOWERS

PARK, STREET & GARDEN TREES + SHRUBS **Leaves** *pp. 58–67* | **Flowers** *pp. 68–73*

LC ▮ Manna Ash *Fraxinus ornus*

A medium-sized deciduous tree (H to 25 m) that is native to southern Europe and south-western Asia and was brought to Britain & Ireland *c.*1710 and planted by a Dr Uvedale of Enfield. **Features:** BARK grey; smooth. BUDS end-bud pinkish-brown to grey-brown, with a dense covering of short grey hairs (*cf.* Ash). LEAVES pinnate; 5–9 leaflets; leaflet-margins finely toothed wavy. FLOWERS white, in dense clusters at the end of some branches. FRUIT long, slender winged seeds.

J F M A M J J A S O N D

LEAFLET-MARGINS
wavier than in Ash

ASH (*p. 298*)
BUDS black

BUDS pinkish/grey-brown

LC ▮ Narrow-leaved Ash *Fraxinus angustifolia*

A tall deciduous tree (H to 30 m) naturally occurring in the Mediterranean and into Asia. Introduced in 1815 with two subspecies and a cultivar commonly planted: ssp. *angustifolia* has hairless leaves; ssp. *oxycarpa* has leaves with hairs on the underside near the midrib. In autumn leaves are yellow or green, although striking red in the 'Raywood Ash' cultivar. Narrow-leaved Ash is very similar to Ash – primary differences detailed below. **Features:** BUDS brown. LEAFLETS typically 7–11; narrower; leaflet-margins more coarsely toothed.

J F M A M J J A S O N D

LEAFLETS
much
narrower
than those
of Ash

BUDS brown

LEAF-MARGIN
coarsely toothed

▮ Tree-of-heaven *Ailanthus altissima*

A medium-sized deciduous tree (H to 15 m) native to China and Taiwan and introduced via seed sent to the Chelsea Physic Garden in 1751. Planted in parks and gardens, especially in London, it can reproduce from suckers and can be invasive. **Features:** LEAVES pinnate (L up to 90 cm); 15–40 pointed-oval leaflets (L 7–15 cm) with 1–6 large teeth at the base. FLOWERS ♂ and ♀ flowers usually on separate trees; small; greenish white in spikes (L 10–20 cm); ♂ flowers with 'rancid peanut' aroma. FRUIT reddish brown seeds with distinctive twisted tips.

J F M A M J J A S O N D

LEAFLETS 15–40

340

Escallonia *Escallonia rubra* var. *macrantha*

An evergreen shrub (H to 3 m) that is native to Argentina and Chile. Introduced in 1847 and recorded as forming hedges in the County Kerry and the west of England at the turn of the 20th century, it has now become naturalized in hedges, on sea cliffs, roadsides and waste ground. **Features:** LEAVES arranged oppositely; oval (L 2–5 cm); tapering to the base; tip ± pointed; leaf-margin toothed towards the tip; underside dotted with sticky glands. FLOWERS pink to red; trumpet-like (L ± 15 mm); 5 petal-lobes. FRUIT broad egg-shaped capsule (L ± 8 mm).

J F M A M J J A S O N D

LEAVES base tapered; tip slightly pointed at most

LEAF UNDERSIDE with glands

Weigelia *Weigela florida*

A deciduous shrub with arching stems (H to 3 m) that is native to China and eastern Asia. Introduced in 1845 it has escaped from gardens and become naturalized in many waste spaces and in hedges and on roadsides and railway banks. **Features:** LEAVES arranged alternately; oval (L 4–6 cm); usually with a long-pointed tip and rounded base; underside densely hairy on veins; leaf-margin finely-toothed. FLOWERS pink to red; trumpet-like (L 30–35 mm) with 5 petal-lobes. FRUIT curved cylinder (L 20–25 mm).

J F M A M J J A S O N D

LEAVES base rounded; tip a drawn-out point

Flowering quinces *Chaenomeles* spp.

Open ± spiny shrubs. **Chinese Quince** *C. speciosa* and **Japanese Quince** *C. japonica* hybridize (**Hybrid Quince** *C. ×superba*), with the hybrids showing mixed and intermediate characters of the parents. FLOWERS red, orange, pink or white (L 35–45 mm). FRUIT apple-like; yellowish green.

J F M A M J J A S O N D

CHINESE   JAPANESE

	Chinese	Japanese
Form	H to 3 m	H to 1 m
Twigs	smooth	warty
Leaves	L 40–100 mm; fine teeth	L 25–60 mm; coarse teeth
Fruit	± 40 mm	50–65 mm

FRUIT

PARK, STREET & GARDEN TREES + SHRUBS   Leaves *pp. 58–67* | Flowers *pp. 68–73*

Judas-tree *Cercis siliquastrum*

Small deciduous tree (H to 14 m) native to eastern Europe and western Asia and introduced by 1596 by the herbalist John Gerard. Leaves heart-shaped; wider than long. Flowers bright pink; pea-like; in clusters; growing straight from the bark of the trunk.

J F M A M J J A S O N D

Katsura *Cercidiphyllum japonicum*

Small to medium-sized deciduous tree (H to 30 m) native to China and Japan. Introduced in 1865 and often planted in gardens and parks for their stunning autumn colours. Distinctive oval to heart-shaped leaves have an aroma of burnt toffee as the leaves fall in autumn.

J F M A M J J A S O N D

Black Mulberry *Morus nigra*

Small tree (H to 10 m) originating in Asia and first introduced to Britain & Ireland by 1548 and widely planted (mainly in southern England) for its large, black raspberry-like fruit. Heart-shaped leaves and distinctive fruit (*inset*).

J F M A M J J A S O N D

Indian Bean Tree *Catalpa bignonioides*

Medium-sized deciduous tree (H to 20 m) native to the southeastern USA. Introduced in 1726. Large heart-shaped leaves (L to 25 cm); white trumpet-shaped flowers with yellow spots inside (*left*); distinctive pendent, long, thin bean-like pods (*right*).

J F M A M J J A S O N D

Fruit *pp. 74–77* | **Twigs** *pp. 78–85*　　　　　　　　　　PARK, STREET & GARDEN TREES + SHRUBS

Dove Tree *Davidia involucrata*
Medium-sized deciduous tree (H to 20 m) that is native to China. Brought to Britain & Ireland in 1902 by the plant hunter Ernest Wilson, after a dangerous journey in which his boat sank in rapids. Flowers highly distinctive; purple, in a cluster; surrounded by 2 large white bracts which flutter in a breeze and look like a flapping dove or handkerchief.　　J F M A M J J A S O N D

Cider Gum *Eucalyptus gunnii*
Tall, fast-growing tree (H to 30 m; growth up to 2 m a year) with distinctive shedding bark (*inset*). It was found in Tasmania by Engish botanist Sir J. Hooker in 1840 and introduced to Britain & Ireland shortly after. Decorative young leaves are round and stalkless but become long and willow-like after 4 years.

J F M A M J J A S O N D

LEAVES

LEAVES

Pagoda Tree *Styphnolobium japonicum*
Large deciduous tree (H to 25 m) that is endemic to China and brought to Britain & Ireland by 1753 by James Gordon, a London nurseryman. Leaves pinnate (L to 25 cm) with 7–17 oval leaflets with pointed tip and creamy white, pea-like flowers.

J F M A M J J A S O N D

Honey Locust *Gleditsia triacanthos*
Tall deciduous tree (H to 30 m) that is native to northern and eastern USA. First recorded during the latter half of the 17th century in Fulham, London, it is planted for its deep gold autumn colour. Twigs sharply spiny; leaves pinnate (L to 20 cm) with 14–32 oblong to spear-shaped leaflets (*cf.* False Acacia *p. 300*).

J F M A M J J A S O N D

PARK, STREET & GARDEN TREES + SHRUBS

Magnolias *Magnolia* spp.

A range of species from Asia and the Americas, magnolias are trees (H to 20 m) and shrubs (H to 3 m) with large fragrant, often tulip-like, flowers. Some species are evergreen, but in those that are deciduous the flowers usually appear in the early spring, before the leaves.

J F M A M J J A S O N D

Californian Lilac *Ceanothus* spp.

A range of over 50 species of nitrogen-fixing shrubs (H to 3 m) some of which are widely planted as garden shrubs or hedges. The vibrant blue flowers are usually tiny and fragrant. The seeds are small, hard nutlets which may depend on fire to trigger germination.

J F M A M J J A S O N D

Forsythia *Forsythia* spp.

These shrubs (H to 2 m) are renowned for their bright yellow flowers, which brighten gardens in the late winter/early spring. Named after William Forsyth (1737–1804), a Scottish botanist and a gardener at the Chelsea Physic Garden. He was also a founding member of the Royal Horticultural Society.

J F M A M J J A S O N D

Witch Hazel *Hamamelis* spp.

Deciduous shrubs (H to 5 m) with oval leaves (L 5–15 cm) that have a wavy margin. The species is widely planted in gardens for its distinctive 'spider'-like yellowish flowers, which are winter-flowering and appear alongside the previous year's maturing fruit.

J F M A M J J A S O N D

344

Further reading

Chinery, M. (2011). *Britain's Plant Galls*. Princeton WildGuides.

Elwes, H.J. & Henry, A. (1906). *The Trees of Great Britain & Ireland*. Privately printed, Edinburgh.

Evelyn, J. (1662). *Sylva, or, a Discourse of Forest-Trees, and the Propagation of Timber in his Majesties Dominions: As it was delivered in the Royal Society on the 15th Day of October 1662.*

Eversham, B. (2021). *Identifying British Elms* Ulmus. Wildlife Trust for Beds, Cambs & Northants.

Gruffydd, B. (1987). *Tree Form, Size and Colour*. Taylor & Francis.

Kibby, G. (2020). *Mushrooms and Toadstools of Britain & Europe, Volume 1* (3rd Edition). Geoffrey Kibby.

Lonsdale, D., (ed.) (2013). *Ancient and other veteran trees: further guidance on management.* The Tree Council.

Loudon, J.C. (1838). *Arboretum et Fruiticetum Britannicum; Or, the Trees and Shrubs of Britain*. Printed for the Author.

Mitchell, A. (1974). *Field Guide to the Trees of Britain and Northern Europe*. Collins.

Rackham, O. (2000). *The History of the Countryside*. Weidenfield & Nicholson.

Rich, T.C.G., Houston, L., Robertson, A. & Proctor, M.C.F. (2010). *Whitebeams, Rowans and Service Trees of Britain and Ireland*. BSBI.

Sell, P.D. & Murrell, G. (2018). *Flora of Great Britain and Ireland, Volume 1*. Cambridge University Press.

Stokes, J., Rodger, D. & Miles, A. (2004). *The Heritage Trees of Britain and Northern Ireland*. Constable.

Stokes, J. & Hand, K. (2004). *The Hedge Tree Handbook*. The Tree Council.

Stroh, P.A., Walker, K.S., Humphrey, T.A., Pescott, O.L. & Burkmar, R.J. (2023). *Plant Atlas 2020: Mapping Changes in the Distribution of the British and Irish Flora*: Princeton University Press.

Strouts, R.G., & Winter, T.G. (1994). *Diagnosis of Ill Health in Trees*. HMSO.

Strutt, J.G. (1830). *Sylva Britannica; or, Portraits of Forest Trees, Distinguished for their Antiquity, Magnitude, or Beauty*: Published for the author by Longman, Rees, Orme, Brown and Green.

White, J.E.J. (1994). New tree species in a changing world: *Arboricultural Journal*. 18:2, pp. 99–112. AB Academic Publishers.

White (1996). Trees for community woodland and ornamental Plantings in Britain. In *Landscape Plants* (ed. Thoday, P. & Wilson, J.). *Proc. Inst. Hort. Conf.* 2nd–4th April 1995. pp. 31–39. Cheltenham and Gloucester College of Higher Education.

Wolton, R. (2024). *Hedges*. Bloomsbury.

Appendix I

Measurements of optimal short shoot leaves* of elm species after Sell (2018) and Eversham (2021)   *See p. 203*

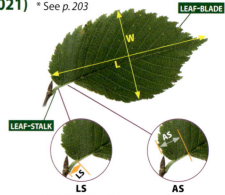

NOTE Sequential numbers for taxa included in the book (*pp. 202–208*) and the 62 species in Eversham (2021) are given. European White-elm (Eversham **22**) is excluded from the table. The table follows the current thinking on genetic grouping. Orange tints denote those taxa that, in the species accounts, are presented outside of their genetic group and instead are covered in the text in morphological groups for pragmatic use in the field.

L – leaf length (from base of blade to tip)
W – leaf width at the widest point
LR – Leaf length: width ratio
LVC – Leaf vein count (on longest side)
AS – length of leaf-base asymmetry
LS – length of leaf-stalk (from join with twig to where it first joins the leaf-base)

LS LEAF-STALK LENGTH the length between the stalk base and the lowest part of the connected leaf-blade

AS LEAF-BASE ASYMMETRY the distance between the base of the lowest side and the base of the higher side

BOOK	EVERSHAM	English name	Scientific name	L Leaf length	W Leaf width	LR Length/Width ratio	LVC Longer side vein count	AS Basal asymmetry	LS Leaf-stalk length
		'WYCH' ELMS							
1	61	Northern Wych Elm	*Ulmus glabra*	10–18cm	4–9cm	2–3	17–27	3–13mm	3–6mm
1a	60	Southern Wych Elm	*Ulmus scabra*	8–17cm	4–11cm	1·6–2·1	16–26	2–8mm	3–7mm
1b	50	Moss's Elm	*Ulmus mossii*	10–15cm	6–9cm	1·5–2·0	16–26	2–8mm	3–7mm
1c	62	Camperdown Elm	*Ulmus camperdownii*	14–18cm	8–11cm	1·6–1·8	18–25	11–14mm	6–8mm
1d	55	Exeter Elm	*Ulmus exoniensis*	8–11cm	6–8cm	1·3–1·5	17–19	7–9mm	5–10mm
		'FIELD' ELMS							
3	40	English Elm	*Ulmus procera*	6–9cm	4–7cm	1·2–1·6	10–16	3–7mm	5–9mm
3	56	Atinian Elm	*Ulmus proceriformis*	5–8cm	3–6cm	1·1–1·6	10–13	1–4mm	3–7mm
4	41	Corky-barked Elm	*Ulmus gyrophylla*	7–12cm	4–7cm	1·2–1·8	15–19	4–8mm	5–11mm
11	59	Dutch Elm	*Ulmus hollandica*	7–14cm	4–8cm	1·5–1·8	13–21	4–13mm	6–13mm
12	53	Huntingdon Elm	*Ulmus vegeta*	8–16cm	5–9cm	1·5–2·0	12–22	6–18mm	9–16mm
13	47	Laxton Elm	*Ulmus crassa*	8–12cm	3–6cm	2–3	21–24	4–7mm	6–9mm
14	32	Cut-leaved Elm	*Ulmus incisa*	5–7cm	3–4cm	1·7–2·2	12–17	2–6mm	7–10mm
15	52	Davey's Elm	*Ulmus daveyi*	5–10cm	5–7cm	1·5–1·7	15–22	6–9mm	6–10mm
–	58	Guernsey Elm	*Ulmus insularum*	6·5–8·5cm	4·5–6·0cm	–	–	–	8–12mm
16	46	Fat-leaved Elm	*Ulmus obesifolia*	5–9cm	3–5cm	1·2–1·9	14–20	2–9mm	4–9mm
17	57	Scabrid Elm	*Ulmus scabrosa*	7–12cm	4–7cm	1·5–2·0	10–20	4–8mm	6–8mm
18	1	Cornish Elm	*Ulmus cornubiensis*	2–4cm	1·5–2·0cm	1·5–2·0	9–12	0–2mm	1·5–6mm
19	3	Goodyer's Elm	*Ulmus sativa*	3–6cm	2–4cm	1·5–2·0	9–15	1–4mm	4–11mm
20	4	Jersey Elm	*Ulmus wheatleyi*	5–7cm	3–5cm	1·2–1·7	11–17	2–8mm	6–11mm
21	2	Small-leaved Elm	*Ulmus stricta*	5–7cm	3–4cm	1·5–2·2	11–16	0·5–3mm	5–8mm
22	51	Western Elm	*Ulmus occidentalis*	6–11cm	4–7cm	1·4–1·7	11–20	2–8mm	3–9mm

APPENDIX 1 | MEASUREMENTS OF ELMS

BOOK	EVERSHAM	English name	Scientific name	L Leaf length	W Leaf width	LR Length/Width ratio	LVC Longer side vein count	AS Basal asymmetry	LS Leaf–stalk length
23	24	Peninsula Elm	*Ulmus peninsularis*	5–8cm	3–4cm	1·5–2·3	12–17	3–8mm	6–10mm
24	8	Smooth-leaved Elm	*Ulmus rasilis*	3–5cm	1·5–2cm	1·6–1·9	9–13	0–4mm	4–8mm
25	20	Long-toothed Elm	*Ulmus longidens*	5–7cm	3–4cm	1·5–2·0	12–17	2–5mm	7–10mm
26	12	Assington Elm	*Ulmus serrata*	4–7cm	2–4cm	1·4–2·0	10–16	1–4mm	5–9mm
27	27	Essex Elm	*Ulmus pseudelegantissima*	4–7cm	3–4cm	1·5–2·0	9–12	2–4mm	4–9mm
5	54	Bassingbourn Elm	*Ulmus platyphylla*	7–13cm	5–8cm	1·3–2·1	9–16	2–9mm	6–11mm
6	48	Woodland Elm	*Ulmus cantabrigiensis*	6–10cm	4–6cm	1·5–2·0	12–17	1–4mm	6–11mm
7	43	Hayley Elm	*Ulmus crenata*	5–10cm	3–6cm	1·5–2·0	11–19	1–6mm	7–10mm
8	42	Dark-leaved Elm	*Ulmus atrovirens*	6–9cm	3–5cm	1·7–2·0	12–16	2–6mm	9–11mm
9	44	Madingley Elm	*Ulmus madingleyensis*	5–11cm	3–6cm	1·5–2·0	13–20	2–7mm	5–12mm
10	45	Sacombe Elm	*Ulmus pseudobovata*	5–7cm	3–4cm	1·8–2·0	11–20	2–8mm	3–9mm
28	15	Bonhunt Elm	*Ulmus acutissima*	6–10cm	3–4cm	2·0–2·5	15–19	3–8mm	8–13mm
29	19	Fat-toothed Elm	*Ulmus obesidens*	6–9cm	4–6cm	1·4–1·8	15–20	5–14mm	6–9mm
30	23	Wedge-leaved Elm	*Ulmus cuneiformis*	8–11cm	3–5cm	2·0–2·5	17–21	3–9mm	7–14mm
31	29	Long-tailed Elm	*Ulmus longicaudata*	6–10cm	3–5cm	1·5–2·3	13–22	3–10mm	5–9mm
32	26	Narrow-crowned Elm	*Ulmus multidentata*	5–8cm	3–5cm	1·8–2·2	12–17	2–8mm	5–12mm
33	13	Prominent-toothed Elm	*Ulmus prominentidens*	5–8cm	3–4cm	1·5–2·0	12–18	–8mm	5–10mm
34	30	Chaters' Elm	*Ulmus chaterorum*	6–10cm	3–5cm	1·3–1·9	13–16	2–7mm	4–9mm
35	18	Burred Elm	*Ulmus serratifrons*	6–10cm	3–4cm	1·4–2·3	13–20	3–8mm	9–14mm
36	16	Jagged-leaved Elm	*Ulmus longidentata*	6–9cm	3–5cm	1·7–2·0	13–22	2–6mm	6–10mm
37	35	Pebmarsh Elm	*Ulmus serratula*	6–9cm	4–5cm	1·6–2·1	13–16	4–8mm	9–13mm
38	33	Tall Elm	*Ulmus alta*	5–9cm	3–7cm	1·5–2·0	11–16	3–8mm	7–11mm
39	36	Pointed-leaved Elm	*Ulmus acuminatissima*	5–9cm	3–5cm	1·5–2·0	15–18	3–9mm	7–11mm
40	34	Narrow-leaved Elm	*Ulmus oblanceolata*	5–10cm	3–5cm	1·7–2·3	11–20	2–8mm	6–13mm
41	10	Cambridge Elm	*Ulmus minor*	5–10cm	3–5cm	1·7–2·5	10–21	2–10mm	7–14mm
42	7	Luffenham Elm	*Ulmus angustiformis*	4–7cm	2–3cm	2·0–2·7	13–18	1–4mm	7–9mm
43	17	Hatley Elm	*Ulmus sylvatica*	6–10cm	3–4cm	1·7–2·7	13–22	2–9mm	7–11mm
44	6	Anglo-Saxon Elm	*Ulmus anglosaxonica*	4–7cm	2–4cm	1·5–2·0	11–15	1–4mm	4–8mm
45	5	Curved-leaved Elm	*Ulmus curvifolia*	5–7cm	2–4cm	1·9–2·2	12–19	0–4mm	5–10mm
46	38	Pale-leaved Elm	*Ulmus asymmetrica*	5–8cm	3–5cm	1·5–2·0	14–17	5–15mm	10–20mm
47	49	Large-toothed Elm	*Ulmus prionophylla*	6–9cm	3–5cm	1·6–2·4	13–21	2–6mm	6–9mm
48	25	Rhombic-leaved Elm	*Ulmus rhombifolia*	5–9cm	3–6cm	1·3–1·9	9–17	1–6mm	7–11mm
49	28	East Anglian Elm	*Ulmus diversifolia*	4–7cm	3–4cm	1·3–2·1	11–16	1–5mm	4–9mm
50	21	Leathery-leaved Elm	*Ulmus coriaceifolia*	5–7cm	3–4cm	1·7–2·2	12–17	2–6mm	7–10mm
51	39	Coritanian Elm	*Ulmus coritana*	5–7cm	3–4cm	1·7–2·2	12–17	2–6mm	7–10mm
52	14	Round-leaved Elm	*Ulmus pseudocoritana*	5–8cm	4–5cm	1·4–2·0	12–17	3–8mm	5–10mm
53	9	Dwarf-leaved Elm	*Ulmus microdon*	5–7cm	3–4cm	1·5–2·9	12–17	1–6mm	6–12mm
54	11	Plot's Elm	*Ulmus plotii*	4–6cm	2–4cm	1·3–2·3	8–14	4–10mm	1–6mm
55	31	Midland Elm	*Ulmus elegantissima*	5–7cm	3–5cm	1·7–2·0	8–12	2–6mm	5–10mm
56	37	Sowerby's Elm	*Ulmus sowerbyi*	6–8cm	3–4cm	1·8–2·1	13–16	3–8mm	9–14mm
57	–	'Tasburgh' Elm	–	–	–	–	–	–	–

Appendix 2

Taxonomic list of trees and shrubs mentioned in the text

Colours refer to species' status: ■ Native; ■ Archaeophyte; ■ Neophyte; ▨ Hybrid (see *p. 8* for definitions)

GYMNOSPERMS

Ginkgo	**Ginkgoaceae**
Ginkgo	*Ginkgo biloba*
Monkey-puzzle	**Araucariaceae**
Monkey-puzzle	*Araucaria araucana*
Podocarp	**Podocarpaceae**
Plum-fruited Yew	*Podocarpus andinus*
Cypress	**Cupressaceae**
Dawn Redwood	*Metasequoia glyptostroboides*
Coast (Coastal) Redwood	*Sequoia sempervirens*
Giant Redwood	*Sequoiadendron giganteum*
Japanese Red-cedar	*Cryptomeria japonica*
Swamp Cypress	*Taxodium distichum*
Sawara Cypress	*Chamaecyparis pisifera*
Lawson's Cypress	*Chamaecyparis lawsoniana*
Monterey Cypress	*Cupressus macrocarpa*
Smooth Arizona Cypress	*Cupressus glabra*
Leyland Cypress	*Cupressus ×leylandii*
Common Juniper	*Juniperus communis* ssp. *communis* ssp. *hemisphaerica* ssp. *nana*
Western Red-cedar	*Thuja plicata*
Northern White-cedar	*Thuja occidentalis*
Nootka Cypress	*Xanthocyparis nootkatensis*
Yew	**Taxaceae**
Yew	*Taxus baccata*
Pine	**Pinaceae**
Deodar	*Cedrus deodara*
Atlas Cedar	*Cedrus atlantica*
Cedar of Lebanon	*Cedrus libani*
Eastern Hemlock-spruce	*Tsuga canadensis*
Western Hemlock-spruce	*Tsuga heterophylla*
Noble Fir	*Abies procera*
Grand (Giant) Fir	*Abies grandis*
European Silver-fir	*Abies alba*
Caucasian Fir	*Abies nordmanniana*
Douglas Fir	*Pseudotsuga menziesii*
European Larch	*Larix decidua*
Japanese Larch	*Larix kaempferi*
Hybrid Larch	*Larix ×marschlinsii*
Sitka Spruce	*Picea sitchensis*
Norway Spruce	*Picea abies*
Macedonian Pine	*Pinus peuce*
Weymouth Pine	*Pinus strobus*
Bhutan Pine	*Pinus wallichiana*
Maritime Pine	*Pinus pinaster*
Black Pines	*Pinus nigra* ssp. *nigra* ssp. *laricio*
Scots Pine	*Pinus sylvestris*
Dwarf Mountain Pine	*Pinus mugo*
Monterey Pine	*Pinus radiata*
Lodgepole Pine	*Pinus contorta*

ANGIOSPERMS

MAGNOLIDS

MAGNOLIALES

Magnolia	**Magnoliaceae**
Magnolias	*Magnolia* spp.
Tulip-tree	*Liriodendron tulipifera*

LAURALES

Laurel	**Lauraceae**
Bay	*Laurus nobilis*

MONOCOTS

ASPARAGALES

Asparagus	**Asparagaceae**
Butcher's-broom	*Ruscus aculeatus*
Cabbage-palm	*Cordyline australis*
Spanish-dagger	*Yucca gloriosa*

EUDICOTS

RANUNCULALES

Barberry	**Berberidaceae**
Barberry	*Berberis vulgaris*
Darwin's Barberry	*Berberis darwinii*
Thunberg's Barberry	*Berberis thunbergii*
Gagnepain's Barberry	*Berberis gagnepainii*
Chinese Barberry	*Berberis julianae*
Mrs Wilson's Barberry	*Berberis wilsoniae*
Hedge Barberry	*Berberis ×stenophylla*
Oregon-grape	*Mahonia aquifolium*

PROTEALES

Plane	**Platanaceae**
Oriental Plane	*Platanus orientalis*
American Plane	*Platanus occidentalis*
London Plane	*Platanus ×hispanica*

BUXALES

Box	**Buxaceae**
Box	*Buxus sempervirens*

SAXIFRAGALES

Currant	**Grossulariaceae**
Flowering Currant	*Ribes sanguineum*
Black Currant	*Ribes nigrum*
Red Currant	*Ribes rubrum*
Downy Currant	*Ribes spicatum*
Mountain Currant	*Ribes alpinum*
Gooseberry	*Ribes uva-crispa*

APPENDIX 2 | TAXONOMIC LIST

Katsura — **Cercidiphyllaceae**

Katsura — *Cercidiphyllum japonicum*

Witch-hazel — **Hamamelidaceae**

Witch Hazel — *Hamamelis spp.*

CELESTRALES

Spindle — **Celastraceae**

Spindle — *Euonymus europaeus*
Evergreen Spindle — *Euonymus japonicus*
Large-leaved Spindle — *Euonymus latifolius*

MALPIGHIALES

Willow — **Salicaceae**

Aspen — *Populus tremula*
Grey Poplar — *Populus ×canescens*
White Poplar — *Populus alba*
Black Poplar — *Populus nigra*
  Lombardy Poplar — 'Italica'
Hybrid Black Poplar — *Populus ×canadensis*
  'Marilandica'
  'Robusta'
  'Serotina'
  'Serotina Aurea'
Western Balsam-poplar — *Populus trichocarpa*
Balm-of-Gilead — *Populus ×jackii*
Eastern Balsam-poplar — *Populus balsamifera*
Hybrid Balsam-poplar — *Populus ×hastata*
Grey Willow — *Salix cinerea*
  Rusty Sallow — ssp. *oleifolia*
  Grey Sallow — ssp. *cinerea*
Goat Willow — *Salix caprea*
  ssp. *caprea*
  ssp. *sphacelata*
Eared Willow — *Salix aurita*
Purple Willow — *Salix purpurea*
Bay Willow — *Salix pentandra*
Osier — *Salix viminalis*
White Willow — *Salix alba*
Crack-willow — *Salix ×fragilis*
Almond Willow — *Salix triandra*
Eastern Crack-willow — *Salix euxina*
Olive Willow — *Salix elaeagnos*
Weeping Willow — *Salix ×sepulcralis*
Golden Willow — *Salix alba* var. *vitellina*
Willow hybrid — *Salix ×reichardtii*
Willow hybrid — *Salix ×multinervis*
Broad-leaved Osier — *Salix ×smithiana*
Silky-leaved Osier — *Salix ×holoserica*
Tea-leaved Willow — *Salix phylicifolia*
Dark-leaved Willow — *Salix myrsinifolia*
Creeping Willow — *Salix repens*
  var. *repens*
  var. *fusca*
  var. *argentea*
Downy Willow — *Salix lapponum*

Woolly Willow — *Salix lanata*
Whortle-leaved Willow — *Salix myrsinites*
Mountain Willow — *Salix arbuscula*
Dwarf Willow — *Salix herbacea*
Net-leaved Willow — *Salix reticulata*

FABALES

Pea — **Fabaceae**

Judas-tree — *Cercis siliquastrum*
Honey Locust — *Gleditsia triacanthos*
False Acacia — *Robinia pseudoacacia*
Spanish Gorse — *Genista hispanica*
Petty Whin — *Genista anglica*
Dyer's Greenweed — *Genista tinctoria*
  ssp. *tinctoria*
  ssp. *littoralis*
Hairy Greenweed — *Genista pilosa*
Gorse — *Ulex europaeus*
Western Gorse — *Ulex gallii*
Dwarf Gorse — *Ulex minor*
Hybrid Gorse — *U. ×breoganii*
Broom — *Cytisus scoparius*
  ssp. *maritimus*
Spanish Broom — *Spartium junceum*
Laburnum — *Laburnum anagyroides*
Scottish Laburnum — *Laburnum alpinum*
Pagoda Tree — *Styphnolobium japonicum*

ROSALES

Rose — **Rosaceae**

Rowan — *Sorbus aucuparia*
Hupeh Rowan — *Sorbus hupehensis*
Wild Service-tree — *Sorbus torminalis*
Orange Whitebeam — *Sorbus croceocarpa*
Common Whitebeam — *Sorbus aria*
Rock Whitebeam — *Sorbus rupicola*
Swedish Whitebeam — *Sorbus intermedia*
English Whitebeam — *Sorbus anglica*
Grey-leaved Whitebeam — *Sorbus porrigentiformis*
True Service-tree — *Sorbus domestica*
Lancastrian Whitebeam — *Sorbus lancastriensis*
Irish Whitebeam — *Sorbus hibernica*
Scannell's Whitebeam — *Sorbus scannelliana*
Arran Whitebeam — *Sorbus arranensis*
Arran Service-tree — *Sorbus pseudofennica*
False Rowan — *Sorbus pseudomeinichii*
Menai Strait Whitebeam — *Sorbus arvonicola*
Stirton's Whitebeam — *Sorbus stirtoniana*
Llangollen Whitebeam — *Sorbus cuneifolia*
Ley's Whitebeam — *Sorbus leyana*
Welsh Whitebeam — *Sorbus cambrensis*
Thin-leaved Whitebeam — *Sorbus leptophylla*
Least Whitebeam — *Sorbus minima*
Llanthony Whitebeam — *Sorbus stenophylla*

APPENDIX 2 | TAXONOMIC LIST

Devon Whitebeam	*Sorbus devoniensis*
Watersmeet Whitebeam	*Sorbus admonitor*
Somerset Whitebeam	*Sorbus subcuneata*
Margaret's Whitebeam	*Sorbus margaretae*
Bloody Whitebeam	*Sorbus vexans*
Rich's Whitebeam	*Sorbus richii*
Cheddar Whitebeam	*Sorbus cheddarensis*
Twin Cliffs Whitebeam	*Sorbus eminentoides*
Gough's Rock Whitebeam	*Sorbus rupicoloides*
Observatory Whitebeam	*Sorbus spectans*
Leigh Woods Whitebeam	*Sorbus leighensis*
Willmott's Whitebeam	*Sorbus wilmottiana*
Bristol Whitebeam	*Sorbus bristoliensis*
White's Whitebeam	*Sorbus whiteana*
Round-leaved Whitebeam	*Sorbus eminens*
Doward Whitebeam	*Sorbus eminentiformis*
Herefordshire Whitebeam	*Sorbus herefordensis*
Green's Whitebeam	*Sorbus greenii*
Symonds Yat Whitebeam	*Sorbus saxicola*
Evans' Whitebeam	*Sorbus evansii*
Ship Rock Whitebeam	*Sorbus parviloba*
Medlar	*Mespilus germanica*
Juneberry	*Amelanchier lamarckii*
Brideworts	*Spiraea* spp. + cultivars
Cherry Laurel	*Prunus laurocerasus*
Portugal Laurel	*Prunus lusitanica*
Bird Cherry	*Prunus padus*
Rum Cherry	*Prunus serotina*
Wild Cherry	*Prunus avium*
Dwarf (Sour) Cherry	*Prunus cerasus*
Japanese Cherry	*Prunus serrulata*
Blackthorn	*Prunus spinosa*
Cherry Plum	*Prunus cerasifera*
Wild Plum	*Prunus domestica*
Wild Apple (Crab Apple)	*Malus sylvestris*
Domestic Apple	*Malus domestica*
Chinese Quince	*Chaenomeles speciosa*
Plymouth Pear	*Pyrus cordata*
Wild Pear	*Pyrus pyraster*
Cultivated Pear	*Pyrus communis*
Hawthorn	*Crataegus monogyna*
Midland Hawthorn	*Crataegus laevigata*
Cockspurthorn	*Crataegus crus-galli*
Large-sepalled Hawthorn	*Crataegus rhipidophylla*
Wild Cotoneaster	*Cotoneaster cambricus*
Bullate Cotoneaster	*Cotoneaster rehderi*
Hollyberry Cotoneaster	*Cotoneaster bullatus*
Franchet's Cotoneaster	*Cotoneaster franchetii*
Himalayan Cotoneaster	*Cotoneaster simonsii*
Entire-leaved Cotoneaster	*Cotoneaster integrifolius*
Wall Cotoneaster	*Cotoneaster horizontalis*
Firethorn	*Pyracantha coccinea*
Asian Firethorn	*Pyracantha rogersiana*

Buckthorn	**Rhamnaceae**
Alder Buckthorn	*Frangula alnus*
Buckthorn	*Rhamnus cathartica*
Californian Lilac	*Ceanothus* spp.
Sea-buckthorn	**Elaeagnaceae**
Sea-buckthorn	*Hippophae rhamnoides*
Elm	**Ulmaceae**
European White-elm	*Ulmus laevis*
Wych Elm	*Ulmus glabra*
Field Elm	*Ulmus minor*
Fig	**Moraceae**
Black Mulberry	*Morus nigra*
Fig	*Ficus carica*

FAGALES

Southern Beech	**Nothofagaceae**
Roble	*Nothofagus obliqua*
Rauli	*Nothofagus alpina*
Beech	**Fagaceae**
Beech	*Fagus sylvatica*
Turkey Oak	*Quercus cerris*
Evergreen (Holm) Oak	*Quercus ilex*
Cork Oak	*Quercus suber*
Pin Oak	*Quercus palustris*
Red Oak	*Quercus rubra*
Scarlet Oak	*Quercus coccinea*
Pedunculate (English) Oak	*Quercus robur*
Sessile Oak	*Quercus petraea*
Hybrid (Rose) Oak	*Quercus ×rosacea*
Lucombe Oak	*Quercus ×crenata*
Sweet Chestnut	*Castanea sativa*
Sweet Gale	**Myricaceae**
Bog-myrtle	*Myrica gale*
Walnut	**Juglandaceae**
Walnut	*Juglans regia*
Black Walnut	*Juglans nigra*
Birch	**Betulaceae**
Hazel	*Corylus avellana*
Turkish Hazel	*Corylus colurna*
Filbert	*Corylus maxima*
Hornbeam	*Carpinus betulus*
Common Alder	*Alnus glutinosa*
Grey Alder	*Alnus incana*
Italian Alder	*Alnus cordata*
Red Alder	*Alnus rubra*
Green Alder	*Alnus viridis*
Common × Grey Alder	*Alnus ×hybrida*
Silver Birch	*Betula pendula*
Downy Birch	*Betula pubescens* ssp. *pubescens* ssp. *celtiberica* var. *fragrans* (ssp. *tortuosa*)
Himalayan Birch	*Betula utilis*
Paper Birch	*Betula papyrifera*
Dwarf Birch	*Betula nana*

APPENDIX 2 | TAXONOMIC LIST

MYRTALES

Myrtle — **Myrtaceae**
Cider Gum — *Eucalyptus gunnii*

Willowherb — **Onagraceae**
Fuchsia — *Fuchsia magellanica*

SAPINDALES

Maple — **Sapindaceae**
Field Maple — *Acer campestre*
Sycamore — *Acer pseudoplatanus*
Norway Maple — *Acer platanoides*
Silver Maple — *Acer saccharinum*
Cappadocian Maple — *Acer cappadocicum*
Red Maple — *Acer rubrum*
Ashleaf Maple — *Acer negundo*
Horse Chestnut — *Aesculus hippocastanum*
Indian Horse Chestnut — *Aesculus indica*
Red Horse Chestnut — *Aesculus ×carnea*

Simaroubaceae
Tree-of-heaven — *Ailanthus altissima*

Anacardiaceae
Stag's-horn Sumach — *Rhus typhina*

MALVALES

Mallows — **Malvaceae**
Small-leaved Lime — *Tilia cordata*
Common (European) Lime — *Tilia ×europaea*
Large-leaved Lime — *Tilia platyphyllos*
Silver Lime — *Tilia tomentosa*

Mallows — **Thymelaeaceae**
Spurge-laurel — *Daphne laureola*
Mezereon — *Daphne mezereum*

CARYOPHYLLALES

Tamarisk — **Caryophyllaceae**
Tamarisk — *Tamarix gallica*

CORNALES

Hydrangea — **Hydrangeaceae**
Mock-oranges — *Philadelphus spp.*

Dogwood — **Cornaceae**
Dogwood — *Cornus sanguinea*
Red-osier Dogwood — *Cornus sericea*
White Dogwood — *Cornus alba*
Cornelian Cherry — *Cornus mas*

Nyssaceae
Dove Tree — *Davidia involucrata*

ERICALES

Heather — **Ericaceae**
Rhododendrons — *Rhododendron ponticum*
Yellow Azalea — *Rhododendron luteum*
Labrador Tea — *Rhododendron groenlandicum*
Strawberry-tree — *Arbutus unedo*

AQUIFOLIALES

Holly — **Aquifoliaceae**
Holly — *Ilex aquifolium*
Highclere Holly — *Ilex ×altaclerensis*

ESCALLONIALES

Escalloniaceae
Escallonia — *Escallonia rubra var. macrantha*

DIPSACALES

Moschatel — **Adoxaceae**
Common Elder — *Sambucus nigra*
Red-berried Elder — *Sambucus racemosa*
American Elder — *Sambucus canadensis*
Guelder-rose — *Viburnum opulus*
American Guelder-rose — *Viburnum trilobum*
Asian Guelder-rose — *Viburnum sargentii*
Wayfaring-tree — *Viburnum lantana*
Chinese Wayfaring-tree — *Viburnum veitchii*
Laurustinus — *Viburnum tinus*
Wrinkled Viburnum — *Viburnum rhytidophyllum*

Honeysuckle — **Caprifoliaceae**
Weigelia — *Weigela florida*
Snowberry — *Symphoricarpos albus*
Hybrid Coralberry — *Symphoricarpos ×chenaultii*
Coralberry — *Symphoricarpos obiculatus*
'Doorenbos Coralberry' — *Symphoricarpos ×doorenbosii*

GARRYALES

Garryaceae
Spotted Laurel — *Aucuba japonica*

LAMIALES

Olive — **Oleaceae**
Forsythia — *Forsythia spp.*
Lilac — *Syringa vulgaris*
Wild Privet — *Ligustrum vulgare*
Garden Privet — *Ligustrum ovalifolium*
Ash — *Fraxinus excelsior*
Manna Ash — *Fraxinus ornus*
Narrow-leaved Ash — *Fraxinus angustifolia*

Plantain — **Plantaginaceae**
'Hedge Hebe' — *Veronica ×franciscana*
Narrow-leaved Hebe — *Veronica salicifolia*
Lewis's Hebe — *Veronica ×lewisii*

Figwort — **Scrophulariaceae**
Butterfly-bush (Buddleia) — *Buddleja davidii*

Trumpet Vines — **Bignoniaceae**
Indian Bean Tree — *Catalpa bignonioides*

SOLANALES

Nightshade — **Solanaceae**
Duke of Argyll's Teaplant — *Lycium barbarum*
Chinese Teaplant — *Lycium chinense*

351

Appendix 3

Non-tree species mentioned in the text

BACTERIA
Frankia alni — —

OOMYCETES
Phytophthora austrocedri — —
Phytophthora lateralis — —
Phytophthora ramorum — —

LICHENS
Hypogymnia physodes — —
Lobaria pulmonaria — Tree Lungwort
Platismatia glauca — —
Pseudocyphellaria crocata — —
Teloschistes flavicans — Golden-hair Lichen

FUNGI
Amanita muscaria — Fly Agaric
Auricularia auricula-judae — Jelly-ear Fungus
Cronartium ribicola — White Pine Blister Rust
Daedaleopsis confragosa — Blushing Bracket
Encoelia fascicularis — Spring Hazelcup
Gymnosporangium clavariiforme — Tongues of Fire
Hericium corralloides — Coral Tooth
Hericium erinaceus — Bearded Tooth
Hydnellum aurantiacum — Orange Tooth
Hymenoscyphus fraxineus; formerly *Chalara fraxinea* — 'Ash dieback'
Hyphodontia sambuci — Elder Whitewash
Hypocreopsis rhododendri — Hazel Gloves
Inonotus obliquus — Chaga
Laccaria laccata — Deceiver
Lactarius aspideus — Willow Milkcap
Lactarius helvus — Fenugreek Milkcap
Lactarius rufus — Rufous Milkcap
Laetiporus sulphureus — Sulphur Polypore
Leccinum holopus — Ghost Bolete
Melampsora amygdalinae — a rust fungus
Melampsora capraearum — a rust fungus
Melampsora epitea — a rust fungus
Melampsora reticulatae — a rust fungus
Paxillus involutus — Brown Rollrim
Phellinus hippophaeicola — Sea-buckthorn Bracket
Phellinus igniarius — Willow Bracket
Piptoporus betulinus — Birch Polypore
Piptoporus quercinus — Oak Polypore
Puccinia graminis — a stem rust
Reticularia lycoperdon — False Puffball
Rhytisma acerinum — Tar Spot
Russula laccata — Willow Brittlegill
Russula ochroleuca — Ochre Brittlegill
Sarcodon glaucopus — Greenfoot Tooth
Schizophyllum amplum — Poplar Bells
Taphrina betulina — 'Witches' broom'
Taphrina carpini — 'Witches' broom'
Taphrina johnsonii — Aspen Tongue

Taphrina padi — Bird Cherry Pocket Gall
Taphrina pruni — Pocket Plum Gall
Tuber aestivum — Summer Truffle
Tuber uncinatum — Autumn, or Burgundy, Truffle

BRYOPHYTES: Mosses
Isothecium myosuroides var. *myosuroides* — Slender Mouse-tail Moss
Thuidium tamariscinum — Common Tamarisk-moss

BRYOPHYTES: Liverworts
Plagiochila punctata — Spotty Featherwort
Plagiochila spinulosa — Prickly Featherwort
Scapania gracilis — Western Earwort

PLANTS: Pteridophytes
Hymenophyllum wilsonii — Wilson's Filmy-fern

PLANTS: Flowering Plants
Allium ursinum — Wild Garlic
Cephalanthera longifolia — Narrow-leaved Helleborine
Dryas octopetala — Mountain Avens
Epipogium aphyllum — Ghost Orchid
Eriophorum angustifolium — Common Cottongrass
Goodyera repens — Creeping Lady's-Tresses
Hyacinthoides non-scripta — Bluebell
Lathraea squamaria — Toothwort
Linnaea borealis — Twinflower
Melampyrum cristatum — Crested Cow-wheat
Moneses uniflora — One-flowered Wintergreen
Narthecium ossifragum — Bog Asphodel
Neottia cordata — Lesser Twayblade
Neottia nidus-avis — Bird's-nest Orchid
Primula elatior — Oxlip
Pyrola media — Intermediate Wintergreen
Sambucus ebulus — Dwarf Elder
Trientalis europaea — Chickweed Wintergreen
Viscum album — Mistletoe

ARACHNIDS: Spiders
Hybocoptus decollatus — —
Hyptiotes paradoxus — Triangle Spider
Verrucosa arenata — Triangle Orbweaver

ARACHNIDS: Mites
Aceria campestricola — A leaf-gall mite
Aceria tenella — A leaf-gall mite
Aculus laevis — A mite
Eriophyes convolvens — A gall mite
Eriophyes tiliae — A blister mite

INSECTS: Coleoptera
Agelastica alni — Alder Leaf Beetle
Ampedus rufipennis — Red-horned Cardinal Click Beetle
Anaplodera sexguttata — Six-spotted Longhorn
Byctiscus populi — Aspen Leaf-rolling Weevil

352

APPENDIX 3 | NON-TREES MENTIONED IN TEXT

Chrysomela populi	Red Poplar Beetle or Creeping Willow Beetle
Cryptocephalus decemmaculatus	10-spotted Pot Beetle
Curculio nucum	Nut Weevil
Dorcus parallelipipedus	Lesser Stag-beetle
Dorytomus filirostris	a bark beetle
Dorytomus ictor	a bark beetle
Dorytomus longimanus	a bark beetle
Erotides cosnardi	Cosnard's Net-winged Beetle
Eucnemis capucina	False Click Beetle
Family Eucnemidae	False Click Beetle
Gelechia hippophaella	Seathorn Groundling
Gnorimus nobilis	Noble Chafer
Gnorimus octopunctatus	a chafer
Gnorimus variabilis	Variable Chafer
Hypebaeus abietinus	Moccas Beetle
Lettura quadrifasciata	Four-banded Longhorn
Limoniscus violaceus	Violet Click Beetle
Mycetophagus piceus	Hairy Fungus Beetle
Platycis minutus	Small Net-winged Beetle
Prionychus ater	Darkling Beetle
Pyrrhalta viburni	Viburnum Leaf Beetle
Rhagium mordax	Black-spotted Longhorn
Rhynchaenus populi	a leaf-mining weevil
Saperda carcharias	Large Poplar Longhorn
Saperdo populnea	Small Poplar Borer
Scolytus multistriatus	an elm bark beetle
Scolytus scolytus	an elm bark beetle
Sinodendron cylindricum	Rhinoceros Beetle
Smaragdina affinis	Short-horned Leaf Beetle

INSECTS: Hemiptera: Aphids

Aphis fabae	Black Bean Aphid
Aphis sambuci	Elder Aphid
Chaitophorus capreae	Pale Sallow Leaf Aphid
Drepanosiphum platanoidis	Common Sycamore Aphid
Dysaphis aucupariae	Wild Service Aphid
Eucallipterus tiliae	Lime Aphid
Myzus padellus	an aphid
Pemphigus spyrothecae	Poplar Spiral Gall Aphid
Periphyllus testudinaceus	Common Periphyllus Aphid
Rhopalosiphum padi	Bird Cherry Oat Aphid

INSECTS: Diptera

Callicera rufa	Pine Long-horned Hoverfly
Contarinia petioli	Aspen Petiole Gall Midge
Eristalinus sepulcralis	Small Spot-eye
Eristalis arbustorum	Plain-faced Dronefly
Eristalis nemorum	Stripe-faced Dronefly
Eristalis tenax	Drone Fly
Harmandiola globuli	a gall midge
Iteomyia capreae	a gall midge
Iteomyia major	a gall midge
Phytomyza ilicis	Holly Leaf Miner
Rabdophaga cinerearum	a gall midge
Rabdophaga salicis	a gall midge
Taxomyia taxi	Yew Gall Midge

INSECTS: Lepidoptera: Butterflies

Apatura iris	Purple Emperor
Callophrys rubi	Green Hairstreak
Celastrina argiolus	Holly Blue butterfly
Favonius quercus	Purple Hairstreak
Gonepteryx rhamni	Brimstone
Nymphalis polychloros	Large Tortoiseshell
Pyronia tithonus	Gatekeeper
Satyrium pruni	Black Hairstreak
Satyrium w-album	White-letter Hairstreak
Thecla betulae	Brown Hairstreak

INSECTS: Lepidoptera: Moths

Abraxas grossulariata	Magpie
Abraxas sylvata	Clouded Magpie
Agrotis clavis	Heart and Club
Agrotis puta	Shuttle-shaped Dart
Angerona prunaria	Orange Moth
Archiearis notha	Light Orange Underwing
Argyresthia sorbiella	—
Argyresthia pruniella	Cherry Fruit Moth
Cacoecimorpha pronubana	Carnation Tortrix
Cameraria ohridella	—
Campaea margaritaria	Light Emerald
Catocala sponsa	Dark-Crimson Underwing
Cerura vinula	Puss Moth
Coleophora coracipennella	—
Coleophora vibicella	Large Gold Case-bearer
Colocasia coryli	Nut-tree Tussock
Cosmia diffinis	White Spotted Pinion
Cyclophora linearia	Clay Triple-line
Cyclophora porata	False Mocha
Cydalima perspectalis	Box Moth
Ectoedemia heringella	—
Ectoedemia intimella	Black-spot Sallow Pigmy (White-spot Sallow Dot)
Ectropis crepuscularia	The Engrailed
Enarmonia formosana	Cherry-bark Moth
Epicallima formosella	a concealer moth
Epione vespertaria	Dark Bordered Beauty
Eriogaster lanestris	Small Eggar
Furcula bicuspis	Alder Kitten
Furcula furcula	Sallow Kitten
Gelechia hippophaella	Seathorn Groundling
Hemithea aestivaria	Common Emerald
Hydriomena ruberata	Ruddy Highflyer
Laothoe populi	Poplar Hawkmoth
Lasiocampa quercus	Oak Eggar
Malacostria	Lackey
Mimas tiliae	Lime Hawkmoth
Opisthograptis luteolata	Brimstone
Ourapteryx sambucaria	Swallow-tailed Moth
Pareulype berberata	Barberry Carpet
Phyllocnistis xenia	Kent Maze-miner
Phyllonorycter oxyacanthae	Common Thorn Leaf-miner
Protolampra sobrina	Cousin German
Ptilodon cucullina	Maple Prominent
Saturnia pavonia	Emperor Moth

353

APPENDIX 3 | NON-TREES MENTIONED IN TEXT

Sesia apiformis	Hornet Clearwing
Sphinx ligustri	Privet Hawkmoth
Sphinx pinastri	Pine Hawkmoth
Spilosoma lutea	Buff Ermine
Stauropus fagi	Lobster Moth
Stigmella suberivora	Holm-oak Pigmy Moth
Swammerdamia passerella	—
Synanthedon formicaeformis	Red-tipped Clearwing
Synanthedon tipuliformis	Currant Clearwing
Thera juniperata	Juniper Carpet
Trisateles emortualis	Olive Crescent
Venusia cambrica	Welsh Wave
Watsonalla cultraria	Barred Hooked-tip
Xanthia gilvago	Dusky-lemon Sallow
Xestia baja	Dotted Clay
Yponomeuta cagnagella	Spindle Ermine
Yponomeuta evonymellus	Bird Cherry Ermine
Zeuzera pyrina	Leopard Moth

INSECTS: Hymenoptera

Andrena ferox	Oak Mining Bee
Andrena thoracica	Cliff Mining Bee
Andricus foecundatrix	Artichoke Gall Wasp
Andricus quercuscalicis	Knopper Gall Wasp
Apis mellifera	Honeybee
Bombus lucorum	White-tailed Bumblebee
Bombus pratorum	Early Bumblebee
Bombus terrestris	Buff-tailed Bumblebee
Mesopolobus diffinis	a chalcid wasp
Syndiplosis petiole	Aspen Petiole Gall Midge
Torymus nigritarsus	a chalcid wasp

INSECTS: Hymenoptera: Sawflies

Aproceros leucopoda	Elm Zigzag Sawfly
Arge enodis	—
Cimbex conatus	Large Alder Sawfly
Euura amerinae	—
Euura arbusculae	—
Euura plicaphylicifolia	—
Euura polita	—
Euura purpureae	—
Euura salicispurpureae	—
Euura triandrae	—
Euura vesicator	—
Euura viminalis	—
Euura weiffenbachiella	—
Hemichroa crocea	Banded Alder Sawfly
Nematus pravus	—
Pristiphora luteipes	—
Pristiphora retusa	—
Rhogogaster viridis	Green Sawfly

BIRDS

Acanthis flammea	Redpoll
Bombycilla garrulus	Waxwing
Carduelis carduelis	Goldfinch
Carduelis spinus	Siskin
Chloris chloris	Greenfinch
Coccothraustes coccothraustes	Hawfinch
Columba palumbus	Woodpigeon
Cyanistes caeruleus	Blue Tit
Dendrocopos major	Great Spotted Woodpecker
Emberiza citrinella	Yellowhammer
Erithacus rubecula	Robin
Ficedula hypoleuca	Pied Flycatcher
Fringilla coelebs	Chaffinch
Garrulus glandarius	Jay
Lagopus muta	Ptarmigan
Lagopus scotica	Red Grouse
Loxia scotica	Scottish Crossbill
Luscinia megarhynchos	Nightingale
Oenanthe oenanthe leucorhoa	'Greenland' Wheatear
Pandion haliaetus	Osprey
Parus major	Great Tit
Passer montanus	Tree Sparrow
Phoenicurus phoenicurus	Redstart
Phylloscopus collybita	Chiffchaff
Poecile montanus	Willow Tit
Prunella modularis	Dunnock
Pyrrhula pyrrhula	Bullfinch
Sitta europaea	Nuthatch
Sturnus vulgaris	Starling
Sylvia communis	Common Whitethroat
Tetrao tetrix	Black Grouse
Tetrao urogallus	Capercaillie
Turdus iliacus	Redwing
Turdus merula	Blackbird
Turdus philomelos	Song Thrush
Turdus pilaris	Fieldfare
Turdus viscivorus	Mistle Thrush

MAMMALS

Akes akes	Elk
Apodemus flavicollis	Yellow-necked Mouse
Apodemus sylvaticus	Wood Mouse
Barbastella barbastellus	Barbastelle
Bos primigenius	Wild Ox / Auroch
Capreolus capreolus	Roe Deer
Castor fiber	European Beaver
Dama dama	Fallow Deer
Martes martes	Pine Marten
Megaloceros giganteus	Irish Elk
Meles meles	Badger
Microtus agrestis	Field Vole
Muscardinus avellanarius	Hazel Dormouse
Myodes glareolus	Bank Vole
Myotis bechsteinii	Bechstein's Bat
Oryctolagus cuniculus	Rabbit
Plecotus auritus	Brown Long-eared Bat
Rhinolophus ferrumequinum	Greater Horseshoe Bat
Rhinolophus hipposideros	Lesser Horseshoe Bat
Sciurus carolinensis	Grey Squirrel
Sciurus vulgaris	Red Squirrel
Sus scrofa	Wild Boar
Vulpes vulpes	Fox

Photographic and illustration credits

The production of this book would not have been possible without the help and cooperation of the photographers whose images have been reproduced.

All illustrations are © **Stuart Jackson-Carter** (SJC Illustration).

For readers wanting more information about a specific image, a comprehensive and fully searchable schedule of all of the images is available for download (under 'Resources') from the Princeton University Press Website at: https://press.princeton.edu/ISBN/9780691224169

The following is a complete list of the contributing photographers and the page numbers their images appear on. Those not listed were taken by the author **Jon Stokes** or by **Rob Still**.

John Bingham (*p257*); **Paul D. Brock** (*pp45, 188, 220, 233, 240, 248, 273, 281, 287, 291, 305, 308, 311, 313*); **Suzanne Burgess** (*p237*); **Greg & Yvonne Dean** (*p217*); **Brian Eversham** (*pp204, 205, 206, 208, 209, 210, 211, 212*); **Chris Gibson** (*p194*) ; **William Harvey** (*p45*); **Mark How** (*p150*); **David Kjaer** (*pp45, 97, 141, 188, 215, 217, 220, 277, 279, 299*); **Durwyn Liley & Sophie Lake** (*p9*); **Tim Melling** (*p327*); **Trevor & Dilys Pendleton** (*p198*); **Malcolm Storey** (*p287*); **Andy & Gill Swash** (*pp287, 303*); **David Tipling** (*pp220, 273*); **Nigel Voaden** (*p188*); **Zoonar GmbH / Alamy Stock Photo** (*p224*).

The following contributors are gratefully acknowledged for their work, shared under the Creative Commons licenses noted in square brackets:

Abraham [CC BY-SA 4.0] (*p283*); **AfroBrazillian** [CC BY-SA 3.0] (*p263*); **Agnieszka Kwiecień**, Nova [CC BY-SA 4.0] (*p94*); **Amazone7** [CC BY-SA 3.0] (*p327*); **Andy Morffew** from Itchen Abbas, Hampshire [CC BY 2.0] (*p317*); **Aplants** [CC BY-SA 4.0] (*p343*); **B. Schoenmakers** at Waarneming.nl [CC BY 3.0] (*p273*); Beech woodland on the edge of Windsor Great Park **by Joe** [CC BY-SA 2.0] (*p213*); **Ben Sale** [CC BY 2.0] (*pp119, 175, 198, 257, 273, 313*); **Bernard Dupont** [CC BY-SA 2.0] (*p281*); **Bramfab**, [CC BY-SA 4.0] (*p283*); **Crusier** [CC BY 3.0] (*p121*); **Daderot**, via Wikimedia Commons [CC 0] (*p113*); **Dave Powell**, USDA Forest Service, United States [CC BY-SA 3.0] (*pp95, 114*); **Dellex** [CC BY-SA 4.0] (*pp76, 175*); **Douglas Goldman** [CC BY 4.0] (*p224*); **Famartin** [CC BY-SA 4.0] (*p289*); **Federico Calledda** [CC BY-NC 4.0] https://uk.inaturalist.org/observations/193744727 (*p235*); **Francis C. Franklin** [CC BY-SA 3.0] (*p233*); **Gaëtan Jouvenez** [CC BY-NC 4.0] https://www.inaturalist.org/observations/239367948 (*p235*); **Harald Süpfle** [CC BY-SA 3.0] (*p235*); **Hectonichus** [CC BY 3.0] (*p248*); **Hermann Schachner** [CC 0] (*pp199, 247*); **Hladac** [CC BY-SA 4.0] (*pp75, 185, 277*); **Ilia Ustyantsev** [CC BY-SA 2.0] (*p44*); **J Brandstetter** [CC BY-SA 2.0] (*p263*); **Janet Graham** [CC BY 2.0] (*p273*); **Jerzy Opioła** [CC BY-SA 3.0] (*pp95, 115*); **Katja Schulz** from Washington, D.C., USA [CC BY 2.0] (*p283*); **Krzysztof Golik** [CC BY-SA 4.0] (*p342*); **Krzysztof Ziarnek**, Kenraiz [CC BY-SA 4.0] (*p143*); **Laval University** [CC BY-SA 4.0] (*p289*); **Lazaregagnidze** [CC BY-SA 4.0] (*p343*); **Len Worthington** [CC BY-SA 2.0] (*p189*); **Leonora Enking** from West Sussex, England [CC BY-SA 2.0] (*p309*); **Matthewkwan** [CC BY-NC 4.0] https://www.inaturalist.org/observations/224827050 (*p267*); **Maurice REILLE** [CC Creative Commons Attribution-Share Alike 3.0 Unported license] (*p313*); **Metselaar, J.** [CC BY-SA 4.0] (*p323*); **MOs810** [CC BY-SA 4.0] (*p103*); **MPP Wildlife** [CC BY-NC 4.0] https://www.inaturalist.org/observations/231356966 (*p273*); **Muséum de Toulouse** [CC BY-SA 3.0] (*pp95, 199*); **Nucatum amygdalarum** [CC BY-SA 4.0] (*p124*); **Patrick Clement** [CC BY 2.0] (*pp144, 177, 195, 223*); **Richard Avery** [CC BY-SA 4.0] (*p343*); **S. Rae** from Scotland, UK [CC BY-SA 2.0] (*p103*); **Sally Jennings** [CC BY 2.0] (*p220*); **Sten Porse** [CC BY-SA 3.0] (*pp271, 342*); **Sansum, P. A.** [CC BY-NC 4.0] https://uk.inaturalist.org/observations/20775787 (*p285*); **Syrio** [CC BY-SA 4.0] (*p251*); **Tofts** at Faroese Wikipedia, GFDL http://www.gnu.org/copyleft/fdl.html [CC GNU Free Documentation License, Version 1.2] (*p246*); **Udo Schmidt** from Deutschland [CC BY 2.0] (*p222*); **VirginiaFamartin** [CC BY-SA 4.0] (*p281*); **Yurakuna** [CC 0] (*pp69, 283*); Димитър Найденов / **Dimìtar Nàydenov** [CC BY-SA 4.0] (*p283*).

Index

This index contains the English and *scientific* names of all tree and shrub species mentioned in the species accounts (*pp. 88–344*).

Bold black page numbers indicate species that are afforded a full account.

Italic page numbers indicate species that are mentioned and are depicted with at least one illustration.

Regular text page numbers indicate species that are mentioned in the text only.

A

Abies alba **115**
— *grandis* **114**
— *nordmanniana* 113
— *procera* **113**
Acacia, False **300**
Acer campestre **284**
— *cappadocicum* **289**
— *negundo* **289**
— *platanoides* **288**
— *pseudoplatanus* **286**
— *rubrum* **289**
— *saccharinum* **289**
Aesculus ×carnea **291**
— *hippocastanum* **290**
— *indica* **291**
Ailanthus altissima **340**
Alder, Common **272**
—, Green 274
—, Grey **274**
—, Hybrid 273
—, Italian **274**
—, Red 274
Alnus cordata **274**
— *glutinosa* **272**
— *×hybrida* 273
— *incana* **274**
— *rubra* 274
— *viridis* 274
Amelanchier lamarckii **336**
Apple, Crab **180**
—, Domestic **180**
—, Wild **180**
Araucaria araucana **129**
Arbutus unedo **301**
Ash **298**
—, Manna **340**
—, Narrow-leaved **340**
Aspen **256**
Aucuba japonica **330**
Azalea, Yellow *325*

B

Balm-of-Gilead **264**
Balsam-poplar, Eastern **264**
—, Hybrid **264**
—, Western **264**
Barberry **316**
—, Chinese **317**

Barberry, Darwin's **317**
—, Gagnepain's **317**
—, Hedge **317**
—, Mrs Wilson's **317**
—, Thunberg's **317**
Bay **331**
Beech **214**
Berberis darwinii **317**
— *gagnepainii* **317**
— *julianae* **317**
— *×stenophylla* **317**
— *thunbergii* **317**
— *vulgaris* **316**
— *wilsoniae* **317**
Betula ×aurata *264*
— *nana* **271**
— *papyrifera* **270**
— *pendula* **268**
— *pubescens* **268**
— ssp. *tortuosa* **269**
— ssp. *celtiberica* **269**
— ssp. *pubescens* **269**
— var. *fragrans* **269**
— *utilis* **270**
Birch, Downy **268**
—, Dwarf **271**
—, Himalayan **270**
—, Paper **270**
—, Silver **268**
Blackthorn **171**
Bog-myrtle **314**
Box **313**
Bridewort **329**
Broom **320**
—, Spanish *318*
Buckthorn **193**
—, Alder **192**
Buddleia **328**
Buddleja davidii **328**
Bullace *173*
Butcher's-broom **315**
Butterfly-bush **328**
Buxus sempervirens **313**

C

Cabbage-palm **339**
Carpinus betulus **278**
Castanea sativa **216**
Catalpa bignonioides **342**

Ceanothus spp. **344**
Cedar, Atlas **109**
Cedar of Lebanon **109**
Cedrus atlantica **109**
— *deodara* **109**
— *libani* **109**
Cercidiphyllum japonicum **342**
Cercis siliquastrum **342**
Chaenomeles japonica **341**
— *speciosa* **341**
— *×superba* 341
Chamaecyparis lawsoniana **127**
— *pisifera* **128**
Cherry, Bird **174**
—, Cornelian 304
—, Dwarf **179**
—, Japanese **179**
—, Rum **179**
—, Sour **179**
—, Wild **176**
Chestnut, Horse **290**
—, Indian Horse **291**
—, Red Horse **291**
—, Sweet **216**
Cockspurthorn 186
Coralberry **332**
—, Doorenbos **332**
—, Hybrid **332**
Cordyline australis **339**
Cornus alba **305**
— *mas* **304**
— *sanguinea* **304**
— *sericea* **305**
Corylus avellana **276**
— *colurna* *274*
— *maxima* *274*
Cotoneaster, Bullate **190**
—, Entire-leaved **191**
—, Franchet's **191**
—, Himalayan **191**
—, Hollyberry **190**
—, Wall **191**
—, Wild **189**
Cotoneaster horizontalis **191**
— *rehderi* **190**
— *bullatus* **190**
— *cambricus* **189**
— *franchetii* **191**
— *integrifolius* **191**

356

INDEX

Cotoneaster simonsii — **191**
Crack-willow — *242*
—, Eastern — **244**
Crataegus crus-galli — *186*
— *laevigata* — **186**
— ×*media* — *186*
— *monogyna* — **186**
— *rhipidophylla* — *186*
Cryptomeria japonica — **125**
Cupressus ×*leylandii* — **126**
Cupressus glabra — **125**
— *macrocarpa* — **125**
Currant, Black — **322**
—, Downy — **323**
—, Flowering — **322**
—, Mountain — **324**
—, Red — **323**
Cypress, Lawson's — **127**
—, Leyland — **126**
—, Monterey — **125**
—, Nootka — **128**
—, Patagonian — *88*
—, Sawara — **128**
—, Smooth Arizona — **125**
Cypress, Swamp — **121**
Cytisus scoparius — **320**
— ssp. *maritimus* — **119**

D

Damson — *173*
Daphne laureola — **325**
— *mezereum* — **325**
Davidia involucrata — **343**
Deodar — **109**
Dogwood — **304**
—, Red-osier — **305**
—, White — **305**
Douglas-fir — **112**

E

Elder — **306**
—, American — *307*
—, Red-berried — **307**
Elm, Anglo-Saxon — *211*
—, Assington — *208*
—, Atinian — *205*
—, Bassingbourn — *206*
—, Bonhunt — *209*
—, Burred — *210*
—, Cambridge — *210*
—, Camperdown — *204*
—, Chaters' — *210*
—, Coritanian — *211*
—, Corky-barked — *205*
—, Cornish — *208*
—, Curved-leaved — *211*
—, Cut-leaved — *207*
—, Dark-leaved — *206*
—, Davey's — *207*

Elm, Dutch — *207*
—, Dwarf-leaved — *211*
—, East Anglian — *211*
—, English — *205*
—, Essex — *208*
—, European White — **199**
—, Exeter — *204*
—, Fat-leaved — *207*
—, Fat-toothed — *209*
—, Field agg. — **201**
—, Goodyer's — *208*
—, Guernsey — *346*
—, Hatley — *211*
—, Hayley — *206*
—, Huntingdon — *207*
—, Jagged-leaved — *210*
—, Jersey — *208*
—, Large-toothed — *211*
—, Laxton — *207*
—, Leathery-leaved — *211*
—, Long-tailed — *210*
—, Long-toothed — *208*
—, Luffenham — *211*
—, Madingley — *206*
—, Midland — *212*
—, Moss's — *204*
—, Narrow-crowned — *210*
—, Narrow-leaved — *210*
—, Northern Wych — *204*
—, Pale-leaved — *211*
—, Pebmarsh — *210*
—, Peninsula — *208*
—, Plot's — *212*
—, Pointed-leaved — *210*
—, Prominent-toothed — *210*
—, Rhombic-leaved — *211*
—, Round-leaved — *211*
—, Sacombe — *206*
—, Scabrid — *207*
—, Small-leaved — *208*
—, Smooth-leaved — *208*
—, Southern Wych — *204*
—, Sowerby's — *212*
—, Tall — *210*
—, 'Tasburgh' — *212*
—, Wedge-leaved — *210*
—, Western — *208*
—, Woodland — *206*
—, Wych agg. — **200**
Escallonia — **341**
Escallonia rubra var. *macrantha* — *341*
Eucalyptus gunnii — **343**
Euonymus europaeus — **280**
— *japonicus* — *278*
— *latifolius* — *280*

F

Fagus sylvatica — **214**
Ficus carica — **337**

Fig — **337**
Filbert — *274*
Fir, Caucasian — **113**
—, Douglas — **112**
—, Giant — **114**
—, Grand — **114**
—, Noble — **113**
Firethorn — **334**
—, Asian — *332*
Fitzroya cupressoides — **215**
Forsythia — **344**
Frangula alnus — **192**
Fraxinus angustifolia — **340**
— *excelsior* — **298**
— *ornus* — **340**
Fuchsia — **333**
— *magellanica* — **333**

G

Genista anglica — **320**
— *hispanica* — *317*
— *pilosa* — **321**
— *tinctoria* — **321**
— ssp. *littoralis* — **321**
— ssp. *tinctoria* — **321**
Ginkgo — **338**
— *biloba* — **338**
Gleditsia triacanthos — **343**
Gooseberry — **324**
Gorse — **318**
—, Dwarf — **318**
—, Hybrid — *316*
—, Spanish — *317*
—, Western — **318**
Greengage — *173*
Greenweed, Dyer's — **321**
—, Hairy — **321**
Guelder-rose — **309**
—, American — *307*
—, Asian — *307*
Gum, Cider — **343**

H

Hamamelis spp. — **344**
Hawthorn — **186**
—, Large-sepalled — *186*
—, Midland — **186**
Hazel — **276**
—, Turkish — *274*
—, Witch — **344**
Hebe, 'Hedge' — **328**
—, Lewis's — *328*
—, Narrow-leaved — **328**
Hemlock-spruce, Eastern
—, Western — **111**
Hippophae rhamnoides — **194**
Holly — **302**
—, Highclere — **303**
Hornbeam — **278**

357

INDEX

Horse-chestnuts 290

I
Ilex ×*altaclerensis* 303
— *aquifolium* 302

J
Judas-tree 342
Juglans nigra 196
— *regia* 196
Juneberry 336
Juniper, Common 118
Juniperus communis 118
— ssp. *communis* 119
— ssp. *hemisphaerica* 119
— ssp. *nana* 119

K
Katsura 342

L
Labrador Tea 325
Laburnum 330
— *alpinum* 328
— *anagyroides* 330
—, Scottish 328
Larch, European 106
—, Hybrid 107
—, Japanese 107
Larix decidua 106
— *kaempferi* 107
— ×*marschlinsii* 107
Laurel, Cherry 178
—, Portuguese 178
—, Spotted 330
Laurus nobilis 331
Laurustinus 335
Ligustrum ovalifolium 312
— *vulgare* 312
Lilac 329
—, Californian 344
Lime, Common 295
—, European 295
—, Large-leaved 296
—, Silver 297
—, Small-leaved 294
Liriodendron tulipifera 339
Locust, Honey 343
Lycium barbarum 333
Lycium chinense 333

M
Magnolias 344
Mahonia aquifolium 331
Malus domestica 180
— *sylvestris* 180
Maple, Ashleaf 289
—, Cappadocian 289
—, Field 284
—, Norway 288

Maple, Red 289
—, Silver 289
Medlar 336
Mespilus germanica 336
Metasequoia glyptostroboides 120
Mezereon 325
Mock-oranges 334
Monkey-puzzle 129
Morus nigra 342
Mulberry, Black 342
Myrica gale 314

N
Nothofagus alpina 215
— *obliqua* 215

O
Oak, Cork 225
—, English 218
—, Evergreen 223
—, Holm 223
—, Hybrid 219
—, Lucombe 225
—, Pedunculate 218
—, Pin 224
—, Red 224
—, Rose 219
—, Scarlet 224
—, Sessile 221
—, Turkey 222
Oregon-grape 331
Osier 240
—, Broad-leaved 245
—, Silky-leaved 245

P
Pear, Cultivated 184
—, Plymouth 183
—, Wild 184
Philadelphus spp. 334
Picea abies 105
— *sitchensis* 104
Pine, Bhutan 103
—, Black 98
—, Bristlecone 88
—, Dwarf Mountain 103
—, Lodgepole 102
—, Macedonian 103
—, Maritime 100
—, Monterey 101
—, Scots 96
—, Weymouth 103
Pinus contorta 102
— *longaeva* 88
— *mugo* 103
— *nigra* 98
— ssp. ssp. *laricio* 98
— ssp. ssp. *nigra* 98
— *peuce* 103

Pinus pinaster 100
— *radiata* 101
— *strobus* 103
— *sylvestris* 96
— *wallichiana* 103
Plane, American 283
—, London 282
—, Oriental 283
Platanus ×*hispanica* 282
— *occidentalis* 283
— *orientalis* 283
Plum, Cherry 172
—, 'Shropshire Prune' 173
—, 'Victoria' 173
Plum, Wild 173
Podocarpus andinus 117
Poplar, Black 261
ssp. *betulifolia* 261
—, 'Golden' 262
—, Grey 258
—, Hybrid Black 262
—, Lombardy 260
— 'Marilandica' 262
— 'Railway' 262
— 'Robusta' 262
— 'Serotina Aurea' 262
— 'Serotina' 262
—, White 260
Populus alba 260
— *balsamifera* 264
— ×*canadensis* 262
— ×*canescens* 258
— ×*hastata* 264
— ×*jackii* 264
— *nigra* 261
— *tremula* 256
— *trichocarpa* 264
Privet, Garden 312
—, Wild 312
Prunus avium 176
Prunus cerasifera 172
— *cerasus* 179
— *domestica* 173
ssp. *domestica* 173
ssp. *insititia* 173
ssp. ×*italica* 173
— *laurocerasus* 178
— *lusitanica* 178
— *padus* 174
— *serotina* 179
— *serrulata* 179
— *spinosa* 171
Pseudotsuga menziesii 112
Pyracantha coccinea 334
— *rogersiana* 332
Pyrus communis 184
— *cordata* 183
— *pyraster* 184

Q

Quercus cerris **222**
— *coccinea* **224**
— ×*crenata* **225**
— *ilex* **223**
— *palustris* **224**
— *petraea* **221**
— *robur* **218**
— ×*rosacea* **219**
— *rubra* **224**
— *suber* **225**
Quince, Chinese **341**
—, Hybrid 341
—, Japanese **341**

R

Rauli 215
Red-cedar, Japanese **125**
—, Western **124**
Redwood, Coast **122**
—, Coastal **122**
—, Dawn **120**
—, Giant **123**
Rhamnus cathartica **193**
Rhododendron groenlandicum 325
— *luteum* 325
— *ponticum* **327**
Rhododendrons **327**
Rhus typhina **337**
Ribes alpinum **324**
— *nigrum* **322**
— *rubrum* **323**
— *sanguineum* **322**
— *spicatum* **323**
— *uva-crispa* **324**
Robinia pseudoacacia **300**
Roble 215
Rowan **140**
—, False **300**
—, Hupeh 140
Ruscus aculeatus **315**

S

Salix alba **241**
— var. *caerulea* 241
— var. *vitellina* **244**
— *arbuscula* **252**
— *aurita* **236**
— *caprea* **234**
— ssp. *scaprea* **234**
— ssp. *sphacelata* **234**
— *cinerea* **232**
— ssp. *cinerea* **232**
— ssp. *oleifolia* **232**
— *elaeagnos* **244**
— *euxina* **244**
— ×*fragilis* **242**
var. *decipiens* 242
var. *fragilis* 242

Salix ×*fragilis*
var. *furcata* 242
var. *russelliana* 242
— *herbacea* **254**
— ×*holoserica* **245**
— *lanata* **251**
— *lapponum* **250**
— ×*multinervis* **245**
— *myrsinifolia* **247**
— *myrsinites* **253**
— *pentandra* **239**
— *phylicifolia* **246**
— *polaris* 226
— *purpurea* **238**
— ×*reichardtii* **245**
— *repens* **248**
var. *argentea* **248**
var. *fusca* **248**
var. *repens* **248**
— *reticulata* **255**
— ×*sepulcralis* **244**
— ×*smithiana* **245**
— *triandra* **243**
— *viminalis* **240**
Sallow, Grey **232**
—, Rusty **232**
Sambucus canadensis 307
— *nigra* **306**
— *racemosa* **307**
Sea-buckthorn **194**
Sequoia sempervirens **122**
Sequoiadendron giganteum **123**
Service-tree, Arran **152**
—, German 146
—, True **148**
—, Wild **142**
Silver-fir, European **115**
Snowberry **332**
Sorbus admonitor **158**
— *anglica* **147**
— *aria* **144**
— *arranensis* **152**
— *arvonicola* **154**
— *aucuparia* **140**
— ×*avonensis* **164**
— *bristoliensis* **164**
— *cambrensis* **156**
— *cheddarensis* **162**
— *croceocarpa* **143**
— *cuneifolia* **154**
— *decipens* 165
— *devoniensis* **158**
— *domestica* **148**
— *eminens* **166**
— *eminentiformis* **166**
— *eminentoides* **162**
— *evansii* **168**
— *greenii* **168**
— *herefordensis* **166**

Sorbus hibernica **150**
— *hupehensis* 140
— *intermedia* **146**
— *lancastriensis* **149**
— *leighensis* **164**
— *leptophylla* **156**
— *leyana* **154**
— *margaretae* **160**
— *minima* **156**
— ×*motleyi* 154
— *mougeotti* 146
— *parviloba* **168**
— *porrigentiformis* **147**
— *pseudofennica* **152**
— *pseudomeinichii* **300**
— *richii* **160**
— *rupicola* **145**
— *rupicoloides* **162**
— *saxicola* **168**
— *scannelliana* **150**
— *spectans* **164**
— *stenophylla* **156**
— *stirtoniana* **154**
— *subcuneata* **158**
— ×*thuringiaca* 146
— *torminalis* **142**
— *vexans* **160**
— *whiteana* **166**
— *wilmottiana* **164**
Spanish-dagger **339**
Spartium junceum **318**
Spindle **280**
Spindle, Evergreen 278
—, Large-leaved 280
Spiraea spp. **329**
Spruce, Norway **105**
—Spruce, Sitka **104**
Spurge-laurel **325**
Steeple-bush **329**
Strawberry-tree **301**
Styphnolobium japonicum **343**
Sumach, Stag's-horn **337**
Sycamore **286**
Symphoricarpos albus **332**
— ×*chenaultii* **332**
— ×*doorenbosii* 332
— *obiculatus* 332
Syringa vulgaris **329**

T

Tamarisk **326**
Tamarix gallica **326**
Taxodium distichum **121**
Taxus baccata **116**
Teaplant, Chinese **333**
—, Duke of Argyll's **333**
Thuja occidentalis **128**
— *plicata* **124**
Tilia cordata **294**

INDEX

Tilia ×europaea 295
— platyphyllos 296
— tomentosa 297
Tree, Dove 343
—, Indian Bean 342
—, Pagoda 343
Tree-of-heaven 340
Tsuga canadensis 111
— heterophylla 111
Tulip-tree 339

U

Ulex ×breoganii 318
— europaeus 318
— gallii 318
— minor 318
Ulmus acuminatissima 210
— acutissima 209
— alta 210
— anglosaxonica 211
— angustiformis 211
— asymmetrica 211
— atrovirens 206
— camperdownii 197, 204
— cantabrigiensis 206
— chaterorum 210
— coriaceifolia 211
— coritana 211
— cornubiensis 208
— crassa 207
— crenata 206
— cuneiformis 210
— curvifolia 211
— daveyi 207
— diversifolia 211
— elegantissima 212
— exoniensis 204
— glabra 204
— gyrophylla 205
— hollandica 207
— incisa 207
— insularum 346
— laevis 199
— longicaudata 210
— longidens 208
— longidentata 210
— madingleyensis 206
— microdon 211
— minor 210
— mossii 204
— multidentata 210
— obesidens 209
— obesifolia 207
— oblanceolata 210
— occidentalis 208
— peninsularis 208
— platyphylla 206
— plotii 212
— prionophylla 211

Ulmus procera 205
— proceriformis 205
— prominentidens 210
— pseudelegantissima 208
— pseudobovata 206
— pseudocoritana 211
— rasilis 208
— rhombifolia 211
— sativa 208
— scabra 204
— scabrosa 207
— serrata 208
— serratifrons 210
— serratula 210
— sowerbyi 212
— stricta 208
— sylvatica 211
— vegeta 207
— wheatleyi 208

V

Veronica ×franciscana 328
— ×lewisii 328
— salicifolia 328
Viburnum lantana 310
— opulus 309
— rhytidophyllum 335
— sargentii 307
— tinus 335
— trilobum 307
— veitchii 311
—, Wrinkled 335

W

Walnut 196
—, Black 196
Wayfaring-tree 310
—, Chinese 311
Weigela florida 341
Weigelia 341
Wellingtonia 123
Whin, Petty 320
White-cedar, Northern 128
White-elm, European 199
Whitebeam, Arran 152
—, Bloody 160
—, Bristol 164
—, Broad-leaved 165
—, Cheddar 162
—, Common 144
—, Devon 158
—, Doward 166
—, English 147
—, Evans' 168
—, Gough's Rock 162
—, Green's 168
—, Grey-leaved 147
—, Herefordshire 166
—, Irish 150

Whitebeam Lancastrian 149
—, Least 156
—, Leigh Woods 164
—, Ley's 154
—, Llangollen 154
—, Llanthony 156
—, Margaret's 160
—, Menai Strait 154
—, Mougeot's 146
—, Observatory 164
—, Orange 143
—, Rich's 160
—, Rock 145
—, Round-leaved 166
—, Scannell's 150
—, Ship Rock 168
—, Somerset 158
—, Stirton's 154
—, Swedish 146
—, Symonds Yat 168
—, Thin-leaved 156
—, Twin Cliffs 162
—, Watersmeet 158
—, Welsh 156
—, White's 166
—, Willmott's 164
Willow, Almond 243
—, Bay 239
—, Bedford 242
—, Creeping 248
—, Cricket-bat 241
—, Dark-leaved 247
—, Downy 250
—, Dwarf 254
—, Eared 236
—, Goat 234
—, Golden 244
—, Grey 232
—, Mountain 252
—, Net-leaved 255
—, Olive 244
—, Polar 238
—, Purple 226
—, Tea-leaved 246
—, Weeping 244
—, White 241
—, Whortle-leaved 253
—, Woolly 251

X

Xanthocyparis nootkatensis 128

Y

Yew 116
Yew, Plum-fruited 117
Yucca gloriosa 339